Fire in the Hills

Fire in the Hills

A HISTORY OF RURAL FIRE-FIGHTING IN NEW ZEALAND

Helen Beaglehole

CANTERBURY UNIVERSITY PRESS

To my grandchildren,

Dom, Celia, Elsie, Luke, Thomas and Miles.

First published in 2012
CANTERBURY UNIVERSITY PRESS
University of Canterbury
Private Bag 4800, Christchurch
NEW ZEALAND
www.cup.canterbury.ac.nz

ISBN 978-1-927145-35-7

A catalogue record for this book is available from the
National Library of New Zealand.

Book design and layout: Quentin Wilson, Christchurch
Printed in China through Bookbuilders, owned by APOL

Contents

Acknowledgements

Writing history invariably requires collaboration and assistance, and this book has been no different. Those in the past who were engaged in questions associated with rural fire-fighting left invaluable documents and photos that record and elaborate their work and their commitment.

I was privileged to meet people with experience ranging from the Forest Service to private companies and bush fire forces, many of whom still play important roles: Doug Ashford, Jack Barber, John Barnes, Peter Berg, Jock Darragh, Murray Dudfield, Murray Ellis, Rod Farrow, Don Geddes, Morrie Geenty, Ross Hamilton, Kerry Hilliard, Mike Hockey, Charlie Ivory, Roy Knight, Ian Millman, Syd Moore, Lesley Porter, Warren Reekie, Bill Studholme, Dave Thurley, Gavin Wallace and John Ward. These people gave me sometimes hours of their time, answering innumerable questions and giving me a sense of the passion with which they approached their work. Even where space has limited my opportunities to quote them directly, their narratives underpin and helped focus my understanding.

All interviews, now stored in the Oral History Archive, Alexander Turnbull library, are available for future researchers. Brian Molloy provided useful perspectives on the impact of the early East Polynesian migrations; Robin Cox, Ashley Cunningham, Barry Keating, Gary Lockyer and Grant Pearce willingly responded to what must have seemed at times very elementary questions. The online rural fire history (www.ruralfirehistory.org.nz/5th.htm), developed by Gavin Wallace, has been an invaluable resource.

Wally Seccombe provided photos from his extensive private collection of the Forest Service at work, and offered useful information on a number of the images. Bill Studholme's photos added to my understanding of the 1987 Dunsandel fire. I am also very grateful to Peter Berg, Gerald Hensley and Bill Studholme for their advice on the background needed to contextualise my research, and for commenting on what I have written.

Institutions, too, have played a part. Staff at Archives New Zealand, the Auckland City Library, the National Library and the Oral History Centre all helped as I worked my way through books, articles, files and images. I must also acknowledge the huge research benefit of online material. PapersPast (www.paperspast.natlib.govt.nz) enabled me to scan newspapers in a way that would have been impossible had I been limited to microfiche; the digitising of the *Appendices to the Journals of the House of Representatives* has greatly facilitated my work.

I am hugely grateful to Jock Phillips for suggesting the topic and for starting me on years of fascinating research; to Michael Roche for reading the manuscript, and for his knowledgeable suggestions and comments that sharpened my presentation of the context in which the Forest Service was developing its work in rural fire-fighting. Murray Dudfield, Jock Darragh, Kerry Hilliard and John Ward's comments on the evolving text give me confidence that my interpretation of the history in which they played an integral role is not entirely awry.

I am also very grateful for the financial assistance I have received. Murray Dudfield at the National Rural Fire Authority gave annual grants for three of the years over which I researched and wrote this book, as well as a generous sum towards publication; the New Zealand Forest Owners' Association gave $1000 towards payment for the images that so enhance the text. Doug Ashford at the Forest and Rural Fires Association of New Zealand gave me a travel grant for interviews. In 2009 I received a Ministry of Culture and Heritage New Zealand History Research Fund Award in History. Such financial assistance is always more than welcome, for its vote of confidence in your work as well as for the hard cash!

I'm indebted, too, to the many former Forest Service men who readily offered me access to their collections of memorabilia, publications and photos. These add considerably to our understanding of the past. Ian Millman generously lent a number of old publications that I did not find elsewhere, and Wally Seccombe has amassed a large collection of photos of the Forest Service and its operations in the two last decades or so of its existence.

Canterbury University Press, under publisher Rachel Scott, has produced a book with the high production values and attention to both visual and factual detail that typify their work. They also have dealt patiently with my propensity to continue to change the text up until the last minute. My thanks, too, to editor Tania Tremewan.

Finally, I must thank my family, who have probably heard more than enough about rural fire-fighting. Special thanks go to my husband Tim, who turned manfully to each new draft of each chapter, whose thoughtful comments and insights I always value, and whose moral support is of inestimable value.

Foreword

This book will be of particular interest to many whose livelihood, like mine, is forestry. It is a history of rural fire-fighting and it is intimately bound up with the broader history and development of forest management and policy in this country – although it also appropriately recognises that fire is both a tool and a threat for all sections of the rural community.

As one reads this book one appreciates a little better the sheer majesty of the natural forests that once dominated the landscape, but even more the scale and destructiveness of the forces applied to once and for always remove these. Fire, sometimes accidental but more often deliberate, was the primary instrument of the forests' removal, with some estimates suggesting more than 90 per cent of the cover was removed without any attempt at wood recovery. So successful was this activity that within 50 years of serious settlement commencing, and in a country noted for its grand forests, tree planting was being mooted as the only way to ensure a wood supply for future generations.

If the last 60–100 million years during which New Zealand's forests evolved in relative isolation from the rest of the world were represented by a single day, then the period during which mankind and his fellow travellers grossly modified them is no more than the last one second. The author has explored the development of New Zealand's rural fire strategy and policy over this critical period of one second, and considered the effectiveness of both the process and its implementation.

Undoubtedly the most important moment of the period was in 1919 when the government decided to set up the agency that became the New Zealand Forest Service, which immediately established controls on harvesting and other forms of clearance of the state's remaining native forests, introduced fire-protection measures and quickly ramped up the planting of fast-growing exotic tree species to ensure that future generations had access to a reliable and sustainable source of wood products.

The Forest Service also quickly invested in training and research capability to support the management, protection and utilisation of the quickly growing resource, and fire – as both a tool and a threat – was an important consideration in each of these processes. Such was the calibre of this effort that the Service became identified as the lead organisation on matters associated with rural fire … a situation that persisted for the next six decades.

These were very positive years for rural fire administration – notable for strong support from many rural communities where individuals and small groups joined forces to provide local fire-fighting capability. These volunteers worked with the Forest Service to improve their skills, becoming the first port of call in many small communities for a range of incidents and developing great teamwork and esprit de corps.

In 1987 the Forest Service was broken up into a commercially driven forestry corporation, largely based around the planted forests; a Department of Conservation to manage native forests but with no wood production; and a regulatory, policy and forest facilitation Ministry of Forestry. Suddenly local authorities found themselves thrust into a position of greater responsibility for rural fire, a position they had largely resiled from in earlier times. Rural fire administration now rests within the New Zealand Fire Service, although a separate National Rural Fire Division manages this process. Some 3000 part-time/volunteers still engage, and a well-directed programme of training, research and monitoring supports the structure very ably.

Of course the strength of the country's rural fire-fighting capability has been in its people who, without exception, have striven to achieve results of the highest quality, often despite quite meagre equipment and even poorer information. Not surprisingly, the Rural Fire Service abounds in stories of colourful and determined people. This book brings much of that 'character' to life and in my view is worth reading not just because it the story of rural fire that it so accurately portrays, but also because in many respects the story of rural fire is effectively a microcosm of the history of New Zealand with all the same elements – often overwhelming odds, poor resourcing, a make-do/no. 8 wire mentality and great success.

As Helen Beaglehole concludes: 'Some are driven by a passion for fire and its management, or by the urge for achieving the self discovery that comes with tough physical effort. For others the primary drivers are about contributing to their community's safety or protecting its natural assets. But whatever the motivation … they, their predecessors, and their histories take their place … in the annals of New Zealand's history.'

I am personally delighted to be able to endorse this book as an accurate and important contribution to the history of rural fire administration in New Zealand, but more importantly for its contribution to the record of the settlement and development of rural New Zealand as we now know it.

Peter Berg ONZM, FNZIF, B.For.Sc

Preface

This is a history of rural fire-fighting: a history of how forest and other vegetation fires in rural areas – or wildfires, to give them the American appellation increasingly in vogue – have been prevented, extinguished or controlled. Today fire brigades, rural and otherwise, have a role in fighting vegetation fires. But although Chapter Five traces their history to explain their historically limited role in rural fires, the brigades are not my primary focus. Rather, I am primarily interested in how New Zealanders dealt with the fires that for decades devastated the New Zealand countryside, destroying a timber resource, threatening an increasingly valuable investment in plantation forests, and directly or indirectly jeopardising agricultural lands.

In writing this history I found aspects of my past intricately linked to it. Every summer my family drove across the Kaimai Range from Hamilton to Rotorua and every summer we passed the ragged bush edge and scrub and trunks lying dead on the ground – marks of controlled or uncontrolled fire. My father, a great advocate for breaking in 'waste land', told us about James Fletcher and the work of his firm, about planting pines, and about their future value to the country. We picnicked and swam in the Blue Lake (now Lake Tikitapu), surrounded by *Pinus radiata*; along the roadside we saw the first of the now common fire-danger signs in the shape of a half moon and we enthusiastically read the fire messages that were strung out, word by word.

One of my mother's friends was a daughter of Henry Valder, a founding partner in Ellis and Burnand Ltd, the largest sawmilling operation in the King Country. Her sisters, most unusually for the time, were farming on land they had broken in during the 1930s on the slopes of Mt Pirongia; when I stayed there in the early 1960s they were still digging up stumps from the clearance burns. I grew up lighting fires to cook potatoes in the glowing embers – and just lighting fires. I also knew about making sure the fires were out before I left. In the excited beginning-of-the-school-year rush to the local stationery shop for new books, rubbers and pencils, I bought the coveted wooden rulers inset with small panels of native timber. From the late 1960s I went tramping in forest parks and used the huts, tracks and signposts that the Forest Service developed.

When, some 40 years on, a fortuitous train of events set me writing the history of rural fire-fighting, my childhood experiences seemed to slot easily into parts of the narrative, though I now understand it in different terms. The roadside signs about fire and fire danger, I now know, were part of a public education drive on fire control and prevention; being taught not to leave fires alight was also part of that effort (it owed something, too, to my mother's tramping background). The rulers were to encourage us to realise the beauty and value of New Zealand timber and, by extension, the value of the bush. (I don't ever remember having one of the little leaflets that would help us understand this; we simply liked the way they looked.) My mother's protests against my father's desire to see land broken in and 'improved' drew on a valuing of trees and spaces that reflected early arguments about conservation (though not, I think, its utilitarian aspects of retaining the forest for future logging). Was she, as an early woman tramper, one whom the Forest Service had encouraged to go into the bush, learn its beauty and lessen its destruction in the public education initiative started in 1921 by MacIntosh Ellis? Its aim was to build a public consciousness and appreciation of the bush as a tangible asset, and it was still operative when I began my tramping. Unwittingly we were perhaps both participating in the by then century-old debate about the relative value of trees as a future timber resource, scenic areas, protection bush and agricultural land. By contrast, my father, who loved and planted trees, stolidly maintained that to be productive farmers must burn the bush and break in the land – an argument, I now know, that fits into a wider framework of the imperatives of frontier societies, the practicalities of settlement and the settlers' history, religion and politics.

As I worked through the history of rural fire-fighting, I found two things surprising. First, although fire was, and continues to be, a land management tool, I had not realised the impact of the early burning. For the settlers it was vital for clearing the land. The flush of green grass in its aftermath, too, was a highly valued result. Yet from the beginning of European settlement the bush fires, caused by burning off or carelessness, were destructive and terrifying. My father would have been a small boy when, in 1908, some of New Zealand's most devastating bush fires swept across much of the North Island. The land between the Hokianga and Gisborne was alight, proclaimed one headline. Ships encountered the smoke 160 kilometres off the East Cape. Fires swept towards

Wellington, the smoke making lights necessary at midday. Ships could not leave the harbour. New Plymouth used its foghorn to signal the whereabouts of its harbour entrance. The top of the South Island, too, suffered. The Rai Valley, separated from Nelson only by a range of hills, was badly burned. When in the 1920s my father came as a young man to Hamilton, the huge Waikato swamps were being drained and turned into agriculturally productive land. Smoke and peat fires, which often got out of control, were a part of Hamilton life.

These fires brought considerable financial loss. The accounts syndicated in all newspapers, sometimes for days at a time, describe desperate measures to save household treasures and homes; the choking, blinding smoke; the losses. They hint at the emotional horror. Increasingly photographs captured the fires' terrible beauty as they dramatically modified our landscape. In one extraordinarily intensive decade of burning, 1890–1900, 27 per cent of the existing forest (about 13 per cent of the total land area) was cleared, a deforestation rate four times that of tropical Asian rainforests during the late 1990s.[1] The rate slowed but the loss continued – in 1921 Ellis fulminated against the burning of almost a million hectares of virgin bush over 50 years and lamented the 'useless barren waste' caused by public apathy and indifference.

These clearances continued until the 1930s. I was born in 1946, the year the 'nationally calamitous' Taupo fires raged out of control over thousands of hectares for almost a month and caused uncalculated damage to indigenous and privately owned exotic forests. Yet none of these fires formed part of the narrative on which I was brought up. Indeed, if casual discussion is anything to go by, this gap seems part of a public amnesia that extends to the vast controlled burns over the 1960s, 1970s and into the 1980s when vast clouds of smoke, somewhat reminiscent of atomic clouds, accompanied the burning of thousands and thousands of hectares in preparation for future planting. In Chapters One and Two I discuss the scale and effect of the European burning and the settler mindset in which such events were construed; Chapter Three looks at the regime Ellis established to modify that mindset and halt the destruction of timber and land.

Chapter One also briefly traces the origins, extent and effect of the burning that accompanied Maori settlement. This burning does not strictly fit into a history of rural fire-fighting – we know nothing about how Maori responded to what they, wittingly or unwittingly, had set alight. On the other hand, their fires form part of a narrative about pioneer settlement and landscape modification that reflect, like the European fires, the realities of a frontier society.

Second, I was surprised at the extent to which people, on learning about my research, framed it in terms of the 'vollies' – the volunteer rural fire brigades that over the years have done much sterling work in fighting vegetation fires. What that response seems blind to is the fact that for years the volunteers were underpinned by Forest Service equipment, know-how and training. If my account focuses on the Forest Service, that focus is deserved.

As the efforts to prevent rural fire were initially directed towards conserving a timber resource and maintaining protection of forests to help safeguard arable lands and good water supplies, the State Forest Department became the responsible agency. The Forests Act 1921–22 developed by the enthusiastic MacIntosh Ellis, and its later iterations, established a framework and a *modus vivendi* for how to approach the task; today it continues to inform the management of rural fire. Chapters Four and Six look at the steady growth in the Forest Service's systematic response to rural fire. Increasingly in touch with fire management practices in Australia and the United States, it battled bad roads and meagre budgets to work with recalcitrant millers, farmers and the New Zealand Railway Department – getting them on side, attempting to ensure that they always operated with a regard to fire safety. Huge public education drives sought to develop the general public's notions of fire safety. Increasingly as the exotic plantations developed and matured, self-interest and proximity saw the Forest Service putting out fires. Partly as a deterrent, the costs of doing this were sheeted home to the landowners, and Forest Service staff became as skilled at recovering those costs as they were at fire-fighting. By the 1950s they were seen as the nation's fire-fighters.

From the early 1960s, when the science of fires was still in its infancy, the Forest Service and private companies' controlled burns turned the use of fire as a land clearance tool into an art and contributed enormously to institutional and personal knowledge of fire behaviour and fire-fighting. Debriefs after major fires contributed to this knowledge. The large private forestry companies strengthened their fire protection roles both within and beyond their own forests. The community played a part, whether as honorary forest rangers, or as volunteers in a variety of fire forces trained by the Forest Service and using equipment the Forest Service had written off. Increasingly, too, the Forest Service developed a national co-ordination and facilitation role.

The Forest Service was disestablished in 1987. When the series of bad fires hit Canterbury in the summers of 1987–88 and 1988–89, the Forest Service's national capacity, equipment and manpower were sorely missed. Today the National Rural Fire Authority co-ordinates New Zealand's rural fire management. The New Zealand Forest Owners' Association, working at times alongside the authority, has fire research, education and safety as its top priorities. Three thousand volunteers and part-timers, who belong to some 86 rural fire authorities, themselves often part of the larger Civil Defence framework, do the fire-fighting. The fire authorities also work to educate landowners and the general public about fire safety.

The environment continues to be complex. Humans and their land clearance burns are still responsible for continued unintended burning. Although what is now burnt annually is less than half of what wildfires burnt in 1987, 5000–6000 hectares of land annually still go up in smoke. It is mostly grass and scrubland, but the economic cost is almost $1 million (2008 dollars) annually – and this figure does not include non-market losses from such things as increased flood risk, degraded water quality or impoverished tourist experience. The Rural Fire Authority faces challenging questions. Many consider the risk of intense fires has grown as plantation forests have increased and because significant areas of high country have been returned to the Crown; it, for conservation reasons, uses fire less intensively as a land-management tool. A warming world and concerns about carbon, too, raise questions about the continued use of fire as a silviculture tool. In addition, as the authority seeks to achieve further efficiencies in fire management in terms of both training and organisation, it must consider the effect of such changes on largely unpaid, and at times unsung, volunteer rural fire-fighters.

One of the hugely enjoyable aspects to writing this history has been choosing the illustrations. Although building alterations meant that my access to the Alexander Turnbull Library's collection was limited to its online database, there were rich pickings: working through the Forest Service's vast collection at Archives New Zealand provided a host of possibilities; photographs from the 1950s on introduced me to the talented John Johns who worked for the Forest Service, whose beautiful black-and-white images are held at the McNamara Gallery in Wanganui and can be seen online. I am indebted, too, to ex-Forest Service staff who readily offered me access to their own and other valuable pictures – images that add considerably to our understanding of the past. However, my choice of photos for Chapter Five is sadly limited. The National Library's huge alteration programme made its copying services unavailable to the public and it proved too complicated to get copies of the fascinating images in local brigade histories done elsewhere.

I have changed acres, inches, feet, yards, chains, miles, gallons and other imperial measures to their relevant metric equivalents; in direct quotes I have given the metric figure in brackets. For important sums of money I used the 'general' category of the Reserve Bank's inflation calculator and fourth-quarter values to show today's (2011) monetary values. The online calculator can be found at www.rbnz.govt.nz/statistics/0135595.htm.

There are also issues of terminology. The term 'reforestation' refers to the process of replanting a forest, especially after clear-cutting, in contrast to 'afforestation', which is about creating a new forest where none had existed before, or reforesting areas long deforested. For convenience I have used 'minister' interchangeably with the strictly correct term Commissioner of Forests. Similarly, although the organisation went through a variety of appellations before it became the New Zealand Forest Service, I have used the name Forest Service throughout.

Finally, from the 1970s women were recruited into the Forest Service. The gender-neutral term 'fire-fighters' that I use from that date reflects a changing pattern of language within government agencies that began in that decade. Today, as women are increasingly participating in what was once a male-dominated activity, such gender-neutral language is entirely appropriate.

Chapter One

The burning of New Zealand

Fire is an elemental agent. Like other environmental forces such as lightning strikes, eruptions, erosion, extreme volcanic events and long-term climate change, it has wrought extensive change to landscapes all over the world.[1] New Zealand has been no exception. Pollen and charcoal studies indicate the country suffered significant burning prior to Polynesian settlement. At about 150 AD the Taupo eruption, 'probably the most explosive volcanic eruption in the world in the last 5,000 years', temporarily devastated much of the North Island;[2] by about 450 AD major deforestation had occurred in other areas. But although estimates vary, there is general consensus that at the time of the first substantial Polynesian settlement, around 1100 AD, some 80 per cent of New Zealand's landmass was still clad in bush or scrub (a qualified statement is needed as computer modelling may suggest greater certainty and accuracy than in fact exists).

Those Polynesian voyagers, who gave rise to the distinctive Maori people, and the European settlers who followed them centuries later, modified the landscape dramatically.[3] The fires that both sets of immigrants and their descendants lit have been the greatest agent of change in the country's landscape. But before assessing the extent and damage of this burning, we need to think, first, about the context in which these fires occurred. Both the east Polynesian and European settlers used fire to destroy forest and open land for cultivation and settlement. Their response to their new land was like that of any frontier society. Fires, whether lit by east Polynesian voyagers or European settlers, were an accepted tool for destroying vegetation for cultivation and making the land habitable. Second, whatever the impact of fire on the landscape, its impact on species has been less devastating. As a result of fire, only five or six plant species appear to have become extinct and some species have been listed as threatened or endangered. Other species, like matagouri and the colourful native broom, have flourished in the new habitats the fires (and later top-dressing) created.[4] This sometimes beneficial effect contrasts vividly with the impact of introduced predators and human habitation – 67 animal and bird species now extinct and 2788 threatened or endangered.[5]

The early Polynesian coastal dwellers initially lit small fires, but within about 200 years of initial settlement, clearance was well advanced in both main islands. Drought-prone and flat or rolling areas were destroyed more quickly than wetter hill country. The initial deforestation, and later repeat burning, meant the bush did not recover; in general, tall forest was replaced by seral vegetation, particularly bracken, or tussock grasslands that were even more susceptible to fire.[6] Just before large-scale British and Irish immigration, anthropogenic fires had reduced the total forest area to about 55 per cent of its coverage prior to human habitation.[7]

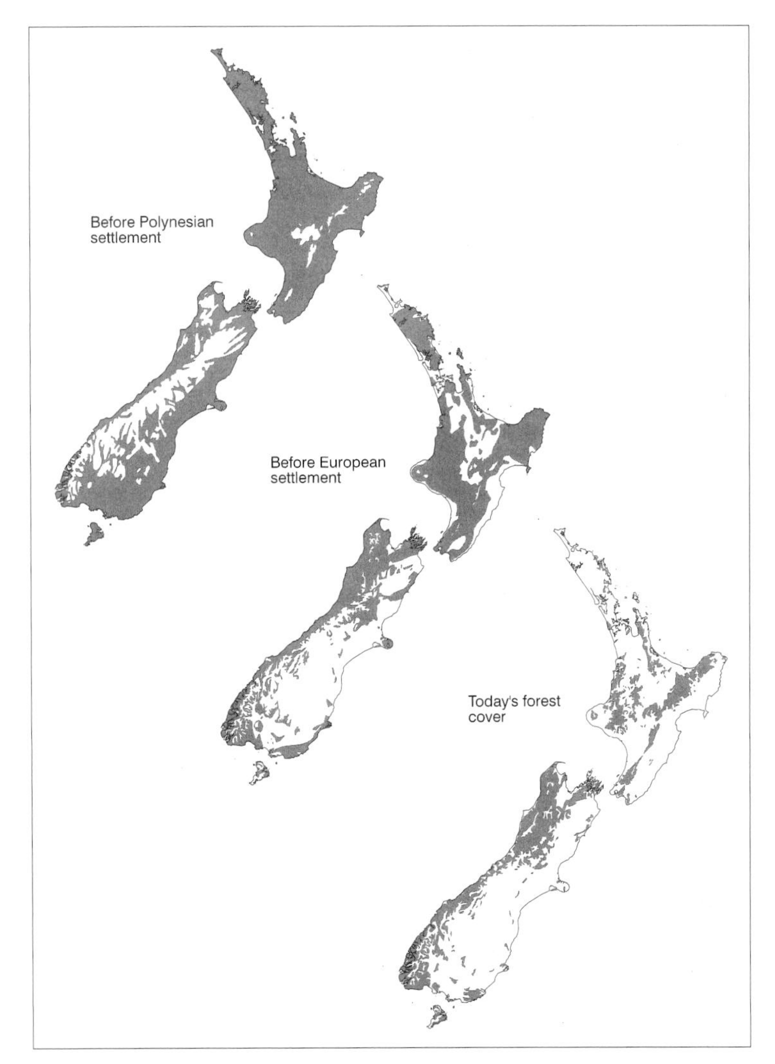

Human impacts on New Zealand's natural forest and tall scrubland from the first Polynesian arrivals to the present.

Fig. 8.5 from Taylor and Smith, *The State of New Zealand's Environment 1997*

The impact of European settlement, with its greater influxes of population and its different patterns of land use, significantly accelerated this devastation. From 1861 to 1881 pasture increased from less than 70,000 hectares to 1.4 million hectares; by 1901, 4.5 million hectares were in grass. In one extraordinarily intensive decade of burning, 1890–1900, 27 per cent of the existing forest (about 13 per cent of the total land area) was cleared, a deforestation rate four times that at which tropical Asian rainforests were being cleared during the late 1990s.[8] Significant burning of indigenous forest only ended in about the 1930s, when the land for an agricultural-based economy had been largely broken in (the term betokens the attitude) for cultivation. But the abandoned billy fire, the cigarette butt casually flicked out the car window, the burn-offs unwisely lit or incompetently managed, indicate attitudes still alive and well at least into the

Buggins clearing, Taranaki, by Colonel Edward Williams, 1864. The wide-open valley show how effectively fire cleared land; the lone chimney, the charred ruins of a house and the skeletal remains of cattle and cartwheels point to its destructiveness.

ATL, C-140-005

1950s and underline the decades it took before the general public realised the importance of 'Keep[ing] New Zealand Green'.

Anthropogenic fire cannot be distinguished from natural fire. However, fire does seem to have been linked to the early Polynesian quest for moa, the now-extinct flightless birds, as an important source of protein and for the bones that were used for ornaments and for fashioning into fish-hooks, harpoon heads and other tools.[9] Moa-hunting or butchering sites have been found primarily in south Taranaki, on the eastern South Island plains, in Central Otago and, to a far lesser extent, in the Hawke's Bay,[10] where the land was covered by the more open and fertile lowland forests that the moa preferred. While the relative roles of long-term climate change and human agency have been extensively debated,[11] studies of pollen counts and carbon dating of charcoal (conducted since the 1950s and with renewed interest at the time of writing) suggest the frequency of fires in the areas where moa were hunted rose dramatically after the Polynesians arrived, that those fires were 'systematic and deliberate', and that they caused 'the majority of forest loss and erosion'.[12] Fire may have been used in hunting moa, though probably dogs, spears or snares were preferred.

In both archaic and classical periods the Polynesians, as their European counterparts would do, deliberately destroyed bush to prepare land for cultivation. In the archaic period, kumara, for instance, may have been grown in the warm micro-climates on Banks Peninsula, and some karaka stands indicate they may once have been planted to provide a valuable food source.[13] Bush may have been fired also to clear travel routes and to deny enemies cover around fortified or settled areas, or (after 1600 AD) to minimise the possibility of fire being used as a weapon against a besieged pa.[14] But whatever the reasons, about 750 years ago the Inland Kaikoura Range was repeatedly burned until it came to be dominated by bracken, tussock and scrub, and fires in Central Otago followed the same pattern.

The extent of the denuding of those landscapes is, in the face of subsequent European landscape modification and agricultural practices, now difficult to appreciate. But in 1844 an early settler who had climbed to the top of the heavily forested Banks Peninsula hills was arrested by what he saw: '[A]n immense plain, apparently perfectly level, stretched away below our feet, extending in a direct line westward at least thirty miles; and, to the southward, as far as the eye could see, backed by a far remote chain of grand snowy summits ... With the exception of one or two groves of black, formal pines of inconsiderable size, this immense plain seemed destitute of timber ...'[15]

In the centuries following their arrival the Maori population expanded. Moa, already vulnerable because of loss of habitat, were hunted to extinction by about 1400, and Maori moved from hunter-gatherer to fisher-gardener. In the 14th and 15th centuries about 80 per cent of the population was concentrated in Northland, Auckland, Bay of Plenty, coastal Taranaki and the North Island East Coast. There, kumara, the only crop that provided adequate yields, could be cultivated on a scale to help support a growing population. Reflecting this population shift, bush clearance was less pronounced in the central North Island, coastal Wellington and Nelson–Marlborough areas, where kumara growing was less reliable.[16]

Elsewhere, from around 1400, fires – like those lit by later Europeans – got out of control and repeatedly destroyed large areas of both lowland and upland beech, totara, matai and kahikatea, which were replaced by bracken and tussock. The central South Island was burned many times, as well as (again) the dry eastern areas of both islands. By 1600 all easily burned forest had been burned off. Loss of bush cover, as European settlers would find, had flow-on effects. The inland South Island basins suffered significant erosion; in the Hawke's Bay fires, along with drought, gales and storms, caused further erosion and loss of bush.[17]

The inland fires that Captain James Cook reported as he sailed the east

***Mount Egmont from the** north of Cooke's Strait, NZ, Natives burning off wood for potato grounds* by Charles Heaphy, 1842.

ATL, C-026-004-b

coast of New Zealand in 1769[18] burned regenerating rather than mature bush. But in the 19th century, having discovered that introduced crops and farm animals fared better in and on soil cleared of virgin bush than on regenerating scrubland,[19] Maori again began using fire for substantial bush clearance, as well as increasing their burning of regenerating vegetation. ' "E" Kuri and his followers', who in Wellington in December 1842 planned to fire an 'extensive tract' so they could plant potatoes,[20] were not atypical. Maori were as keen as any European settler to capitalise on the new markets and opportunities, and in the first half of the 19th century they grew potatoes on a large scale in most districts. Along with pork, these had largely replaced their previous staple diet and were used for trade and export. However, by the late 1860s, as settlers were becoming more self-sustaining and food was more generally available, Maori were badly affected by epidemics and poverty. In the Hawke's Bay (and possibly elsewhere) they were burning less.[21]

The swiddening used in clearing bush for kumara beds showed that Maori cleared the bush systematically, skilfully and effectively. Bush edges were slashed and burned, and kumara were then planted in the ashes. Two to three years later the land would be taken over by dense bracken, a fern that quickly colonised the land after fire and whose rhizomes (fernroot or aruhe) were an important source of carbohydrate for Maori. The kumara beds were temporarily abandoned as the regenerating sites were harvested for bracken, and new ground was burned to prepare new kumara beds.

A wash drawing by Julius von Haast of a fern garden on the slopes above Lake Rotoroa. Maori would have been unable to maintain this garden against regenerating forest without fire.

ATL, A-108-034

The cumulative effect was corrosive. Our man who had climbed the Port Hills noted only scanty remaining patches of bush encircled with 'withered and blackened stumps' and speculated on the impact of successive burning. Trees previously killed or half burned 'in the next summer's fires, readily burn, and dry up and kill those further in. In this manner, the forests are gradually encroached upon and wasted away … we find nothing but stunted fern or grass upon many hills and plains which … were at one time covered with primeval forest.'[22] The continual encroachments destroyed the species at the margins that acted as a buffer; the flammable regenerating scrub and fern that rapidly recolonised the cleared area, and wind drying out the undergrowth, added to the forests' susceptibility to fire.[23]

Moreover, Maori found, as would the European settlers,[24] that the combination of a dry season and high winds could easily spread fires far beyond what was originally intended. In 1840 one commentator reported 'extensive burnt areas' resulting from 'the natives firing indiscriminately any part of the land where they wish to commence planting and letting the fires run into the noble forests, where immense tracts of land are laid bare and timber of immense value wantonly destroyed!'[25] In 1842 William Swainson reported that a small fire that Poverty Bay Maori had started to drive out pigs had set alight an enormous amount of land that burned for three months; Swainson thought that their method of burning would inevitably end in fires well beyond their control.[26] Maori themselves reported in 1847 that fires lit accidentally or on purpose in the Waikato and Waipa counties annually destroyed the fern 'over large tracts of land'.[27]

However, evidence of the incidence and effects of fires tells us little about how Maori construed them or what they did about them. It may be possible to extrapolate some sort of response from the fact that they did evolve rules and customs over time to control access to, and exploitation of, treasures and raw resources that resulted in sustainable land management.[28] But Matt McGlone, an ecological scientist, urges caution:

> Traditional Maori society did practise resource conservation, but only within the context of a landscape they had already depleted and where food shortages were an unpleasantly common fact of life. We make a mistake if we project these attitudes back to a time when the population was small (but expanding) and the animal food resources both large and vulnerable.[29]

So what of the European burning? It should be noted, first, that fire was not the only way in which the European settlers destroyed the bush. In the 1820s a booming spar trade in Northland was already well established. In 1839 Ernst Dieffenbach described kauri logging, by then the mainstay of a major export trade to Australia, the US and Britain, as 'a melancholy scene of waste and destruction'.[30] That destruction would continue. The Kaipara Harbour was only one export port; there 50–100 ships were commonly moored, loading from great floating booms that held about 10,000 logs;[31] logs were also shipped from the Hokianga, Whangape, Herekino, Parengarenga, Kaitaia district, Whangaroa, the Bay of Islands, Whangarei, the Hauraki Gulf ports, Thames, Whitianga and Mercury Bay. When kauri became less accessible (the boom peaked just after the turn of the century) millers switched to kahikatea, and sawmill sites moved south. Yet from the early 1870s to 1910 logging was the major source of employment in the Auckland province. Kauri gum exports, which continued into the early 1950s, reached a zenith of 12,000 tonnes of gum annually about 1900, but after World War I declined with increasing competition from synthetic products and other resins. That L. D. Nathan employed several hundred people in Auckland to clean, grade and pack the gum indicates the scale of the industry at its peak.[32]

Axe and saw also laid waste to the bush. In 1859, 70 sawyers were operating in 'the great bush at Kaiapoi' from where Christchurch got much of its wood.[33] On Banks Peninsula 20 sawmills operated between 1860 and 1900, cutting tens of thousands of cubic feet of the dominant totara and denuding the hills.[34] The mills themselves were not immune from fire. Newspaper reports of sawmills threatened or destroyed by fire were commonplace into the 1920s. But the

A timber milling company in Mangaro Valley, c. 1910. Note the amount of flammable slash left lying about.

NL, 3911½ B/W

clearance of the lower North Island between the 1870s and first decade of the 20th century was 'the biggest such clearance in the country's history'. East of the Tararua and Ruahine ranges, the Seventy Mile Bush (the piece within the Wellington province was the Forty Mile Bush) was cut into from north and south. To the west, sawyers set up successive railheads that penetrated inland and north as far as the country around Ruapehu. Between 1905 and 1907 the lower North Island was, like the northern kauri forests, one of the major saw-milling areas.[35]

Arresting as those facts and figures are, fire remained the major agent of the change, accounting for probably 90 per cent of the European clearance.[36] Until well into the 20th century settlers (a term used into the 1950s) turned to fire as the only way of removing the bush that they saw solely as an obstacle to settlement, to clear land for farming and lay down the ash beds that promoted grass growth. Yet controlled burn-offs too easily, and too often, became uncontrolled fires that wrought havoc. Logs from burning-off operations were left to smoulder; sparks from logging machinery caught and burned, especially in uncleared slash; locomotives showered sparks over the countryside (between 1938 and 1941 sparks from trains caused just over 40 per cent of the fires in state forests[37]).

Well into the next century, and as the countryside was made over, populated and picnicked in, fires casually lit by Public Works Department road gangs, travellers, picnickers and hunters were negligently left alight. Fanned

A milling line at Echo Valley near the Wanganui River bridge at Taumarunui, showing a bush scene with an Aveling Porter locomotive, a steam log hauler and logs in the foreground (1910–1920).

ATL, PAColl-5521-16

by prevailing nor'westers, inoffensive burns became massive conflagrations extinguishable only by a wind change, or by rain. A rubbish fire leapt from stump to stump over the grass paddocks, 'getting a fresh start at every fence it came to; and in a very short time the whole hill face ... seemed to be one sheet

A stationary steam hauler, c. 1900. The threat the industry presented to the bush is obvious.

NL, 29016½

of fire', the *Otago Witness* reported in 1872.[38] The speed seemed a constant surprise: 'One moment a settler would be gazing perhaps somewhat indifferently at a fire raging in the distance. An hour or two later that settler would be fighting desperately to save his own house from the same fire.'[39] Blackened and desolate hillsides, miles of smouldering and charred trunks and logs, and days, even weeks, of great palls of blinding, choking smoke were integral to many settlers' experiences. Newspaper reports, sometimes penned within the fire areas themselves, and sometimes held up for days because of communication delays, further drove the message home. By the 1880s and 1890s most small towns had one, if not two, papers. Well into the 20th century these headlined local fires and syndicated accounts from elsewhere in New Zealand, and from Australia and North America. Given the different North American fire season, fire reports from there ensured a constant fare, providing copy as the Australian and New Zealand accounts went into winter abeyance.

Written records and images provide a clear and dramatic account of New Zealand's European-lit fires and their impact. Newspaper reports of individual blazes began from the early 1840s and the number of reports and the size of the fires they described increased steadily over time. From early in the 20th century the Department of Lands' annual reports reflect the same trend. In the numerous dry seasons the fires burned over huge tracts of country, with devastating effect. Homes, fences, sawn timber, and mills and their wooden railway lines were destroyed. For the many small, uninsured landholders, such losses could mean ruin; those insured were lucky to get back half of the value of what had been destroyed. Nor was the damage limited to the impact on individuals. The loss of a mill (flax as well as timber) affected local employment opportunities; the loss of hay and valuable pasturage forced dairy companies to close; and the timber critical for the colony's homes and buildings, roofing shingles, bridges, culverts, fences and firewood was gone.

Yet it took a surprisingly long time before any systematic approach to preventing these fires began to emerge. The reasons lie primarily in the settler (and government) mindset. That this mindset persisted, despite the damage and loss the fires wrought, is an indication of how strong it was, how deeply embedded in the settler psyche. We turn now to the conflagrations themselves, so we can examine, in Chapter Two, the experience of those fires in the context of settler attitudes to the burning of New Zealand.

From its early days the accidental fires that so marked European settlement meant destruction and loss. As early as 1836 around Tutukaka, 'in a short time the whole forest for miles to leeward was in an awful flame'.[40] The pattern was set – the 'frightful simultaneousness' and the profligate waste and destruction that reduced 'hundreds of acres of our precious timber ... into masses of cinders' would become commonplace.[41]

Not only trees were destroyed. In the South Island from the 1840s the already burnt tussock was further burned as the pastoralists, like their Australian counterparts, sought to boost grass growth or clear stock routes (the fires lit to get a route through from Jollies Pass to the Hanmer plains burned for days[42]). The fibrous litter shed by these tough, wiry grasses remained raw, instead of decaying into humus. In dry seasons this material became particularly flammable, and accidental fires, whether from the burns themselves, from shepherds signalling to each other, or from the fires used in the musters when the going got difficult, were commonplace.[43] In damper districts, bracken and manuka, not tussock, were the predominant growth. Wherever the vegetation became too dense, settlers simply burned 'their way into and through the country'.[44]

Especially in dry years, 'destructive' or 'calamitous' fires increasingly became the standard fare of any newspaper, and it is from these reports that I have largely built up the following accounts.[45] But the often sweeping generalisations and the discrepancies between estimates of the value of property lost suggest we approach the reports sceptically.[46] To some extent papers were seeking headlines and sensation (though sometimes they may have been wanting to elicit charitable responses to calls for relief funds). Nor is it easy for today's researchers to ensure overall accuracy, as the papers themselves recognised, often prefacing their accounts with a cautionary 'It is reported ...' Reporters were compiling minute-by-minute reports as they came to hand from correspondents on the ground and, as the *Taranaki Herald* observed in 1908, '[i]t is difficult to get authentic information from the bush, as every available man is either watching or extinguishing fire'.[47] Men and women fighting to save their property must have had little time or inclination to stop and talk to reporters. Smoke, understandably, could deter the most zealous investigative journalist, and by the time it had cleared properly, and the true stocktake of damage was under way, the correspondents had moved to new topics or new fires.

We also need to understand the headlines from within the context in which they were written. 'Large numbers of settlers burned out' proclaimed the *Evening Post* in January 1898 about a fire in Pahiatua. 'People abandon their houses and flock into town.' Yet apparently only four Pahiatua houses were lost, and three or four other houses in nearby settlements were burned.[48] But even if settlement was sparse, even if 'only' pasture was lost, that loss was someone's livelihood and represented extraordinarily hard work of which they might have felt justifiably proud.

Finally, and perhaps most importantly, even the worst fires cannot be compared with the great, uncontrollable North American firestorms of the time, which swept in massive fronts across thousands of kilometres, annihilating towns and people. The New Zealand fires certainly were terrifying and destructive: they changed the face of the country forever; they jeopardised lives and livelihoods. Fanned by wind, some obviously travelled considerable distances. But in many cases the headlines referring to 'miles of fire' appear, on closer reading, to have been describing not a united front of flame but many localised fires of varying sizes in many often contiguous areas. In the terrible 1908 fires, for instance, the Department of Lands reported in relation to state and Crown land that only 810 hectares of good milling timber in state forests had been lost in the Nelson district; 668 hectares of bush was burned on Crown land in Marlborough; there were no losses in Westland, Canterbury or Southland; and those in Otago had been exaggerated.[49] (Crown losses in Auckland and Wellington were not given.) On the other hand, the attention-grabbing headlines, the vivid accounts of ash 'showering the country like a snow storm', the sun obscured, and a tree blazing 'from top to bottom like a gigantic gas jet' cannot be dismissed.[50] The cumulative total area of forest cover destroyed by fire was enormous. Moreover, the newspapers' hyperbole represents an emotional reality that we have to consider when discussing the settlers' response to fires.

It is clear that the number of fires and their seriousness grew as settlement penetrated into the country and settlers had more to lose. Only three newspapers carried reports of fires in 1859. By January 1863 many papers were reporting frequent fires in the Marlborough district, 'many serious fires' around Mangapai, south of Whangarei, and bush fires raging on the Coromandel gold fields. In Hawke's Bay in early February 'very destructive' fires across the area burned 10,000 posts and 6100 metres of sawn timber and badly damaged the Homewood bush; later that month, in unprecedented heat and dryness, more of the country was afflicted. '[F]earful fires' burned around Masterton and Greytown; fires damaged the big runs in the Nelson area; around Raglan smoke from the fires clouded the sun and the flames were contained only by the swamps and water; choking palls of smoke from numerous fires hung over Russell; and more bush on Banks Peninsula went up in flames.

The following condensed accounts of the fires in the following years show the toll continuing. The 1864 fires on the outskirts of Invercargill, at Riverton and Waikiwi bush, burned large acreages of valuable land and hundreds of yards of wooden tramways – a foretaste of the widespread destruction of bush, buildings, implements, machinery and produce in the future. For two weeks in 1868 'an enormous and destructive conflagration' burned in the bush on the outskirts of Auckland.[51] In 1869 fires (already construed as an inevitable consequence of hot, dry weather) burned parts of the Tararua Range. Banks Peninsula continued to burn – as many as 10,125 hectares were lost in just one fire that year.[52] Further south, settlers in the Clutha, Tuapeka, Waitahuna, Pomahaka and Wyndham districts, burning to produce the desired grasslands, gave little thought to forest protection.[53] In January 1872 big bush fires destroyed a tramway, diggers' huts, stores and other property in the Coromandel, forcing miners to temporarily suspend their work.

Fires all over the Wairarapa were forcing drivers to leave wagons laden with wool in the Waingawa River for safety. Then two boys fooling with fire managed to light 'one of the most devastating fires' in the district. Much of Carterton burned and many small farmers in the district, 'who after the toil of years, [had] managed to create comfortable homesteads', were ruined.[54] Hamilton's sweltering heat in February 1872 was increased by swamp fires, one of which burned for days, destroying the settlers' only wood supply. In December a 'furious' bush fire on Mt Cargill, about 9.5 kilometres from Dunedin, burned for days and destroyed cedar. Other fires, by no means the last, were burning on the peninsula around Saywers Bay and Portobello. In 1873 'wilful and culpable fires' destroyed 8100–12,150 hectares in Canterbury (a loss calculated at £70,000 or almost $8 million in today's money).[55] In January 1876 flames spread over the whole range behind Reefton and burned for about six days, destroying milling companies' tramway shoots and paddocks valued at about £1200 and making 'respiration the reverse of pleasant'.[56] In March bush fires burned strongly for two to three days around Wellington. As late as May that year, when the fire danger would normally have abated, two large fires in Oxford Bush destroyed four sawmills, several houses and railway sleepers before rain came to the settlers' rescue. In 1881 'the entire range of hills in Makara district round to Tinakori Road [was] aglow with flames' and there were fears the fires would continue unchecked until they reached the sea at Island Bay.[57] This would not be the last fire on Wellington's doorstep.

Wet years meant fewer fires. Moreover, drought (and the fires that seemed inevitably to accompany it) in one district did not necessarily mean drought and fires in another – the vagaries of New Zealand's climate caused by its 'long latitudinal range … mountainous terrain, and variations in distance from the sea' saw to that.[58] But the country continued to burn: in 1898 the failure of Francis Guy's family to notice a fire approaching their home until they actually saw the flames[59] suggests the extent to which the smoke and smell of fire were endemic over the summer months. However, from the early 1880s to the great Raetihi fire of 1918 newspaper reports of bush fires indicate that the fires occurred disproportionately in the lower North Island.

The origin of that geographical focus for vast conflagrations lies in the ideologies of land use held by the settlers and their political representatives.

These ideologies were reflected in the economic and development policies originated by the Fox government in 1869–70 and carried on by subsequent Liberal governments. The understandable immigrant urge to maximise the use of such promising land was reinforced by what many perceived as a biblical injunction to use the land well – to leave large areas undeveloped or in the hands of those who were either unable or unwilling to use it was seen as morally wrong. Few wanted large holdings where only a few shepherds worked, or bush inhabited by remnants of Maori. The aim was to own heavily cultivated and closely settled land. People and fields were equated with order and stability, progress and civilisation.[60]

Such beliefs underlay the drive for economic development. This priority, Colonial Treasurer Julius Vogel believed, was critical to the colony's future. The country's most valuable resource was its land, but without more capital and labour, and better communications and transport, colonisation would be slow. Moreover, there was mounting concern about Maori guerrilla action led by Te Kooti and Titokowaru in Taranaki and Hawke's Bay. The aggressive campaigns the Stafford government had pursued against the Maori had been costly and ineffective, and the British government was preparing to withdraw its troops. The new Fox government (including Vogel) voted in June 1869 to demilitarise the North Island frontier and push forward with colonial development.

Vogel's ambitious public works scheme of 1870, designed to increase the population and transform communications, answered economic and military concerns. The imperial government finally agreed (as a sweetener to its withdrawal of its troops) to inject £1 million to be spent over 10 years on immigration and public works. The immigration scheme, very broadly conceived to attract a range of nationalities, was 'to "swamp" the North Island with … settlers who would outnumber Maori and help nullify the risk of rebellion against the Government'.[61] It would also bring in the labour to build the roads and railways and break in the farms. Moreover, roads would allow troops to pacify potential trouble spots, and, as Judith Binney nicely puts it, 'it was assumed (for quite obscure reasons) that Maori employment on building these roads would teach them the "moral benefits" of European living styles and end their "restless ways" '.[62] The work and the presence of men were, Vogel would claim some 20 years later, 'the sole alternative to a war of extermination with the natives',[63] but their value lay also in 'penetration and subjugation'.[64] '[T]he spade and the pickaxe', the 'true weapons of conquest' used in pacifying the Scottish Highlands, had a further clear strategic intent. Roads would lead to leases and land sales.[65]

Nearly 100,000 people were encouraged to leave Britain for New Zealand during the 1870s, and immigrants also arrived from other nations, particularly Norway and Denmark. With much of Canterbury, Nelson and Marlborough

***Tussock Burning*, 1882.** Camped near the Tasman Glacier on his way to attempt the first ascent of Aoraki/Mt Cook, William Spotswood Green painted the hillsides being burnt in a muster.

ATL, A-263-011

and some of the Wairarapa and Hawke's Bay already occupied by extensive sheep runs, many of these settlers went to the new bush frontiers of Woodville, Feilding and Inglewood.[66] There they could own land, independent of the social hierarchies of squire or landlord and, by living off the land, return to the perceived organic relationship with nature and community that the industrial revolution had disrupted.[67] By the mid-1870s the great lower North Island bush clearance was under way.

The Rangitikei–Manawatu block was opened in the early 1870s. Over the same period Scandinavian settlers carved Dannevirke, Ormondville and Norsewood out of the Seventy Mile Bush and mud, and Small Farm Associations fostered further development. In Taranaki, settlement of the interior began with the little townships of Inglewood, Midhurst and Stratford. The King Country became accessible to the European settlers in 1883. Road and rail both occasioned and accompanied settlement. In the Manawatu, for instance, a sawyer's tramway (The Skunk or Coffee Pot), pushing up to the navigable head of the Manawatu River, preceded the establishment of Palmerston North in the early 1870s, in the Papa-i-oea clearing. A survey line – 'a sort of narrow channel cut through the encompassing bush [gave] egress to the earlier and comparatively civilized settlements of Marton and Wanganui' and '[o]ver and round [stumps sticking up in the middle of the line] the coach bumped and pitched, like a ship in a storm'. But within less than a decade rail connected Palmerston North

Burning Bush, Taranaki, by William Stutt, 1856.

ATL: E-453-f-008

An advertisement from the Arthur Yates Seed Company showing a man sowing seed on bush country in the Hokianga. By the 1920s a time-honoured recipe had established the grass that clothed most of the country: 20lb certified perennial ryegrass, 5–8lb Italian ryegrass, 5lb cocksfoot, 2lb white and 1lb red clover, 2lbs crested dog's tail, 2lb Timothy, and lb turnip seed to encourage future animals to fossick among logs.

ATL, Eph-A-Horticulture-Yates-1926-cover

to Wanganui and Foxton. In the 1880s the Manawatu–Wellington line went in;[68] the North Island main-trunk line, a wonder of engineering that linked the district to ports, towns and settlements, was finished in 1908. The opportunities such development offered were almost unimaginable. New railways, along with the 'roads, telegraph and the State' that penetrated the bush, made previously inaccessible forest into potential farmland. Railways were 'the forces of civilisation ... A railway pierces the forest: Roads branch out and penetrate every district ... The railway is ever the greatest pioneer of all. Without it no town can ever develop beyond a certain primitive state ... The railway ... is the way of salvation.'[69]

Other Liberal government policies also contributed to settlement. From the late 1870s the Land Act provided that settlers who cleared the bush, fenced and sowed and built permanent homes could lease or buy the land on favourable terms. Given further impetus from 1891 by the Liberal government's emphasis on settlement (under Minister of Lands John Mackenzie),[70] this scheme, between 1895 and 1914, saw about 648,200 hectares of Crown land leased or sold, of which about 405,000 hectares was in Wellington and Hawke's Bay. There were defaulters.[71] Such assistance, too, encouraged a new mindset about bush country. At a quarter to a half of the price of open land, there was a good reason to look at it more favourably. Breaking it in was extremely hard work but relatively inexpensive – the cost of an axe, bill-hook and grass seed needed for the first crop compared very favourably with that of ploughs, harrows and horses required to break in open country. Moreover, with the bush providing firewood, timber for fences and buildings, and wild pig, deer, native pigeons and even pheasants, families could at least subsist.[72]

There was a symbiotic relationship between farmers and sawmillers and improvements in transport and communications. Sawmilling provided labour opportunities and land clearance; rail and road carried the produce in and out; sawmilling communities spelled markets for horse fodder, dairy produce, vegetables and so on. Later, investment in timber companies offered new opportunities to farmers, who were often the major group of shareholders. Landless men, too, could get by comfortably by cutting roofing shingles, firewood, posts and railway sleepers, even though they had to pay royalties to millers. By the mid-1880s timber and firewood made up 58 per cent of the goods traffic on the Hawke's Bay line, representing a massive southwards shift in the sourcing of these supplies.[73]

The Europeans' recipe for clearing the bush for farmland was simple. 'The best way of cultivating is simply to fell all the timber and run a fire through it when dry, sowing wheat and grass upon the ashes without breaking up the soil or using any means to cover the seed.'[74] These words, however, do not convey the reality of the exhausting and unremitting labour that bush clearance entailed. Initially settlers availed themselves of skilled Maori to assist – at 12 shillings a day ($130 in today's value) for the first clearance, and £2 10s[75] an acre for the second – 'one shilling spent before burning saves five shillings afterwards'.[76] By the late 1860s, as the immigrants themselves developed the necessary skills, that pattern changed. Women and children helped, and young teenagers did men's work. Creepers and scrub were slashed back, then left to dry to open up the bush and provide the dry tinder for the burn. From October trees up to a metre in diameter were felled (bigger ones were left for the fire to deal with) and the air would be filled with the aroma of drying timber.

Then in late summer or early autumn the burns were lit with makeshift kerosene torches or flaming branches. Designed to destroy huge areas of bush, these intentional burns had all the ferocity of a wildfire. Roaring flames leapt from ridge to ridge, and burning leaves and bark were thrown ahead of the fiery red glow of the main fire to start new fires. Seemingly within minutes, rolling black clouds of smoke rising 100 metres and shot with flames would spread across the countryside. Valleys were filled with smoke, trees limbed with fire, and on the ridges a few remaining stumps stood like hairs. At night a glow of continuously changing lights, as if from a town, was visible for kilometres. Men inspecting the burn walked on what logs they could find to avoid traipsing among deep ash so hot that it could singe their trouser legs; they ended the day racked with thirst and blackened by ash. Up to three metres deep, the ash was still warm when wheat or grass was sown. Stumping (sometimes requiring horses) and logging up were done over subsequent years, but stumps could remain half buried, hindering easy cultivation for decades.[77]

Burning had its regional variations. In Taranaki and the central North Island large trees (generally rata) were generally left uncut and, denuded of their foliage, stood guard over bleak, blackened landscapes. Standing timber, too, might be left – higher rainfall than in the Far North, for instance, meant that felled trees could become waterlogged and difficult to burn. In Taranaki burning was delayed to accommodate the very common mahoe, which tended to put out second growth in the area's heavy rainfall. Weather variations within any season also affected practice. Although a wet January might mean a February burn it could also endanger your neighbour's grass, sown conventionally in January. But the techniques of the actual burn were generally agreed. You waited for a day with a steady breeze in the right direction to get a good wide blaze across the section and, when the dew lifted, you set the fires alight. Simple – but a shift in the wind's strength or direction could see things going disastrously wrong.[78]

From the mid-1880s highly skilled contract gangs, which had replaced much of the family labour, used well-honed techniques to fell huge areas.

The construction of shacks with wooden walls and canvas roofs was commonplace in settler society and contributed to fire risk.

NL, 1/1-017831-G

Skilled men (and the boys with them) swinging 1.8-kilogram axes day after day from before daylight to dusk, could each clear 1.6 hectares a week.[79] The very fact of bush clearance and settlers' penetration helped create conditions inimical to the bush. '[T]he cathedral gloom and the damp solitude in which flourished the palm-like nikau and the stately tree-fern are penetrated by the burning sun, and invaded by fierce and parching winds'; the bush was 'denuded of its protection'; trees died and 'each step in the chain of destruction prepar[ed] the way for the next', lamented a contributor in a paper to the New Zealand Institute. Deer, by ringbarking particularly tawa, mahoe, ngaio and whau (a broad-leaved shrub, which as early as 1892 was almost extinct), aggravated the process.[80] Farmers and logging, roading and rail gangs left slash lying partly burnt or unburnt. Wood from one wet season might remain where it fell. A careless match or sparks from a locomotive or a burn could ignite a blaze that would turn to 'fires on the hills to the north, south, east and west – not a log here and a stump there, but hundreds and thousands of fires' that burned with a terrifying rapidity. Almost no year passed without a significant fire being reported somewhere in the country. Frequent dry seasons exacerbated the ever-present threat. There always seemed to be localised burns, but the dry summers of 1879, 1881, 1885, 1886, 1897, 1898 and 1907 saw fires whose scale and destructive force were immense.

The start of a 12,141-hectare fire lit to clear the bush, Pukatora Station, East Coast, 1900. Trees destroyed by an earlier fire lie in the foreground.

NL, 1/2-032845-F

Then came the fires of 1908. In the North Island they lit the pyre that the logging and the settlement policies of the past four decades had built up, and burned huge areas. In the South Island flames spread across parts of Nelson and Canterbury and fires flared in Southland. Even the Department of Lands' annual report fell back on hyperbole and metaphor to describe them. It is to these fires that we now turn.

The summer of 1906–07 had been marked by raging fires in parts of Canterbury and in Southland.[81] By December 1907 settlers would have known they were in for another bad summer. Across the country there had been effectively no rain since 10 October (and there would not be until March).[82] In January 1908 the fires started. In a whole column devoted to fires around the country, the *Hawera & Normanby Star* reported on 11 January that, probably because of a match thrown carelessly among dead manuka, the Waikaia bush (northwest of Gore) had been burning for a week. Unless rain came, its fine timber would be completely destroyed. Fire had cut off five bushmen working in Harringtons Reserve out of Longwood; in the North Island grass fires were sweeping the parched country around the Tararua ranges in the Mt Holdsworth area, and bush and grass fires out of Hastwell had already destroyed a creamery, whare, outhouses and kilometres of fencing, as well as sheep. Eight days later

'The burning-off season in New Zealand: A big bush fire in progress.' 'During the last month,' the *Auckland Weekly News* noted, 'almost the whole of New Zealand has been enveloped in the smoke of innumerable bush fires. In several places the fires spread out into open country, destroying fences and stock, and in one or two cases burning down homesteads and factory buildings. Seen from a little distance the huge column of smoke rising slowly in the air forms a most impressive sight.' Given what was happening, this seems a curiously dispassionate caption.

Auckland Weekly News, 23 January 1908, p. 11

the *Evening Post* reported that welcome rain had checked worrying fires near Greytown, Carterton and Masterton.

It was a temporary respite. Soon new reports were flooding in. By mid-January, with the drought unabated, serious fires had already burned for days around Eketahuna and at Pongaroa, where they had already travelled a long way. 'Sixty miles of fire', shouted the *Evening Post*. '[T]he whole country, except at rare intervals, from Pahiatua to Upper Hutt, appeared to be on fire.' Travellers on the Napier–Wellington line could watch the flames racing the train up old burns on the Rimutaka Incline, observe an extensive fire burning to the west, north of Kaitoke, and note the plainly visible track of another fire that had been 'eating' its way through standing bush around Mungaroa and Upper Hutt.[83] Fire relief parties were battling 'a mass of flames' around Rongokokaka, Nireaha and as far north as Mangamahoe, south of Taihape.

Other districts were also in trouble. In the South Island serious bush and grass fires raged in the Nelson province. Sixteen kilometres of bush were lost at Tadmore, and outbursts of fire in the Sherry, Wangapeka and Motupiko districts destroyed thousands of sheep and long stretches of fencing. In the North Island the smoke from fires raging across the northern districts prevented ships from entering Auckland Harbour, while smoke some 225 kilometres off Cape Maria van Dieman forced another ship to slow its progress down the coast. Further south, country around Raetihi and Ohakune was alight and a huge

'The bane of the New Zealand pioneer settler: Bush fire in progress.' Was this curious image painted in a studio, for photographic reproduction?

Auckland Weekly News, 13 February 1908

fire was burning in the Manawatu Gorge. In Hawke's Bay, just east of Dannevirke, Waipawa was enveloped in dense smoke, Dannevirke itself was threatened, heavy fires were said to be raging in the Norsewood district and Hastings was enveloped in 'stifling smoke'. A possible wind change was the only thing that might save a large portion of the Heretaunga Plains.[84] In Taranaki a strong southerly was carrying fires throughout the district, burning bush, houses, grass, fencing and tramways. New Plymouth itself, in stifling heat, was in semi-darkness all afternoon on 18 January, and the harbour board installed a fog signal at the port – over the past week ships had had difficulty picking the port because of fire-induced fog and heat haze.[85] Four days later the worst was over for the time being, though hot spots continued to ignite.

For much of the Wairarapa that respite was brief. An *Evening Post* extra on 23 January carried news that volunteers, settlers and light rain had largely checked the hill fire that threatened Carterton, and by 24 January the fires around Dannevirke had abated. But fires were still spreading near Eketahuna, threatening houses without water. Without a wind change where would the flames stop? Not until 29 January did lack of smoke beyond Masterton suggest the fires had burned themselves out – and there had been no wind, an absolute godsend.

Elsewhere, each report brought news of fresh fires. In the Nelson province previously unreported fires were sweeping on unchecked. Baigent's mill in Pigeon Valley (above Totaranui) was lost and many men would be out of employment; on land behind the Upper Moutere the smoke was so dense that mustering was impossible and great stock losses were expected; a great pall of smoke hung over the Blenheim district as the Rai Valley and the Marlborough

'The recent devastating bush fires: View in a beautiful bush reserve at Tangoio, Hawke's Bay, which has been completely destroyed.' In this image the particular represents the general.

Auckland Weekly News, 5 March 1908, p. 15

Sounds burned. Nor was the Auckland province spared. Fire in the Rangihau and Kapowai blocks in the Coromandel had been raging for a week, with standing timber and thousands of logs damaged or destroyed. The timber in the dams and the workmen's shanties were lost – without rain, more valuable timber reserves would be threatened.

There was a brief lull – almost like the eye of a storm – then on 12 February reports of new fires in the Wairarapa, between Eketahuna and Pahiatua, gave the first inkling that the fire danger was far from over. Five days later vast areas around Waimarino (just northeast of Turangi) had been swept by fire.

'Recent disastrous bush fires at Kawhia: Sheep smothered by smoke on Mr. Elliot's farm at Oparau.' The paper went on to note the 'great' loss of stock throughout the country. In Elliot's case, from one paddock only 12 sheep out of a flock of 800 were saved.

Auckland Weekly News, 5 March 1908, p. 15

That was still burning and would be fanned into new destruction a few days later. The area around Ohakune and Horopito was ablaze and the road was almost impassable, Raetihi was surrounded by fire, fires were attacking fresh blocks at Mataroa, and in the Ohura valley (south of Taumarunui) settlers were battling a big blaze. By 17 February 'rather bad' fires around Whangamomona (east of Stratford) had created a 'thorough mantle of smoke over the whole coast';[86] fire was practically 'everywhere between the Rangitikei and Kiwitea rivers'. With the 'district in flames', the Kiwitea County Council sat for only two hours so the men could get back to protect their properties. Further south, around Palmerston North, fires were raging in Pohangina and Apiti districts (in the Ruahine foothills) and would still be burning vigorously five days later. The Manawatu Gorge hills were ablaze for kilometres along the tops. Auckland was so enveloped in smoke from bush fires that shipping was delayed, and Raglan was 'completely hidden by smoke' from fires in open grass country farther up the harbour. A couple of days later, around Waingaro in the Far North, only embers remained over many kilometres of bush country; around Dargaville kilometres of the peaty soil (up to three metres deep) of the swamps were burning, while further south the Waiuki district was 'barren of every vestage [*sic*] of green except the swamp'.[87] By 19 February there were bush fires in all directions outside Dargaville and reports of disastrous fires in the Kawhia district.

The southern sections of the Wairarapa were again suffering. Masterton was so enveloped with smoke from fires around Matahiwi and Mikimiki that 'the ranges were entirely obscured, and even close objects of magnitude were

'All that was left of a sawmill recently erected by Mr. H. Armstrong of Oparau, Kawhia.'

Auckland Weekly News, 5 March 1908, p. 15

undiscernible'. Meanwhile, for days (if not weeks) the smoke around Wellington was at times 'so thick that navigation was rendered difficult' and shipping was delayed.[88] Fires in Khandallah's tinder-dry bush threatened homes. Much of Wilton Bush, the only stretch of bush remaining within walking distance of the city, was destroyed – an 'irreparable' loss.[89] The last remnants of bush on the east harbour hills were alight, and though by 22 February the *Evening Post* reported that strenuous fire-fighting had saved 'the magnificent stretch of native bush' behind Days Bay, fires continued to burn to the east, the York and Lowry Bay bush was largely destroyed and the scenic Korokoro bush had been devastated. Continuing strong winds meant the fire danger was not over.

The foothills and plains of Horowhenua were also in trouble by 19 February. At Waikanae mill hands stopped work to help save the town and the swamps; the smoke at Otaki was so dense that 'all views were obscured'.[90] The fires destroyed fences and grass as they moved south towards Paraparaumu, and Paekakariki was full of smoke. Further north, around Shannon, railway sleepers were burned, men were guarding the line, and trucks were removed from the danger zone. Two days later two outbreaks had been contained but the Tokomaru Swamp was alight. Palmerston North and Feilding were enveloped in smoke. Parts of the Taranaki district were again 'full of fire and smoke' as fires continued to burn until 22 February, when the likelihood of rain brought hope for an end.

Areas of the east coast were ablaze. Fires in the Mamaku Ranges swept 'enormous' areas bare; the Rotorua district was in an 'impenetrable haze' of smoke; country near Katikati burned. Fires on Great Barrier Island and Coromandel Peninsula, at Tairua and in Mercury Bay were dampened (briefly) by rain. In the Waikato a number of small fires were burning around Paeroa, the Piako Swamp was on fire and the Maungakawa bush was aflame, threatening the sanatorium just out of Cambridge. 'The smoke pall is resting over practically the whole [Auckland] province, and ships [were] meeting it many hundred miles out to sea.'[91] 'From Hokianga to Gisborne settlers in bush areas are in trouble', reported the *Otago Witness* on 26 February. The entire page it devoted to the reports of fire across the country suggests this was, if anything, an understatement. Moreover, the fires were not simply a rural phenomenon – the charring, the cinders, the smoke and the smell brought the danger to the towns.

Nor was the South Island immune. There were brief reports of further damage behind Nelson. In Reefton fires swept down both sides of the Inangahua Gorge, destroying the town's wooden fluming used for its water supply and creating a severe water shortage. Thick smoke clouds out of Invercargill indicated fires in the northwest.

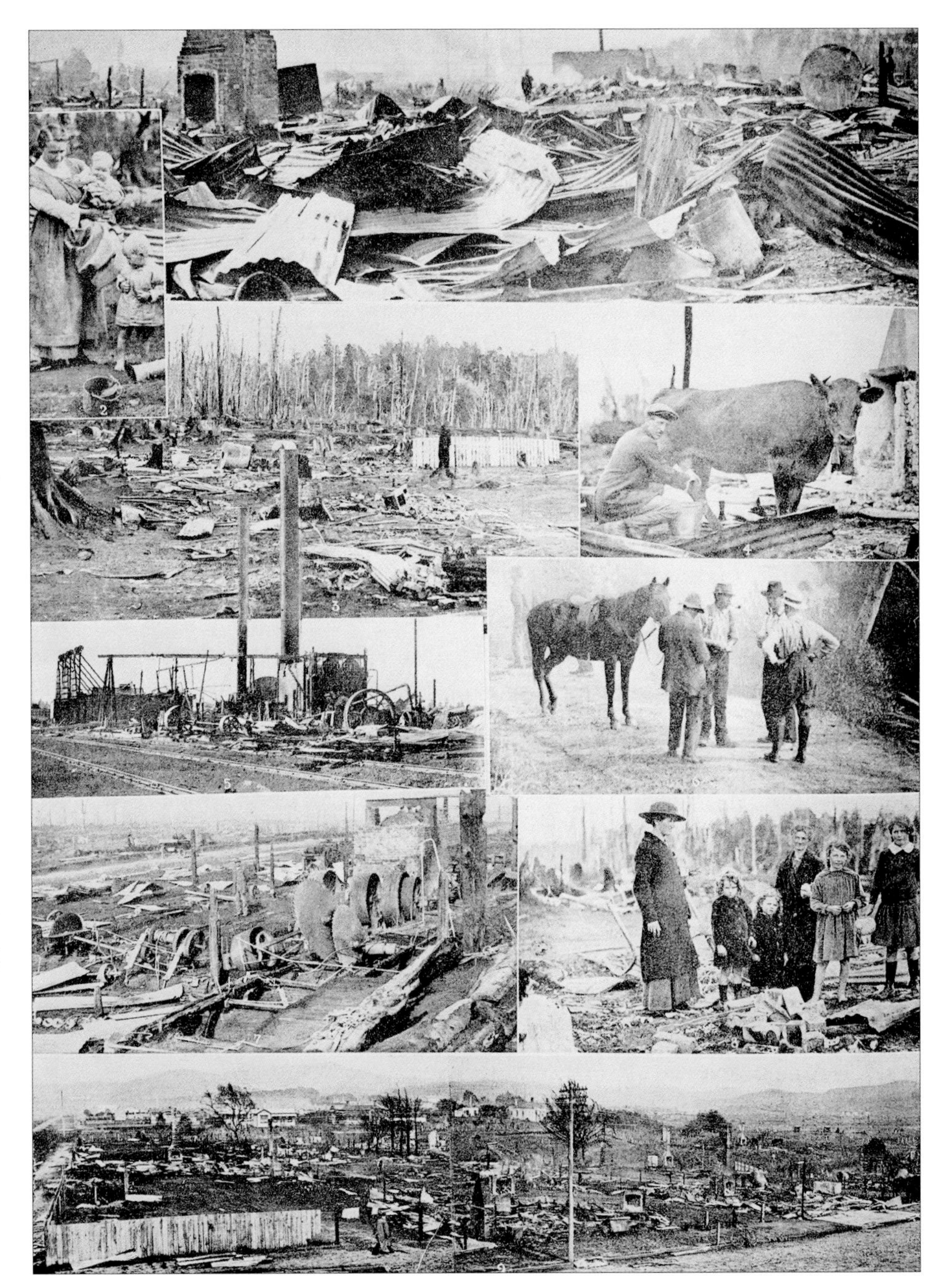

'The King Country disaster: Scenes of destruction in the fire-swept area. The magnitude and heartrending nature of the disaster … cannot be realised by those who have not passed through the area. Mile after mile of open bush country has been swept from end to end, some of the bush having been razed to the ground so effectively as to leave scarcely a stump.' This page of pictures gives some indication of the financial (and emotional) impact of the fires.

Auckland Weekly News, 25 March 1918

Then suddenly, barring a few flare-ups, it was over. It rained in Taranaki; the fires were still smouldering, but unless there was more wind no further trouble was expected. Rain was falling, too, at Ohakune, Mataroa and Taihape. The Piako Swamp fires were out, the Manawatu fires were abating, there was rain (though more was wanted) in Wellington and Otaki and around Feilding. The drought was breaking and the settlers could begin to count their losses.

CHAPTER TWO

A good burn? Changing attitudes from colonisation to 1921

The costs of the 1908 fires, whether to the country as a whole, or to individuals, were substantial. The burnt-out small settlers, particularly in Taranaki and the central North Island, must have watched months of labour burning before their eyes, knowing too that they had little to fall back on. Yet for a long time almost nothing was done to prevent fires from breaking out or to tackle them when they did. Why? As we shall see in Chapter Five, the practical, basic problems of isolated and fragmented settlement, cumbersome and primitive fire-fighting equipment, horse-drawn (at best) conveyances and poor roads all hindered or prevented any easy, effective response. But in terms of root causes those factors are in some ways mere distractions; history would prove such obstacles could be at least partially overcome. One of the critical factors was the settler mentality evinced in the notion of 'a good burn'. Place that concept against another much-evoked concept – 'waste lands' – and the scene is set. Fires, however disastrous at a local or even national level, satisfied a government imperative by achieving several years' work overnight: 'lands ... lying idle' could be sown in grass. In that context, the fires were seen as a 'blessing in disguise ... one of the best things that could have happened to the settlers'.[1]

In the colony and then in the Dominion, fire as a necessary corollary to settlement and civilisation was a long-held and persistent concept. Yet before the end of the 19th century there were suggestions that this attitude might change. The destruction of what was increasingly seen as a highly valuable natural resource was being assessed amid fears of an international timber shortage, and evidence of the downstream effects of deforestation were being drawn to the attention of parliamentarians and the general public alike. However much biblical injunctions to 'settle the land' might excuse the burning, the injunction to 'waste not, want not' was also strong. Aesthetic issues, too, were being raised more vociferously as growing numbers of urban commentators and others deplored the blackened hillsides, the smouldering trunks and the destruction of generations of growth. In addition, the government was developing increasingly valuable exotic forests. The need to respond to all these issues meant that rural fire management began to be addressed.

Let us go back to the 1908 conflagrations to assess the costs to the country, to companies and to individuals. Although for reasons already discussed the reported losses need to be approached sceptically, we can begin to construct a picture of the damage. Critical infrastructure had been damaged when telephone lines and railway bridges and sleepers were burned and when trees fell across railway lines. The value (to both the country and individuals) of the vast areas of bush and timber reserves was irreparably wiped out. Private sawmilling and logging companies lost mills, equipment and miles of tramways.[2] Destruction in the Coromandel, where thousands of logs of standing timber were said to have been damaged or destroyed, where timber in the kauri dams was burned and workmen's shanties were lost, was not unique. The Fairburn Timber Company at Pohui in the Wairarapa was said to have had £3000 (about $500,000 today) worth of stacked timber destroyed; at Warea in Taranaki a mill was destroyed with a similar estimated loss. Lack of insurance, or under-insurance, compounded the financial blows.

Timber output fell. In Taranaki the February fires alone were said to have

'The remains of the New Plymouth sash and door company's sawmill on the Egmont road.'

Auckland Weekly News, 5 March 1908, p. 14

destroyed 12 months of cutting at one mill. In Hawke's Bay, timber output for January 1908 dropped by about two million feet from the previous years, and increased prices on the London market (as well as locally, according to contemporary reports) failed to offset the losses.[3] This trend was reflected in the country's export figures. The amount of timber and its value dropped significantly in 1909; the quantity did not recover until 1911 (although the value of wood exported in 1910 exceeded the 1908 figure).[4] Milling companies employed men to extract what timber they could from the aftermath of fires, but even so the impact on employment would have been severe. In 1906–07, in the Auckland, Hawke's Bay, Taranaki and Wellington districts 209 mills employed 4929 men; 14 mills in Marlborough employed 494 men; and 71 mills in the Nelson district employed 852 men[5] – layoffs and industry downturns would affect a significant number of dependants and workers in the allied trades of contractors, builders and cabinet makers. Similar impacts would have been felt in the gum-digging industry: the 1000 acres (450 hectares) of Mangawhare gum swamp out of Dargaville, said to be valued at £7 ($1077) per acre, were almost completely destroyed.

Flax growers and processing companies also suffered. At Shannon hundreds of fire-fighters battled to save what they could of the highly valuable flax in the huge Makerua Swamp. With 31 mills operating in and around the swamp in 1912, employing up to 800 men (about 250 in the largest mill, Miranui), a lot was at stake for both employer and employee.[6]

Maori lost the flax destroyed by the fires in the Oturei Swamp. Beautiful

'A farm swept by the conflagration on the Egmont road.'

Auckland Weekly News, 5 March 1908, p. 14

and historic carvings were also destroyed, along with the meeting house at Maungakawa in the Waikato. ('The suspicious nature of the Maoris', the *Evening Post* explained, had prevented them 'handing over the treasures to the pakeha for safe keeping.'[7])

Reporters' sweeping and often unquantified generalisations about loss make it difficult to establish the extent to which the individual settler was affected, and the general summary below deflects our attention from the individual heartache, disappointment and distress. However, it is clear the impact of the fires varied from area to area. In north Taranaki, for instance, stock losses were said to be small and only one house was lost; on the east coast Native Affairs Minister James Carroll claimed few stock and little property had been destroyed (though this comment may have been politically motivated). Nevertheless, in the midst of what one man described as one of the worst droughts to hit the country,[8] settlers lost months of work on farms hard-won from the bush. Almost throughout the country stock was burned, fencing was gone and thousands of acres of grass were destroyed. Small, often uninsured farmers (who formed the bulk of those in the worst-hit areas of the North Island) had to sell their remaining stock on a buyer's market or see them starve. The carrying capacity of the land was reduced and, consequently, milk production fell – around Pohui in Taranaki in January 1908 one farmer's weekly output dropped by 454 kilograms and another's by 317.5 kilograms. In some places farmers had difficulty in getting their milk to the factories; in others the creameries were destroyed. Immediate income was jeopardised and the problems of recovery were exacerbated. One commentator saw a silver lining: despite the heavy losses, by October the envisaged butter shortage on the London market would make for good prices. Others were less optimistic, and there were concerns that rain would not come in time to let the pasture recover to carry stock over the winter.

Official figures again provide a useful perspective. The dairy industry was undoubtedly the worst affected: in the 1909–10 financial year the numbers of cattle and dairy cows were still down on the 1906–07 financial year, and the number of pigs (associated with dairy farming) had only just recovered. The number of sheep, however, had risen by some 1500,[9] suggesting the small North Island settler suffered more than the big South Island runholders.

The pattern of losses experienced in the 1908 fires was typical of what had gone before. As early as 1872, for instance, bush fires around Dunedin's North Harbour and Blueskin Road reduced dairying output. In 1878 the *Otago Witness* itemised individual losses from a fire in Pine Hill Bush: Thomas Lockie, 80 cords of wood, 10 acres of bush and fencing; David Young, 40 cords, six acres of bush, a hut and fencing; Benjamin Jeff, 20 acres of bush, a hut and fencing; J. M'Carrow, 30 cords of wood; J. Marshall, ditto; Charles Lawrence,

20 cords. Grass fires around Waipukurau in the same year destroyed so much feed that 4040 stock had to be sold immediately. In the 1897 Wairarapa fires two settlers at Mikimiki (just off State Highway 2, some 15 kilometres north of Masterton) lost fences and feed worth about £250 ($31,000 in today's values). 'Mr Duckett [whose loss was described as serious] is a struggling settler with a family, and as 80 or 90 sheep, the whole of his feed and a lot of his fencing has been destroyed, he is left with nothing. Mr John Campbell estimates his own loss of grass, sheep, totara forest about £1,000.' A year later an *Evening Post* correspondent described the aftermath of fires around Pahiatua, where many settlers were uninsured and homes and possessions had gone. 'In most parts the ground is littered with burnt and blackened logs'; the grass was so burned that where some 50 animals might have grazed, said the *Post*, 'there [was] now not food sufficient for two'.[10]

The papers were full of accounts of heroism in the midst of calamitous events: of men who, well nigh ruined, 'set their teeth at it again';[11] of fighting on 'in suffocating smoke, scorched by the fire, suffering severe burns from the clouds of flying sparks'.[12] Sometimes the speed of a fire's advance would take settlers by surprise, but even with a warning they could do relatively little against the flames. On the big runs, fires were often left to burn, until a natural barrier or a change in the weather intervened. Efforts focused on saving the homesteads, using techniques common to runholder and small settler alike. Recognising the advantages of firebreaks, some settlers cleared around buildings, removing any build-up of debris that might feed the flames; in 1886 grass was burned pre-emptively along the railway lines in Hawke's Bay.[13] Buckets of water were filled and placed strategically; wet, heavy sacks and branches were used to beat out small advance fires. The men defending the 600 hectares of flax in the Makerua Swamp in 1908 developed flails made from fork handles with thongs of stout leather tacked at the end – primitive but effective for attacking smouldering fires – and they blocked the drains to prevent the fires spreading. Farm hands, mill hands, neighbours and communities rallied. Local fire brigades (where they existed) turned out. In Shannon, local volunteers (with a very real interest in saving their employment) formed bucket brigades, fought incipient fires and guarded against further outbreaks until relief parties arrived from Wellington. The many isolated families or couples battled on their own, combating flames, heat, burns and exhaustion. But without rain or a wind change to force the fires to back-burn, the settlers were at the mercy of the elements. Without rain, 'there is no telling when the fires will cease,' the *Evening Post* stated in 1887. That oft-repeated sentiment underlined the settlers' helplessness.

Sometimes outside help arrived. In 1908, facing allegations of official negligence over the destruction in Wilton Bush, the government sent in six artillerymen, and the under-secretary of Crown lands employed labourers to beat out flames among the fallen timber. In some districts, trains assisted the relief effort, taking refugees away from the fires and bringing in the brigades and volunteers. In the 1885–86 fires in Hawke's Bay, thanks to the imaginative thinking of the local railways manager, trains took threatened settlers' furniture away from the danger zone.[14] In 1897, amid fears of 'unparalleled disaster' in the Wairarapa, special trains took sheep to available feed in Wellington; a year later, when some of the railway line had been burned, trains carried only cream consigned to the various creameries and passengers' luggage over the damaged portion to try 'to minimize the heavy losses which have fallen on the farmers in the Forty-Mile Bush'.[15] Nor was the Wellington-Manawatu Railway Company slow to respond to the seriousness of the Khandallah fires that year. Its special train carried 30 permanent artillerymen through a Thorndon veiled in smoke and the innumerable fires around the Crofton Downs Railway Station to Khandallah, and company trains carried the many volunteers to Shannon.

But even outsiders could do little to overcome the fundamental fact that, by today's standards, fire-fighting tools were fairly primitive, fire brigades were not necessarily well equipped, and water, that essential element, was often inadequate or absent. In 1898 the permanent artillerymen railed into Khandallah were armed with about 50 buckets and bundles of slashers. They arrived, amassed their buckets of water, cleared the ground and sat down to await the fire's arrival – without portable pumps and effective ways of carrying water they probably had little alternative. Although local brigades did turn out for fires outside their immediate district and could help protect small towns or specific buildings, the distance was often too great or travel too slow for them to offer timely assistance. In 1879, with fires threatening Carterton, the Masterton brigade, a few kilometres up the road, was expected 'in a few hours'. A special relief train sent in to Waipawa in the 1886 fires achieved 60 kilometres per hour, a phenomenal speed at that time – yet even that was too slow.[16]

Today fire, flood and earthquake insurance provides automatic assistance to those affected by natural disasters. Depending on the scale of the disaster, additional government funding may also be forthcoming. But for our settlers, no matter what they had lost, government assistance was by no means assured and, where it was provided, it was not necessarily helpful. Local fundraising, too, was often inadequate to assist settlers recovering from the destruction and trauma of the fires.

Unusually, in 1886, 1898 and 1907 the government had, under certain conditions, provided settlers with grass seed. The details of a similar offer in

1908 allow us to get a sense of the value of that assistance. In late February 1908 newspapers were voicing concern for those affected by the recent fires, especially small farmers. A relief fund for Taranaki settlers was likely to be set up in New Plymouth. Responding to calls for government assistance, Prime Minister Joseph Ward pledged his intention to give 'practical' assistance to affected settlers – that 'deserving body of people who work early and late in carrying out their important part of the work of the country'.[17] On the basis of local government servants' assessments, grass seed was to be supplied at cost price and interest free for two, three or four years to farmers, whether Crown settlers, freeholders or lease-holders.[18] As various newspapers detailed, cases of exceptional hardship, where settlers had been burned out of their homes, would be considered on their merits; the Railways Department was to carry all materials needed for fencing and building free of charge; and Crown lands, where possible, were to be made available for grazing free of charge.[19] In all, £24,082 was distributed among 647 settlers in the North Island, and £2756 among 67 in the South.[20] The *Evening Post* noted that Wellington, which had borne the 'brunt of the burn', got the highest payment of any district, but that payment was distributed among the greatest number.[21]

How useful the grants were in addressing real need is questionable. In Auckland only 197 of the 296 applications for seed were accepted[22] (perhaps because applicants felt unable to state that they could not finance their own positions, apparently a prerequisite for payment). Moreover, the time over which the relief was available was too short. The Auckland Land Board was arbitrary in its approach to granting rates relief to its tenants; misunderstandings meant that some had not applied for it; and rumour had it that not all could finance the repayments.[23]

Despite the dislike John Mackenzie, Minister of Lands, had for large landowners, they may have profited at the expense of the smaller farmers. We have no information as to whether the grants were averaged out across the various districts or given in relation to individual need. However, on average, recipients in Marlborough got £80 each, Wellington settlers averaged only £41 and the rest even less. In March 1908 the *Wanganui Herald* suggested that the fires had 'resulted in material benefit to [some] individual landowners'.[24] Did the big sheep- and cattle-farming landholders in the South Island have a political leverage unavailable to the small dairy farmers in Taranaki and the King Country? [25]

Moreover, the grant was tightly targeted and as such of limited value. The government wanted to expand settlement and wealth and comfort. Construing dairying as the way forward,[26] it had no scruples in 'practically forcing' seed to be sown early by providing it cheaply.[27] In 1907, when similar relief had been offered to Otago Crown settlers, just over half of them had accepted it.[28] Some may have refused because of perceptions of charity (see below). Others, like their Taranaki counterparts in the reconstruction after the 1886 fires, may not have wanted to tie their reconstruction to dairying. Nor did the relief offer immediate and useful assistance to those who had lost all means of subsistence for months to come or who needed to rebuild homes or restock, and those who had lost timber (cut or uncut) or flax, machinery or equipment got nothing.

But however inadequate, the grants for grass seed were an example of a government recognising the need to provide at least some assistance (and the advantage in doing so). Before that, in the absence of any adequately funded programme or bureaucratic structure to help meet large-scale crises, public subscriptions offering assistance to disaster victims were a vital part of colonial life.[29] The subscriptions were funded in various ways. Concerts were long used to raise money. More common were the relief funds – and here the newspapers' detailed coverage of the fires' progress and the losses suffered, and unabashed soliciting, proved their worth. In January 1897 the *Evening Post* reported the thousands of pounds of damage the fires had already done, the fears that whole settlements were likely to be annihilated, and the fact that the Masterton citizens had set up a relief committee: '[I]t is a condition of things that should at once stir the citizens of Wellington to the desire to help their fellow-colonists in distress that we are proud to think is one of their most prominent characteristics.' The Masterton mayor was yet to gain general support for an organised relief fund, but 'there is already abundant evidence that dire misfortune has overtaken many of the most deserving class in this community – struggling settlers who are fighting hard to wring a livelihood from, and make their homes upon, the bush lands of this Province. He gives twice who gives quickly.'[30] People from all walks of life responded generously, in smaller or greater amounts, and not just once. A year later the *Evening Post* set up a new subscription list for the 1898 fires. Subsequent articles listed the donors and their contributions, which ranged from 2s 6d ($17) to £1 ($144). By 21 January £10 10s 6d had been amassed. A month later the total had grown to £61.[31]

The problem is obvious: the scale of some of the crises defied local effort. In 1897, with fires raging in the Wairarapa at Mikimiki, Hastwell and other districts north of Masterton, at least £2000 was needed to meet just the most pressing needs, and it was unlikely that a fraction of that could be raised.[32] The establishment of relief funds depended on local leaders identifying a crisis, defining the need, identifying sponsors, arranging for the list to be publicised and deciding if amalgamating with other relief subscriptions in the district was appropriate. Leaders unsurprisingly varied in their ability to carry out those roles. Moreover, working out who should get assistance was not necessarily straightforward. Some committees delegated someone to travel to the affected areas to assess damage; in the aftermath of the 1886 Taranaki fires the deserving may have had to approach the appeal committee themselves. Whether assistance was better given in kind or in money, and whether the donations should be averaged out among recipients or given in proportion to need were questions that had to be decided.

There would have been little help for settlers who construed such assistance negatively. How many did so is not easy to establish. However, in the 1880s, 'help from a properly sponsored subscription was probably not looked on as a demeaning charity, but as partaking in a system of mutual support, in which one was sometimes a receiver and sometimes a giver'.[33] Nineteen petitions were handed to the House asking for relief for bush fire damage in 1907; in 1908, although the *Evening Post* considered that those who had suffered would be 'averse to anything like charitable assistance', it also recognised that 'some will welcome Government loans'.[34]

In the aftermath of the 1918 Raetihi fires the government's response was different again. In this context, local initiative was essential because initially the government simply directed that the Department of Lands would work as an agent for men wanting work sowing grass seed. Immediately after the fires, on 19 March 1918, the *Ohakune Times* launched relief efforts. The Caledonian Society established a fund, a Ladies' Relief Committee was set up, and a relief concert was held. By 6 April over 200 people had contributed and, with gifts ranging from a sheep to £10, the Waimarino Fire Relief Association eventually raised £17,000. In May the government stepped up, giving £500 for relief to the most needy and agreeing to match locally raised relief money pound for pound up to

'Recent disastrous bush fires at Kawhia, Auckland: The remains of Mr. McIntosh's homestead at Oparau.' The paper reported that, in the face of the advancing fire and seeing that his place was doomed, Mr McIntosh had buried a few valuables, including the piano. The photo shows the recovered piano on a wagon.

Auckland Weekly News, 5 March 1908, p. 15

A devastated Raetihi. The disaster 'originated in the smouldering fires of recent bush-burning' that were fanned by winds and spread over much of the King Country, with Raetihi the worst hit. The paper reported injuries, deaths, and that many had sheltered in the river beneath the bridge.

Auckland Weekly News, 28 March 1918

£10,000. It would also provide 35-year loans at 5 per cent interest. Government cash grants finally amounted to £21,000 and loans of £6170 were made.[35]

But perhaps the sheer horror of the fires was even worse than the economic loss. Newspapers often touched on the suffering of livestock: the cows that got stuck in smouldering reclaimed swamp; the scorched pigs; the cattle, sheep and horses blinded by the smoke, bewildered and terrified as the flames bore down on them. Cows that had survived the 1898 fires in parts of the Wairarapa stood unmilked, lowing in pain; elsewhere lay 'a heap of between 20 and 30 carcasses of sheep, all singed and scorched, and close by nine dead cows. In a creek a few chains away about a dozen more cows lay where they had perished from the smoke and heat.'[36] The smoke – thick, stifling and ubiquitous – obscured familiar landscape, adding to the distress for both animals and humans. To breathe in the hot smoke was to experience the 'choking sensation of having a rasp sticking somewhere in your thorax', of air 'heated and thickened to a degree sufficient to crack the leathern lungs of a blacksmith's bellows'.[37] The blanketing smoke blinded those fighting the fires and those fleeing them.

As the flames advanced, people took desperate expedients. One man put his family down a well all night; families dug pits for furniture or, like Francis Guy's mother, put household essentials and a few other prized possessions in milk cans in a newly sown paddock so that, when the fire passed, they had something with which to make a meal as well as some treasures intact.[38] Some women fled for the green bush (thought not to burn). Shallow creeks and culverts provided

refuge – though culverts also acted as chimneys for the smoke, which billowed through. 'It was only by bathing their faces continually in water running under their feet that they were able to bear the awful pain,' the *Star* reporter wrote of one group of refugees from the 1908 fires around Taumarunui, who huddled in a culvert for eight hours before being rescued.[39] And when the settlers returned, they returned to desolation.

Yet, two years after the 1908 fires, the rubbish was gone, unprofitable land was under cultivation[40] and, reported the Department of Lands in one of many

Settlers saving the sewing machine and what clothing and pieces of furniture remain after the Raetihi fire swept through the town. At the time the cost of the loss of sheep, cattle, sawmills and timber stocks, productivity, insurances and additional costs were estimated at £214,500.

ATL, 7-A13274

Mandars and Bradley's 'Bush' at Whananaki' after a fire, 1907.

AJHR, 1907, C.–4, Photo no. 4 by H. P. Kavanagh, chief wood exporter.

such statements, minimal effort had achieved major clearance of land that now could soon be brought to full profit.[41] Those views were still held by the majority, despite the devastation of the fires and the heartache and distress they must have caused – and would cause again. Where did such views come from? We need to look at the complex web of practical considerations, tradition, other settler experience and attitudes to the bush that explain why alternative views had little traction for so long.

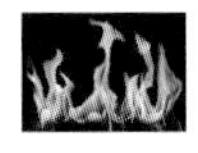

This photo of bullocks hauling logs in the Wanganui district in the 1920s illustrates yet another desolate post-fire landscape. Some contemporary writers and painters found parallels between the fire-blackened stumps in otherwise stark land and the landscape of World War I trenches.

NL, 125078½

Early promotional literature provided intending settlers with a picture of a land whose majestic trees and luxuriant growth were undeniable evidence of a fertile and rich soil, a land waiting for development.[42] Although initially spurred on by such notions and by immigrant imperatives of carving a home and a living out of the land (again the metaphor is instructive), notions of majesty and plenty foundered as settlers battled to create farms from the often impenetrable and seemingly interminable forest that covered much of the country.[43] Only a few species of timber would have seemed to have much value; settlers' immediate needs were for food and marketable produce. For many, burning was perhaps the only practical response. Most of the trees burned well (though beech in high-rainfall areas proved remarkably resistant to fires). A good burn, as settlers were frequently reminded from the late 1860s, cleared the bush cheaply and quickly and the ash fertilised grass seed, producing succulent green grass and giving it a good start.[44] (The fires' intense heat changed the amounts and availability of minerals, such as phosphorus, nitrogen and ammonia in the sterile soil. We now know that those minerals are water soluble and would be leached out over time.[45]) Where fires destroyed state forest, the Department of Lands reported, areas were sown in grass and eagerly sought after.[46] The 1908 fires and the resulting early sowing of seed, said Minister for Lands Robert McNab, had 'renewed the whole face of the country [giving] a magnificent flush of grass'.[47]

Land recovering after being burnt off, Northland, c. 1910. Stumping up could go on for decades.

NL, 1/1-010823-G

As in Australia and North America, therefore, fire made such a vital contribution to the developing frontier economy that it conferred on settlers almost a folk-right to burn. Accidental burns were merely an inevitable part of the progress to a greater good. But the concept of burning to fertilise and settle land went well beyond immediate settler practicalities. It was deeply rooted in the traditions British settlers brought with them. From the late 1600s, if not before, settled land was burned and pared (which involved burning small heaps of soil from spent ground to create fertiliser that was then raked over the ground) to improve the soil; one 19th-century New Zealand correspondent considered this practice to be so common in Ireland and in the west of England that he barely needed to describe it.[48]

In the 18th-century agricultural revolution fire achieved a new prominence. Burning transformed vast amounts of forest, moor, heath and peat into new stretches of farmland. By the late 18th or early 19th century, enclosures, scientific crop rotation and high-yield farming (relatively speaking) had reduced and restricted the need to 'broadcast' fire for pasture improvement, but many who would become New Zealand's immigrants would have played an active part in burning off rank autumn and spring growth and fields after harvest.[49]

Fire, already a tool of land husbandry, was also associated with the settler experience. Arguably New Zealand's rural fire and land clearance history might have been different had the country been settled not by predominantly British stock, who lacked 'forestry traditions', but by continental Europeans, who had them.[50] Under a feudal order with restricted mobility, Europe from the 1300s had preserved or retained much forest for its valuable by-products. By the late 19th and early 20th centuries European forests were smaller and more intensively cared for and there are few records of fire.[51] But many New Zealand settlers had direct or indirect links with North America and Australia, frontier societies where fire played an integral role. High mobility and relative lack of social order saw wildfires accompanying the American continent's expanding frontier, and these fires, whether 'of natural, accidental or incendiary origin [became] part and parcel of a complex frontier of violence and waste'.[52] (In 1825 a fire near New Brunswick destroyed 800,000 hectares of forest, killing 160 people. In Wisconsin in 1871 a 'storm of fire' came 'with the howling of a tornado [and] rained down on the doomed village [of Peshtigo] like flaming missiles shot from unseen artillery'; over 1150 people died and 23 towns and hamlets were destroyed. There were huge fires on the Great Plains during this period. In 1879 in West Texas thousands of square kilometres of land were burned; in 1888 and 1889 towns in North and South Dakota were virtually annihilated; in 1894 a four-day fire in Texas cleaned out over 400,000 hectares, while a chain of fires in northern Idaho forests killed about 700 people. In

1910 a fire in Nebraska was logged travelling at 9.5 kilometres an hour.[53])

In Australia small farms were carved out of bush with axe and firestick. Fires frequently escaped, and in dry seasons these might cover thousands of hectares. Victoria, ravaged by fires about every 10 years, was further devastated in the series of fires that, over 1896 to 1901, destroyed huge areas of its own hinterland; similar destruction was experienced in New South Wales and Tasmania. The extent of coverage in New Zealand's papers varied but over the years the accounts of families caught in blazes, of children covered with wet rags, of fires that lit up the countryside for miles and destroyed thousands of acres, of people 'prostrated with shock', of 'doomed districts' and 'gallant fire-fighters ... straining every nerve to save homes and property' took up many column centimetres.[54] In mirroring (on a different scale) New Zealand settlers' own experience it is likely that such reportage helped validate the notion that burning was inevitable and acceptable.[55]

Attitudes within government reinforced and reinvigorated this notion. In 1874, reflecting the social Darwinism popular at the time, MP John Sheenan considered that the loss of the bush was subject to 'the same mysterious law which appears to operate whenever the white and brown races come into contact – and by which the brown race sooner or later, passes from the face of the earth'.[56] According to MP George Tribe, overseas experience showed that 'wherever a country becomes populated, the timber has to give place to the utilization of the soil for the support of the people'; his colleague William Gibbs from Collingwood suggested settlers were 'compelled' to cut down the forest.[57] To be successful New Zealand needed to expand settlement – and the bush was 'detrimental to progress' (1904);[58] 'the passing of the forest' might be regretted but 'little could be said' if useful grazing eventuated (1928).[59] From this perspective the bush was doomed, an inevitable casualty in the march of progress.

The loss of indigenous forest through burning, accidental or otherwise, was therefore a necessary corollary to settlement and part of a natural process – a mindset aided perhaps by the official statements on the inevitability of forest fires, especially in kauri forests.[60] But the use of, and attitudes to, fire goes perhaps deeper. Many immigrants must have been in England in the 1830s when peasants burned property to demonstrate against changes in their rural world and protest against servility and helplessness.[61] In New Zealand they may have also turned to fire to satisfy a deep psychological need to defy a dark, alien and ubiquitous forest so totally different from the tended landscape at Home. Bush, that 'stronghold of nature', was to be 'assaulted' and 'subjugated'[62] and men became men in doing so.

There were dissenting voices. From early settlement newspapers across the country drew attention to the need to contain bush fires and prevent the destruction of native forest. As early as 1844 the *New Zealand Gazette and Wellington Spectator* suggested banning fires over summer near cleared or occupied land, and the *Nelson Examiner and New Zealand Chronicle* reported that locals were demanding a committee be established in Waimea East to bring those who had lit the recent fires to justice. A correspondent to the *Otago Witness* expounded the need to restrain the 'culpable carelessness and micheviousness [*sic*] of travelling diggers and others' while not 'interfer[ing] with the necessary firing of bush for the purposes of clearing land for settlement, nor with the ordinary operations of agriculture'.[63]

The attempts of the four provincial governments to legislate against bush fires were reported, sometimes at length. By the mid-1870s these issues were being discussed in a new and wider framework. As anxiety about the downstream effects of deforestation mounted, the forest was increasingly viewed as a potentially valuable resource. This perspective was heightened by concerns about the rate of destruction of the New Zealand bush, fears of an international timber famine and high costs of imported wood, disquiet about the displacement of indigenous flora and fauna, and a growing appreciation of the bush's aesthetic values. These issues were raised within Parliament and beyond. Yet initiatives to protect the state forest resource from fire reflected a general political unwillingness to address the question and, broadly speaking, were inadequate and under-resourced. Not until 1921 were the first moves towards an effective fire regime, covering both indigenous and native forests, developed by MacIntosh Ellis, a bold outsider who used a raft of new approaches to address the settler 'fell and burn'[64] mentality. Previous initiatives to protect the state forest resource from fire reflected a general political unwillingness to address the question and were inadequate and under-resourced.

In May 1841, with reports that the kauri forest, and with it the navy's supply of spars, was being destroyed, Lord John Russell, the Secretary for War and the Colonies, appointed Captain William Cornwallis Symonds as conservator of forests. In early November various measures to prevent the loss of the forest were announced: those illegally cutting, burning and grazing could be prosecuted and a £5 reward was offered for information leading to conviction. But when Symonds died he was not replaced, and Hobson's measures to prevent the destruction of kauri on land that Maori had sold to the Crown were no deterrent as they could not be enforced effectively.[65]

Lack of political will and inadequate legislation and enforcement – the factors that determined the fate of the first attempt to conserve the bush ('conserve' did not have the contemporary meaning of preserving, but of ensuring a steady supply for future need) – continued to dog such efforts. Neither the Waste

Lands Act 1849 nor its 1862 amendment contained provisions relating to fire, and although it was perhaps technically possible that the provisions relating to trespassing on Crown lands could be used to prosecute against fire damage,[66] there is no evidence they were. (The newspapers do report prosecutions – though I have been unable to establish under what law these were taken – and these generally languished, as others would subsequently, under the weight of claim and counter-claim or lack of evidence.) In the mid-1860s the Canterbury,[67] Otago[68] and Auckland[69] provincial councils passed bush fire ordinances. Only Otago's limited provisions (quite inappropriate for addressing the fires already sweeping grass land and bush[70]) got vice-regal approval. The more stringent Australian legislation provides an interesting contrast. In 1865 Victoria passed legislation requiring 20 yards (18 metres) of bush be cleared prior to settlement; a year later it appointed a commission to identify ways of 'preserv[ing] the forests as far as possible in Victoria';[71] South Australia's 1867 amendment to its Bush Fires Act placed stringent restrictions on burning.[72]

New Zealand continued to lag behind. MP Thomas Potts, enthusiastic tree planter, pioneer conservationist and amateur naturalist, was appalled by the monetary loss and loss of 'protection forest', particularly in the South Island. In 1869 he finally managed to get the government 'to ascertain the condition of the Forests of the Colony' to see how they could be preserved. Provincial officials were circularised with six questions for this purpose. In their replies, all indicated concerns about the future of a potentially valuable extractive industry and implicated fire among the main causes of the bush's destruction. Southland's Commissioner of Lands referred to the need to address 'the wanton destruction' of the timber reserves; in Taranaki the deputy superintendent was, presciently, aware of the effect of fires 'almost impossible to arrest' as clearing extended into the back country; Canterbury's chief surveyor, only too conscious of the fires sweeping large portions of Banks Peninsula, saw 'nothing to prevent the total destruction of bush on the Peninsula'. Yet Potts' initiative, and further pleas for a royal commission by MP Charles O'Neill (who pointed out that a great part of New Zealand's forests had been destroyed by 'wilful and culpable fires'[73]), produced no action. Not until 1873 were the returns analysed by James Hector, geologist, explorer and '*de facto* adviser to the government of the day on scientific and technical matters'.[74] He estimated that some eight million acres of bush had been lost between 1830 and 1873. As to the bush that remained, he considered that most was valuable as protection forest and that the small amount of commercially valuable forest was best conserved through private ownership. He did not comment on fire, its possible causes, or the frequency with which it had been cited as a destructive agent. Nor did he see a role for central government in preserving indigenous forest: differing conditions in different parts of the country, he concluded, suggested that local legislation, not a general measure, would be more appropriate.[75]

Julius Vogel, unsurprisingly, took a quite different view. His 1873 'Conservation of Forests Bill (An Act to provide for the Preservation and Growth of Timber on Crown Lands)' identified a clear and primary role for government in promoting – and, in a minor way, addressing – the fire issue by requiring licensees to prevent the spread of fire. However, this bill was never introduced; Sir John Cracroft Wilson's 1874 Grass and Forest Fires Prevention Bill was withdrawn at its second reading in favour of further legislation proposed by Vogel in 1874.[76] The 'New Zealand Forests Act (An Act to provide for the Establishment of State Forests, and for the application of the Revenues derivable thereof)' marked the beginning of what would become a huge state enterprise. But, as contemporary commentators pointed out, it contained no practical measures for preventing forest fires, nor for fire-fighting, and the real need – to extend the provisions to all forests, not just those set aside as state forests – had not been met;[77] nor did the regulations designed to prevent 'the danger and spread of fire in State Forests' appear to have been promulgated.[78] Meanwhile the act came to symbolise tensions between centralist and provincial factions. Many saw it as an unwelcome instance of state intervention and a way of promoting forestry against settlement. An attempt at repeal failed, there was a change in government, and the new administration, committed to financial retrenchment, did not vote any money for forestry.[79] Private citizens were voicing their concerns about forest management and fire prevention, but it was not the time for innovative policies.[80]

Eleven years later, in 1885, the State Forests Act was passed. It sought to facilitate the development, management and use of state forests to the best advantage, conserving them for future use and preserving them for climatic and protection reasons. Thomas Kirk, who would become founding professor of botany at Victoria University College and whose 1884 report on the country's forests and state of the timber industry influenced the act's provisions, was appointed chief conservator. His rapidly published new regulations, 'sufficiently elastic to cover the whole country', were also, imaginatively and innovatively, drawn up to give those involved 'a direct interest … in diminishing the risk of fire'.[81] He set penalties large enough to act as a deterrent for lighting fires negligently in a state forest as well as outside fires that spread to or threatened state forests – for the first time, though in a limited way, fires outside state forests were brought into the regulatory framework. Kirk's provisions on trespass in the forests and licensing of sawmillers in the summer months represented further attempts to deal with the careless, irresponsible use of fire.[82] He set up the State Forest Department to administer, manage and enforce the regime.[83]

But Kirk had barely the time over the next two years to report an almost total absence of fire in state forests, where rangers were patrolling, before the old arguments were rehashed and the tables turned again. Particularly in the heavily bushed North Island, politicians saw bush as an obstacle to settlement rather than a revenue source. There were concerns that progress and settlement might be slowed if large amounts of land were locked up in forest. The balance of power of large landowners might alter with new ownership patterns, and conserving bush that was inevitably doomed to destruction was seen as being of dubious value.[84] In 1887 the State Forest Department was disbanded, its employees were laid off, and the department's functions devolved to the Crown Lands Department.[85] (The act was not repealed until 1908, however, and its provisions remained in force until then.) Crown Lands, reported its secretary, would continue to set aside reserves for timber conservation and climatic reasons, and to preserve streams and water sources, but with its resources already stretched, it would focus on the established nurseries and plantations, not add to its responsibilities.[86]

We need now to look more closely at the context in which these developments were taking place – at the commercial, climatic and aesthetic values by which the New Zealand bush was increasingly perceived as having. These views gained currency as the 19th century progressed; from the second decade of the 20th, they were used in developing a public consciousness about rural fires and their damage and would help make fire and its prevention 'everybody's responsibility'.

Timber, playing 'a more or less conspicuous part',[87] was vital to the colony. Without it much of the colony's infrastructure – its houses, barns, fences, pit-struts, railways, bridges and culverts – would not exist. Until the advent of plastic and synthetic materials in the 1950s, cement offered the only – and often unsuitable – substitute. No wonder that some politicians early on warned against profligate use of timber inducing scarcity. In 1873, noting the extent of bush already lost to the colony, and as a non-specialist weighing only slow-growing natives against fast(er)-growing exotics, James Hector reported to the Colonial Secretary on how Australians and Europeans developed, managed, felled and gained revenue from state forests. He attached a memorandum setting out the considerable gains that within 30 years would purportedly accrue from planting exotics,[88] and a report by a Captain Inches Campbell Walker, then conservator of forests in the Madras Conservancy in India.[89]

In 1874 Vogel appointed Campbell Walker as conservator of state forests. The mindset Campbell Walker brought reflected the mid- to late European

H. J. Mathews, chief forester 1896–1908. The title was a misnomer, reflecting how little officials knew about forestry at the time. Mathews' contribution should not be denigrated, but essentially he was a nurseryman hired to plant trees.

ANZ, AAQA, 6506, Box 28, 10/42, neg. 7855

rationales behind 'scientific forestry' – 'the conservation of tracts of natural forest into blocks of quality timber trees of similar ages, capable of being worked in rotation, with the aid of working plans'.[90] Private plantations of exotics already existed, and the 1896 Timber Conference, raising a spectre of an international timber famine, gave rise to a Forestry Department within the Department of Lands and Survey (and its various subsequent manifestations). The Forestry Department, under chief forester H. J. Mathews, who held the post until just before his death in 1909, worked primarily on developing nurseries of exotics and secondarily on afforestation. Three nurseries of exotic trees were originally established; by June 1903 at least 10 were operating – Dumgree, Gimmerburn, Hanmer Springs, Tapanui, Dusky Hill, Eweburn, Naseby Survey, Conical Hill, Starborough and Rotorua. There were also four plantations: Whakarewarewa, Waiotapu, Ruatangata and Somes Island; later annual reports also mention plantations at Tekapo and Puhipuhi.

As better knowledge and greater appreciation of the qualities of New Zealand woods contributed to an ever-growing rate of extraction, concerns with the milling life of the indigenous resource grew. Exacerbating such unease was that the 'infinitesimal' private and government replanting was far less than what was being lost through 'unintentional fire',[91] particularly given the feared international timber famine.[92] By 1907, following by no means the last estimate of the likely life of the native timber resource, officials were envisaging that domestic need would have to be met through imports. In 1918, by which time the US and Canada were apparently cutting more sustainably, Edward Phillips Turner, then responsible for the newly created Forestry Department (an autonomous unit within the Department of Lands and Survey and therefore not bound by its land development goals), pointed out that 'true patriotism, concerning itself more with the future of the country than the present, demands the application of such measures as will ensure a sustained supply for ourselves and those that will succeed us'.[93]

The need to prevent deforestation and its downstream effects also pointed to the desirability of maintaining a balance between the needs of the settlers and those of the burgeoning logging industry. In 1874 Campbell Walker had argued for the value of indigenous forest as protection forest, preventing the otherwise inevitable 'disastrous results, both in the shape of the deterioration of the climate, dangerous floods and landslides and drying up of springs and sources of rivers'.[94] From 1881 reports of the Crown Lands Department and its later iterations show that land deemed unfit for agriculture was progressively put aside to conserve timber, contribute to favourable climatic conditions and preserve healthy streams and watersheds.

The perception of the value of such protection forest, which would remain an aspect of Forest Service policy, had gained steady currency by 1910. Citing overseas studies and experience, administrators pointed to the critical relationship between forest and climate in reducing extremes of temperature, inducing rain, retaining dew, absorbing water and avoiding erosion, preserving water supplies, keeping waterways open for navigation and avoiding the costs of downstream flooding – all factors with major implications for productivity. New Zealand had to be saved from the fate of other countries.

In 1909 Leonard Cockayne's lengthy and lavishly illustrated report, *Our Native Forests*, was laid before Parliament. Over 118 pages in length, it discussed the character of the bush and the co-dependence of the plants within an eco-system, described sawmilling practices, argued for the necessity of preserving a proportion of native forest to avoid the evils of deforestation, and evaluated the rates of replanting since 1896 and the ability to meet future demand.[95] Photos of flood damage and denuded landscapes in New Zealand were placed alongside those of hundreds of kilometres of Chinese land ruined by deforestation and Canadian lakeshores destroyed by clearing. *The Evils of Deforestation*, a booklet by J. P. Grossman, director of the School of Commerce at Auckland University College. This included some of the photos used in departmental reports, was intended for a wider audience.

The value of the forests had for some time been assessed in more than monetary terms. If newspapers are indicators of popular sentiment, even the earliest reports indicate some aesthetic appreciation of the bush – and sadness, disgust, even anger at its loss. '[S]ome distance up the Purakanui Road bold rocks [tower] over the one side, while on the other the land falls away in deep precipitous ravines clothed in trees ... seemingly barren rock once produced these veteran forest kings,' wrote the *Otago Witness* in 1872.[96] Colonial gothic the descriptions may have been, but they captured a mood. Future New Zealanders, opined the *Evening Post* in 1898, would have hard things to say about those who 'have been responsible for the wanton destruction of the good gift of noble forests'. Such vandalism had 'cast from us a great gift that can never be recalled'.[97] Departmental comment was blunter. The loss of subalpine bush around Lake Wakatipu was 'not simply a blot on the district, it is a slur on the colony at large', reported Thomas Kirk in 1886.[98] In 1902 'The Garden', a particularly valuable patch of rare alpine vegetation around Ball Hut in the Southern Alps, was destroyed – an act of 'vandalism', chief forester Mathews wrote, arguing for reserves where plants and animals would not be 'exterminated'.[99] In 1903, after the Land Act 1882 had seen little land set aside for scenery, the Scenery Preservation Act gave 'real impetus for forest preservation'.[100] It enabled land to be taken for permanent historic, scenic or thermal reserves. Fines of up to £100 could be imposed for lighting fires, and any damage had to be paid for. Although the needs of settlement remained the priority, the act did allow aesthetic values to be expressed within a government context. In 1911 Edward Phillips Turner, then Inspector of Scenic Reserves, was one of the more outspoken public servants:

> We have treasures in our scenic reserves which in years to come will be thought priceless by our successors ... It is largely our scenery that makes this Dominion one of the most delightful countries in the world ... no nation with purely utilitarian ideals ever reached real greatness in its highest sense. Our unimaginative settler who protests against the reservation of a very small percentage (and that generally poor land) of the country for scenic and like purposes would be astonished at the stupidity of the Londoner, Berliner, and Parisian for not cutting into allotments the beautiful parks of their cities ... In England pieces of fern land have been bought solely to preserve their distinctive plant-covering. In Canada one reserve alone of 35,000 square miles has been made.
>
> If at present a large number of adults in this country do not appreciate its beauties, then we must try to teach our children to do so.[101]

Edward Phillips Turner, Inspector of Scenic Reserves, secretary and later Director of the Department of Forestry. E. H. F. Swain wrote of him: 'A brave life; a true soul; an eye for the skies; a carriage distinguishing him in the forests of the world through which he trudged. Such was Edward Phillips Turner, gallant gentleman, surveyor, forester.'

ANZ, AAQA, 6506, Box 27, 902.1, Directors-General, 15921

What did these different perspectives on the value of bush mean for fire prevention and protection in the state's ever-expanding forestry assets? Climatic and protection concerns saw government continuing to reserve huge tracts of land clothed with indigenous forest. By 1895, over 456,000 hectares were designated as state forest. In 1900 alone the government added over 281,000 hectares of land from the Tararua and Ruahine ranges, and extending towards East Cape – the main backbone of the North Island mountains. The land was acquired as protection forest, especially to preserve the vegetation at the headwaters of the major rivers. By 1902 over 1.03 million hectares of indigenous forest and plantations of exotics were under reserve; by 1919 the state owned, under a variety of acts and designations, 4.24 million hectares, or 15.9 per cent of the country's 26.65 million hectares in total (though millable timber covered only 2.08 per cent[102]). As we shall see in Chapter Three, a huge exotic planting programme in the late 1920s and 1930s would significantly extend state holdings. The bush's perceived ability to regenerate itself and create new forests at a lower cost than planting exotics added a new economic imperative to preserve.[103] Attitudes to fire and fire prevention and protection practices did change, but on the whole were focused on state rather than private lands.

In the decade or so after the abolition of the State Forest Department (1887) the fire prevention measures the government took, even on its own lands, were paltry. The three rangers in the North Island continued the impossible task of patrolling the valuable and flammable kauri forests to stop trespass and guard against fire. The fires, and the consequent monetary loss, continued. A conservative Liberal MP Richard Monk, a landowner, farmer of some repute, educationist and speaker of Maori, estimated that one group of fires in 1885 lost the country an extraordinary £1.29 million through its effects on timber value and subsequently on costs such as wages and freight.[104] Although under the Land Act 1892 lessees or licensees of lands held for pastoral purposes were charged with preventing timber or bush on the licensed land from being burned, such provisions did not strengthen the government's ability to prevent fire or prosecute for starting fires.

The 1898 Act to Regulate the Kauri-gum Industry aimed to prevent unlicensed digging and limit access to the diggings. (In spite of their obvious self-interest in preventing fires, gum-diggers, along with other marginalised social groups such as Maori and swaggers, were persistently seen as responsible for them. Reading the debates on the legislation today it is easier to identify xenophobia rather than a concern with fire as the driving force behind the legislation. The 'Austrians', some politicians argued, were merely in the country to 'denude' the bush in a systematic and organised way';[105] their 'sole object was to earn quickly and easily' the £4 or £5 weekly 'with which to return to their native land.[106]) To point out, as Monk did, the futility of the measures given the capacity of fires to spread, was to waste one's breath. The legislation fell far short of the considered, specific and practical provisions he had proposed in an 1886 publication to address the problem of fires in the kauri forests.[107] George Perrin, reporting to government a year after Monk's publication, detailed forward-looking provisions for a Fire Act: 'the most urgently needed of all measures for forest reform in New Zealand'.[108] The report sank without trace. Settlement remained the major imperative; even Monk's provisions were drawn up with a view to avoiding 'unreasonably interfer[ing] with the arrangements incident to settlement'. For its part, the Department of Lands and Survey continued to accept fires as inevitable and, particularly in the northern kauri forests, unstoppable. It did not, therefore, address the question of fire prevention.

In 1908 the State Forests Act (which repealed the 1885 act and its 1886 amendment) was enacted to reserve, control and manage state forests. It gave the governor power to regulate against destruction within the forests (s.15(h)), as well as, specifically, to prevent 'the danger and spread of fire (s.15(i)) and trespass and to regulate access (s.17) – factors seen to have an impact on fire.

Infringements could incur a fine of up to £50 – almost $8000 in today's values. Under this act the Forestry Department was split into two divisions. The Forest Conservation Division was to care for, inspect and deal with the remaining indigenous forests. Its staff consisted of local commissioners, timber experts, Crown land rangers and conservators of state forests, who were responsible for different parts of the Crown's holdings under the various fragmented legislative arrangements. In the Afforestation Division, under the chief forester and superintendent of forest nurseries, were assistant foresters and nurserymen.[109] The act, which reflected similar approaches to many Australian states, was represented as a revamping of New Zealand's forest legislation and administrative structures to deal with the concerns about the over-exploitation and destructive burning of forests while a wood-hungry world was fearing an international timber shortage.[110]

Efforts to manage exploitation, to conserve native trees for future milling and to prevent the burning of indigenous forest were limited to the northern kauri forests. In the exotic nurseries and plantations, fire preventative practices had been introduced even before the 1908 act. By 1900 firebreaks had been established and cleared (by horse and plough or harrow)[111] and, as the trees matured, sheep were grazed to reduce the otherwise labour-intensive hand-clearing. (Puhipuhi plantation in 1907, however, had to be cleared by hand because the ground was so steep.[112])

Of course there was no guarantee of immunity from dangers outside the nurseries and plantations. In 1903 a careless match ignited the tussock of a neighbouring run and fire spread to the Tekapo plantation, destroying about 20 hectares. Access roads might form effective breaks but the Public Works Act prevented them from being closed, and plantations and nurseries were endangered by visitors' and roadmen's insouciance with fire. The considerable time forestry staff had to devote to watching for fires was to the obvious detriment of their other work. Keeping people out of the area was the only way staff thought fires could be prevented. In 1906 under-secretary Marchant was quite prepared to challenge the act, close the Conical Hill road 'and fight out the matter if anyone attempts to trespass'.[113] At Waiotapu plantation just out of Rotorua access to 'the Sights' and geothermal activity made such moves impossible and fire signs were erected at all relevant turnstiles.[114]

By 1911 fire prevention was recognised as 'one of the most difficult problems' facing the Afforestation Division, and a wider range of responses was being developed. By then some 58 kilometres of firebreak roads had been formed in South Island plantations and more were being constructed using 'continental ideas' on size.[115] Breaks in planting wide enough for traffic ran along ridges to allow ready access to fires from all directions. Along the all-important boundary breaks, fire-resistant trees were planted and the ground kept clear of inflammable debris so that encroaching fires could be easily beaten out.[116] Yet such measures were 'altogether inadequate', commented Leonard Cockayne in the report of the Royal Commission on Forestry in 1913. He recommended 40-metre-wide belts of fire-resistant trees (possibly poplar, but he also suggested trialling New Zealand fuchsia and wineberry) on the outsides of the plantations, backed by firebreaks 20 metres wide and installed before the nurseries were established.[117]

By 1913 there were nearly 12,400 hectares of exotics in plantations and nurseries. The size of this investment made issues of fire prevention and control imperative. The level of precautions taken varied widely outside the big plantations and nurseries. At Waiotapu and Whakarewarewa plantations, which extended over some 4856 hectares, internal and external firebreaks of 30 metres (or at least 20 metres if the terrain precluded the wider break) had been created, and adjoining public roads and breaks were kept clear by burning or grazing. Sheep were grazed to help offset an annual cost of tenpence for every acre planted. Blocks of planting did not exceed 600 acres (243 hectares).[118] By 1917 Waiotapu had boundary breaks of up to 100 metres, which were disced and ploughed and could be broadened if necessary. Where fernland adjoined the boundary, the ground up to 5 metres away was cleared of debris. In the South Island R. G. Robinson, the superintending nurseryman, found that considerable heat was needed to ignite the fallen autumn foliage of *Populus deltoides* along firebreaks, and that fire could be quickly extinguished, so the poplar was planted accordingly.[119]

The width of firebreaks was also subject to terrain and budgets. After the 1916 Hanmer Springs fire, work intensified on strengthening the weak spots in the plantation firebreaks. Where the tussock had to be chopped out manually from the rocky, hilly countryside, the cost of a 40-metre break represented 'almost a prohibitive figure and in such cases a lessening in width is the only alternative'.[120] However, there, as elsewhere, grazing sheep tended to be profitable and helped to offset costs, even if it was not a trouble-free venture. Dogs were worrying the sheep on the Whakarewarewa plantation in 1919. On the Dusky and Conical Hill plantations, Robinson's successor, J. D. Buchanan, hoped to solve the problems of managing the 965 sheep by getting a returned serviceman as a shepherd. He was also facing a winter feed shortage and could not sell off sheep because the market had slumped.[121]

Attention was also turning to fire control. By 1913 at Waiotapu and Whakarewarewa something of a fire protocol had been developed. During the dry season an officer was on hand at all times and other men were employed as watchers to strengthen the capacity for a rapid strike. The firebreaks allowed quick access to the fires; there were (unspecified) fire-fighting tools and woollen fire-fighting clothes; and drinking water was to be carried so the men could go on fighting.[122] But such a regime did not allow for a rapid response to fires in the weekend; nor did it overcome the difficulty of maintaining a watch along a boundary of some 40 kilometres in length. For the 1914 summer a well-equipped force was on constant duty. Three rangers at Whakarewarewa and two at Waiotapu patrolled boundaries in the summer months. Their accommodation had a phone connection through to the officer in charge. At Whakarewarewa one house served as a fire depot with fire-fighting tools for 10 to 12 men; the two other huts were positioned to overlook several kilometres of boundary. In dry weather, men took turns at staying over on Sundays – an old Health Department kitchen and dining shelter was converted for their accommodation – so they were available as an immediate fighting force.

Fire depot, South Island plantation, 1914. By then the Conical Hill plantation had 11 of these depots and others were being installed at Hanmer Springs and in Central Otago stations. The clearly marked galvanised iron boxes stored 10 handled sack fire-beaters, two pipe torches, five shovels, two small axes, four buckets, one water-bag, three pannikins, one oil-bottle and waste. The sacks were impregnated with antipyrene to make them less flammable; the nurseryman was considering getting wire ones. All employees knew where the depots were and how to fight any outbreaks.

AJHR, 1914, C.–4, 1B

The isolated and medium-sized South Island plantations did not warrant such appointments, but at Conical Hill, 11 fire-fighting depots were established at accessible points in the forest, and similar measures were being taken at Dusky Hill and the Central Otago and Hanmer Springs plantations.[123] In 1917 firebreaks and fire-ranger duties at Whakarewarewa cost £270. There are no comparable figures for the other plantations. Enforcement, at last, was also beginning to be taken seriously. In 1913 the Department of Lands, reporting convictions for relatively minor fires that had spread to state forests, hoped the fines might be a deterrent. Actually securing convictions would continue to present difficulties, but those convictions signalled an advance on Mathew's frustration in 1907: 'ample powers are provided to deal with most offenders – what is wanting is *enforcing* of such laws as already exist'.[124]

From 1909 to 1920 there were only two relatively minor fires within New Zealand's plantations, and fires outside the plantations were either prevented from entering or extinguished.[125] In 1913 superintending nurseryman Robinson commented on the ease of establishing efficient fire protection. He said the results of sheep grazing on the firebreaks 'greatly surpass[ed] our most sanguine expectations', and made the breaks effective and economical.[126] In 1915, after a forestry conference in Scotland and a tour of the US and Europe, he reported that '[i]n no place so visited was such adequate provision against the prevention and spread of fire in artificially-raised plantations noticed as in New Zealand'.[127] Such comments suggest a sense of pride in what had been achieved. His overseas visit also marks a developing characteristic of the department: it was open to ideas from overseas; its staff were reading books and absorbing information in the start of a long and productive process of contributing their ideas, gleaned from reading and experience, for improving fire protection.[128]

The New Zealand Forestry League was established in 1916 as a lobby group to promote state forestry, the preservation of forest areas and the interests of forestry, and to take on a broad public educational role.[129] Along with the newly formed National Society for the Protection and Preservation of New Zealand Forests and Bird-life, it helped disseminate ideas to a wider public. At this time it must have seemed as if fire, at least within exotic forests, could be managed and controlled. Perhaps, too, it was even possible to take on a wider responsibility. Overseas a huge amount of work was going into 'the judicious conservation of natural forests'.[130]

But how far did those enthusiasms and beliefs extend into New Zealand settler society? From 1895 to 1920 forest fire losses averaged 40,000 hectares per annum – a combined loss of about 1.01 million hectares of virgin bush, with 'North Auckland, Thames, the central backbone of the North Island, the Nelson and Marlborough Provinces and the eastern slopes of the Southern Alps' reduced to barren waste – a loss of £1 million (almost $182 million today) annually.[131] The extent of settler (and government) carelessness and indifference that allowed this level of destruction comes across strongly in Phillips Turner's reports on the scenic reserves. By 1916 about 114,000 hectares had been set aside to 'form a beautiful and increasingly valuable asset'.[132] But that value remained secondary to the need for settlement: the Scenery Preservation Board regularly assured the government that land useful for settlement would not be

Tangarakau River at the junction with the Wanganui River. Scenes such as this quiet stretch of river, or of dramatic gorges or great stands of bush, accompanied the Scenery Preservation Board's reports.

AJHR, 1912, C.–6

withheld for aesthetic purposes, while 'Farmer Bill' Massey's policies meant settlement paid scant regard to the intrinsic values of the bush.

Year after year Phillips Turner battled against a 'spirit of destruction'. He wanted belts of cleared land around the reserves, drought-resistant grasses sown, and fencing to keep cattle out, to prevent the bush from drying and becoming more flammable.[133] Recommendations in 1910 to prohibit burning near reserves or, in 1911, for special rangers with prosecution powers, fire notices in *Land Guides* (quarterly publications with information on Crown lands open to selection) and daily press warnings about burning on Crown property – all small, cheap measures that could save forest worth thousands of pounds[134] – went ignored by government and settlers alike. The latter, wanting the land to make a profit, not a home, remained

> indifferent to, or hostile to, forest preservation. If one points out to them what has been the result of deforestation in other countries, one is considered a faddist. The evil effects of deforestation in the back districts have not yet been sufficiently serious to impress them.[135]

Settlers continued to burn, completely insouciant as to 'where the fire would end', Phillips Turner reported after the 1918 Raetihi fires. In 1919, and briefly chief officer of the new State Forest Service, he dismissed the notion that fires were inevitable: 'Fire protection is simply a matter of staff and the adoption of correct methods … there is no more reason for accepting [forest] fires as unavoidable' than for considering city fires unavoidable.[136] Clearly, however, restrictive provisions could go only so far. If the fires were to be stopped, if the vast timber resource was to be safeguarded, that settler mentality had to be addressed.

CHAPTER THREE

'A practical, constructive, and well-ordered' fire policy and its implementation 1919–27

In 1919 Leon MacIntosh Ellis, long remembered as 'that florid-cheeked Canadian in shirt sleeves',[1] was appointed director of the new, professional and stand-alone State Forest Service. He arrived in 1920, with an impressive background. He had a degree in forestry; his work experience ranged from 'dirt' forestry jobs with Canadian timber companies to six years in forest management, protection, utilisation, silviculture and economics in the James Bay and Canadian Pacific Railway; he went with the Canadian Forestry Corp to France during World War I; and just prior to his New Zealand appointment he had worked briefly as an advisory forestry officer in the Board of Agriculture in Scotland.[2] A 'big picture' man, blunt, energetic and forceful (or, as he himself termed it, cleanly aggressive[3]) and prepared to trample over public-service conventions (characteristics that could also be construed as being bold to the point of reckless[4]), he arrived well versed in the mission of Americans Gifford Pinchot and Bernhard Fernow to have the US Forest Service manage its forests on scientific lines and to achieve the 'wise use' of forest resources for 'the greatest good for the greatest number for the longest time'.[5]

Ellis wanted to make New Zealand self-sufficient in wood. He had a clear vision of the value of the indigenous forests as a national asset and of their role as both protection forest and a commercial asset waiting to be enhanced by new approaches in silviculture and in processing and manufacturing. He argued that the state, by far the biggest holder of forested land in the country, should facilitate the resource's efficient use[6] by adopting a conventional, sustained-yield forest management plan, administered by a competent state forest service with the powers needed to protect, administer and manage all Crown forest lands and national parks. But while Ellis initially envisaged the role of exotic forestry, planted on unproductive land, as primarily supplementing what was seen as

Leon MacIntosh Ellis, a kinder, younger photo than many others. After his sudden departure from New Zealand (basically because of dissatisfaction over remuneration) Ellis took up private consulting in Australia. His death in 1941 went largely unrecognised here.

ANZ, AAQA 6506, Box 28, folder 11/2, 901.1, M1289

rapidly diminishing indigenous forest, his later perception that planting *Pinus radiata* would involve a small cost for a potentially huge economic return, led, under his stewardship, to a vast expansion of state exotic forestry.

Ellis's Forests Act 1921–22 established the direction and *modus vivendi* for state forestry in New Zealand long after he resigned in 1928. Initially drawn up to establish the principal legal framework by which the indigenous resource would be protected and managed,[7] it provided for a Forest Service with appropriately technically skilled staff, housing, adequate salaries and salary scales and other civil service machinery; a forest products laboratory; and two schools of forestry. However, by 1925 renewed concern about a rapidly diminishing indigenous forest resource had Ellis turning to large-scale planting of exotics. That year, and envisaging a 121,406-hectare planting programme to be accomplished within a decade, representing a tenfold increase on then current planting rates,[8] he reported that 44,726 hectares had been planted that year, a 'British Empire record' in both extent and formation cost.[9] Even more grandiose plans followed two years later when he recognised the potential for a pulp and paper industry. The act would be increasingly used to develop and protect an ever-expanding exotic forest and forestry industry. It was largely in those forests, and in developing that industry, that the Forest Service developed and honed its fire-fighting skills.

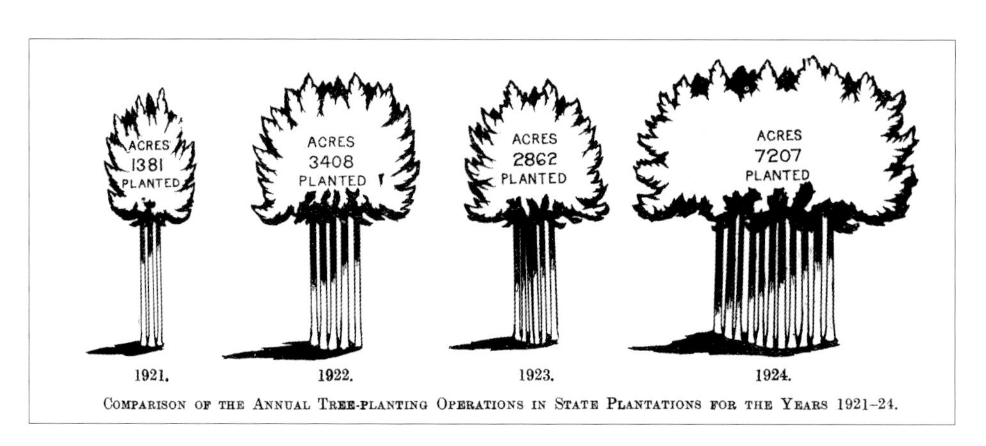

COMPARISON OF THE ANNUAL TREE-PLANTING OPERATIONS IN STATE PLANTATIONS FOR THE YEARS 1921–24.

Comparison of the annual planting operations in state forests for the years 1921–24. A considerable amount of this planting was in the cobalt-sick lands in the central North Island where there was no competition from agriculture.

AJHR, 1924, C.–3, p. 2

When Ellis arrived, protecting indigenous bush was a pressing concern. He immediately compiled a report called *Forest Conditions in New Zealand, and the Proposals for a New Zealand Forest Policy.*[10] This left him well placed to pronounce authoritatively on the constant onslaught bush faced from insects, deer, possums, wild pigs, goats – and fire.[11] In 1921 he voiced his concern to his minister, Sir Francis Dillon Bell, in a typically direct fashion:

> During the past generation two and half million acres [1.01 million hectares] of virgin timber-land has been destroyed, and in its place is useless barren waste ... During this year the orgy of destruction was maintained and over 50,000 acres of State woodland went up in smoke. Your Director is appalled at the apathy and indifference displayed at this wanton decimation ...[12]

Good timber being destroyed, and fires were being lit that had a real potential to become unstoppable infernos. The recent history of fires in Victoria, Australia, alone illustrate the difficulties of containing, let alone suppressing, fire. In 2009, after years of pervasive drought, fires whose energy was 1500

'Puffer Valley [in the Akatarawa region] swept by fire', January 1928.

ANZ, AAQA 6506, folder 8/30, 435, H701

times that of the atom bomb dropped on Hiroshima wiped out whole communities and killed 173 people;[13] the 1983 Ash Wednesday fires killed 75 people and destroyed over 2000 homes; the 1939 Black Friday fires killed 71 people and a quarter of the state was burned. In the first decades of the 20th century fires not extinguished in their initial stages became almost impossible to put

'Fire in Raurimu township, 1925.' Whatever the need to clear the land, inhabitants of small bush towns must have faced the fire season with some anxieties.

ANZ, AAQA 6506, Box 21, folder 8/30, 435, 2306

out. Prevention was the only way to safeguard the forest resource and this was the primary focus of Ellis's fire policies and practices; it was also the aim of the provisions in the Forests Act.

In 1922, aiming for 100 per cent fire prevention within state forests within a few years, Ellis pointed out that '[t]his is the first year in the history of New Zealand that any conscious and planned effort has been made to deal with the arch enemy of destruction and waste in its forest resources'.[14] However one reads Ellis's target (was it a policy goal rather than a pragmatic assessment?) his Forests Act 1921–22 moved away from a punitive and restrictive approach. (I use the word 'his' advisedly. Ellis was vital to its conception and drafting and in persuading politicians and public alike that this 'practical' approach was needed.[15]) The act's strength, Ellis considered, lay in its implementation:

> Effective control [of fires] is not so much a matter of statute or regulation, although adequate fines and penalties are desirable and necessary, as it is a matter of public forest consciousness and appreciation of the forest as a tangible asset … It is only by co-operation in the best sense of the word between the forest officers, settlers, travellers, hunters, and loggers that successful fire-control

'**Millions of trees for farmers** produced by modern nursery practice and present-day machine methods by State Forest Service.' Planting on a vast scale was encouraging settlers to realise trees as a 'tangible asset' within their everyday lives. Eucalypts were being planted for fencing timber, *Pinus insignus* for shelterbelts. Leaflets and posters fanned 'the widespread interest' in forestry. In 1922, 627,950 trees were despatched to farmers, public bodies, school committees and soldier settlers; others wanted advice or seeds.

AJHR, 1925, C.–3, facing p. 4

> will be attained. This is a matter profoundly sensed by the Service, and in its fire patrols made every effort has been to secure an actual and real bond of sympathy between the public and the responsible fire-patrol officers.[16]

Ellis was tackling the settler mentality about burning full on. However, he did not neglect the practical measures needed in fire prevention and suppression. The means he determined to use, how these were implemented and with what success we shall now investigate more closely.

As the 1920s approached, New Zealand's fire protection and prevention practices lagged behind those in North America and in some Australian states. In the US, forest fire protection on unreclaimed frontiers was neither attempted nor thought desirable: 'Many frontiersmen lived in a high fire regime and wanted to keep it; the rest knew that the inexorable evolution from settlement to civilisation would obviate the need' for other than traditional practices.[17] But although there was a general indifference to saving public land where life or property was not endangered, by the 1880s social mores and state-initiated fire codes helped regulate burning undertaken for land-management purposes. In the south, neighbours all burned off at the same time; on the Dakota grasslands plough-lines and firebreaks were regular precautions; and in Minnesota all prairie burning had to be officially supervised. As early as 1833 Maryland deemed railroads to be legally liable for damage caused by trains, companies used spark-arresters on steam engines, section gangs patrolled the lines, and rights-of-way were maintained. Federal government was first involved in 1905 when the Bureau of Forestry (soon to become the Forest Service) was formed within the Department of Agriculture, with professional staff who had training and scientific knowledge. Within five years fire control was seen as a 'fundamental obligation'.[18] A year later new legislation gave the secretary of agriculture the ability to agree to co-operate with states in organising and maintaining a system of fire protection on any private or state lands, with federal funding for up to 50 per cent of the costs. In 1924 those provisions were broadened to include all forests and critical land in state or private ownership.[19]

North America also invested in public education to facilitate prevention. By about 1920 logging operators in Washington state were required to develop and implement detailed fire plans for areas of high hazard in camps, around machinery and near railroads and logging operations. These forced them to address the nature and causes of fire and to take preventative measures. (The plans themselves were seen as vital in saving thousands of dollars.)[20] Forest associations, some officially recognised and supported, spent considerable sums annually on lecturing on the value of forests and their role in water conservation and climate. Fire Prevention Day was initiated in 1914; by 1920 it had been expanded to include forest fires and was exploiting new advertising techniques. Some publications were aimed at children; increasing mobility saw programmes being taken widely afield by car and train; the new 'moving projection' provided 'a combination of salesmanship and entertainment reminiscent of the patent medicine drummers, peripatetic revivalists and travelling circuses of the old frontier'.[21] In an innovative move that combined education with assistance, some states organised boys aged between eight and 18 into bands of Forest Scouts, who could report or put out fires. Posters and slogans – such as 'One tree can make a million matches, one match can destroy a million trees' – were developed.[22]

In Canada fewer measures were in place because conflict with farmers wanting to burn led to the withdrawal of fire rangers who had been attempting to protect small communities. However, by 1921 a number of provinces had fire wardens with powers of arrest and they could co-opt males between 19 and 60 to fire-fight (the assistance, mostly willingly given, was paid). Fire-fighting appliances and field telephones improved efficiency. A raft of restrictive measures did provide for practice to vary between states. Ontario JPs could summarily convict and impose a fine of between £10 and £24 on anyone found not helping to extinguish a fire. Permits to burn and compulsory spark-arresters on engines, on railways and logging machinery were required. British Columbia had forward-looking fire protection provisions. Its Forest Act 1911 established closed seasons when burning required a ranger-approved permit, given on inspection and on conditions relating to firebreaks and assistance for dealing with fires in high winds or following any dry spells that had occurred after the permit had been issued.[23] Permit holders were also responsible for any damage done. The act required millers to fit spark-arresters on their machinery and carry fire-fighting tools; fire laws and regulations were posted in camps; even portable mills were inspected. The Forest Service itself was organised for maximum fire-fighting efficiency. Rangers were specially trained. Duties for all officers were carefully identified, including their responsibility for their own fire-fighting equipment and replacing any loss. Water caches were developed and maintained. Mobility within forests was maximised by trails and railroads that had to be kept clear; and patrolmen's cabins built in the forests ensured an immediately available fighting force. Extensive public education resulted in emergency lists of locals who could help or provide transport; stores kept food supplies; tool caches were developed. Police reported fires; road gangs were trained on immediate suppression techniques. Fire signs were posted in schools and other public buildings, school talks were given, and special signs were printed for activities such as berry-picking (which drew the general public to the woods in greater numbers than usual).[24]

In Australia the eastern states were yet to suffer the widespread fires of 1926 in which many lost their lives, and Australian fire protection was rather less developed and more haphazard than in North America. However, by the start of the 20th century, and 20 years in advance of New Zealand, at least some Australians were no longer accepting fire as a fact of life. Forestry was developing in all states and some forestry authorities were taking on responsibility for their areas. Moreover, as water catchment areas increasingly extended into bushland, water authorities had clear imperatives to train staff in fire control.[25] In the eastern states bushfire brigades began to be developed;[26] slightly later, farmers, levied on the number of sheep they owned, had to organise themselves into these brigades, nominate leaders and get the special equipment.[27] Some states had also developed restrictive provisions. Victoria banned fires within 90 metres of a state forest and imposed penalties for lighting fires; Queensland required those exploiting state forests to protect them; New South Wales also had obligations on preventing the spread of fire;[28] and by 1921 the foresters of the Western Australian Forests Department had powers of arrest and could impress help for fire-fighting.[29]

But if the above shows that in international terms New Zealand was slow in taking protection and prevention measures, it did not mean that other countries had resolved their fire problems. Lack of equipment and funding hampered the Australian forestry and water catchment authorities' training.[30] In Washington state between 1918 and 1922, logging operations, despite their fire plans, were responsible for starting 770 fires (154 per annum on average) of 'a more or less serious nature'. The industry suffered a total loss of US$2.7 million and the fires were said to be growing more serious each year.[31]

Ellis's energy and involvement with shaping and directing an autonomous forestry department and employing specialised staff, against the trend at that time, were parallelled by his vigour and energy in drawing on overseas experience and practice to help forestry flourish in New Zealand. His first-hand acquaintance with Canadian management policies may have added edge to his oft-repeated message: 'The effective protection of the national forest estate against its arch enemy, fire, is the keystone of successful forestry.'[32] He made prevention an

Looking back along the main Rimutaka Range after fire. Undated (c. 1928). From a forestry perspective – indeed, from the country's perspective – timber that should have been managed on a sustained-yield basis, or been part of a protection forest, was now valueless.

ANZ, AAQA 6505, Box 21, folder 8/30, 435, 701

'**Flood damage, Taranaki** district: Destruction follows denudation of forests on steep hillsides.' Images like these, which appeared in the department's annual reports every couple of years, continued to drive home the message: trees were valuable as both timber and protection.

AJHR, 1935, C.–3, facing p. 26

integral part of Forest Service activities; and its fire protection measures, both legal and practical, were enhanced. Perhaps most far-reaching was a huge and ongoing public education initiative, involving a diversity of strategies to enlist the general public's interest in trees and forests and their protection. His reports and recommendations to his minister employed a blunt, uncompromising and direct tone uncommon in the New Zealand public service and speak much of the man and his energy.

Ellis arrived in New Zealand in 1920 and the Forests Act 1921–22 was passed in February 1922. Earlier relevant legislation was repealed; the 'new, clean ordinance'[33] that Ellis had wanted established a uniform approach over all Crown lands. The act certified that the governor-general could proclaim Crown lands as permanent or provisional state forests (all existing provisional state forests were deemed permanent), and that status could be revoked only by agreement within both the House of Representatives and the Legislative Council. Moreover, private lands could be bought for state forests, or taken under the Public Works Act. The impetus behind the Forests Act was primarily commercial, but its effect was that more land could gain fire protection.

The provisions for fire protection and prevention, which reflected both industry input and advice from British Columbian and Western Australian forest services,[34] gave Ellis the 'full and adequate powers' he had sought for the

New Zealand department in dealing with fire.[35] The act substantially strengthened the earlier restrictive provisions. If anybody lit, or helped anyone else light, a fire in or within 46 metres of a state forest and the forest was damaged, that person could be summarily convicted for up to three months' jail or fined £50 (over $4000 in today's values). The same penalty applied to those who left without putting a fire out properly. The fine remained the same as in the 1908 act, but the provisions around it had been tightened and the ability to fine no longer required the governor-general's agreement. Fifty pounds, too, could be paid (on the minister's approval) to anyone apart from forestry officers who gave information leading to a conviction. Any male aged over 16 who was living or working within an eight-kilometre radius of a fire outbreak in a state forest or a fire district (discussed below) could be impressed for fire-fighting duties. They were paid but anyone who refused to comply was summarily fined £25, with the onus resting on them to prove they had not refused. With the minister able to appoint locals (excluding principal agents, millers or loggers) as honorary rangers, who were then deemed to be Forest Service officers, the capacity to police the forests was increased. Finally, the governor-general had the power to regulate, to prohibit or suppress the right to light fires in or near state forests.

But the most innovative aspect of the fire protection measures, apparently adopted at a late stage from British Columbian provisions,[36] was preventative. Aware of settler imperatives and needs, Ellis did not ban fire as a land management tool. Rather he used the concept of fire districts to control lighting of fires during the danger season. The minister, on the recommendation of the Director of Forestry and the district's Land Board, could gazette land in or outside state forests, whether in Crown or private ownership, as a fire district. Once gazetted and with relevant landowners informed, access during closed seasons was prohibited except on a Forest Service officer's written permit, and no burning of any sort was allowed unless the precautions the officer prescribed were followed.

The rest of Ellis's tenure was taken up with developing his department as an effective organisation, and implementing policies and practices that would set the nature and tone of the agency's relationship with its employees, the public and the international forestry community until its disestablishment in 1987, and would influence the practice of subsequent agencies. Through communications (continued long after he had left) with his counterparts in Australia and North America, on fire issues, equipment and the regime he had instituted, Ellis put the Forest Service and its fire policies and practices on the international map. He put it in touch with international ideas. Measures to maximise his men's effectiveness included subscribing to international forestry journals, and drawing foresters' attention to relevant articles on fire policies and prevention practices. In the first of the increasingly common exchanges and training opportunities a forester was sent to Victoria in the mid-1920s.[37] (Perhaps it was that forester who brought back the Victorian Commissioner of Forests A. V. Galbraith's little booklet, with its forthright messages on 'the scourge' of fire, the value of protecting forest and the need to spend adequately to ensure good protection.[38]) Ellis expected his conservators to comment on policy and its implementation – agreeing with him was not mandatory – and he ensured that good operational practice was reported across the Service. Moreover, the flexibility he gave to his management to respond to local situations, to juggle items within budget and to act legally without reference to him[39] introduced the Forest Service to a degree of decentralisation unusual in the public sector at the time.

Ellis also moved rapidly to implement the Forests Act's fire provisions. Thirty-three honorary rangers were promptly appointed in the Wellington conservancy to work alongside Forest Service staff, warning against lighting fires and trespassing, and advising trampers, tourists and others on what they could – and could not – do in forests, and as a way of enlisting their interest in and appreciation of forestry.[40] Annual appointments of honorary rangers swelled their ranks; by 1927 there were 97. Their approach to 'their onerous duties in and out of season'[41] and their contributions in supervising otherwise unsupervised, remote districts were gratefully acknowledged each year.

The Forest Service also appointed fire patrolmen and co-operative patrolmen. In the Wellington conservancy, as early as 1921, fire patrolmen were appointed over the summer months, essentially as back-ups for hard-pressed rangers. The patrolmen focused on back roads and logging operations' railway lines rather than the bush, which was little used by hunters and campers. They posted fire signs in the most conspicuous places. They were to keep diaries and report every two weeks on expenses incurred, where they had travelled, fires seen (including those in fire districts), visits made, weather conditions and likely fire risk. Their work had a strong educational element; they had to attend any burns and advise locals, roadmen and Public Works gangs burning slash about good burning practice – don't burn on windy days; do burn in the afternoon when, it was thought, there is likely to be less wind than in the morning.[42] The men were also to assist rangers in developing firebreaks and other fire-prevention work. Pay was 13 shillings ($65 today) a day, for seven days a week; an additional three shillings daily was paid as a horse allowance if the men had their own mount – or 10 shillings a month for using their own bikes. Actual travelling costs when away from home could be reimbursed.[43] It was a system built on mutual self-interest. Patrolmen were often farmers who had a genuine interest in helping to minimise fire danger in forests near their farms,

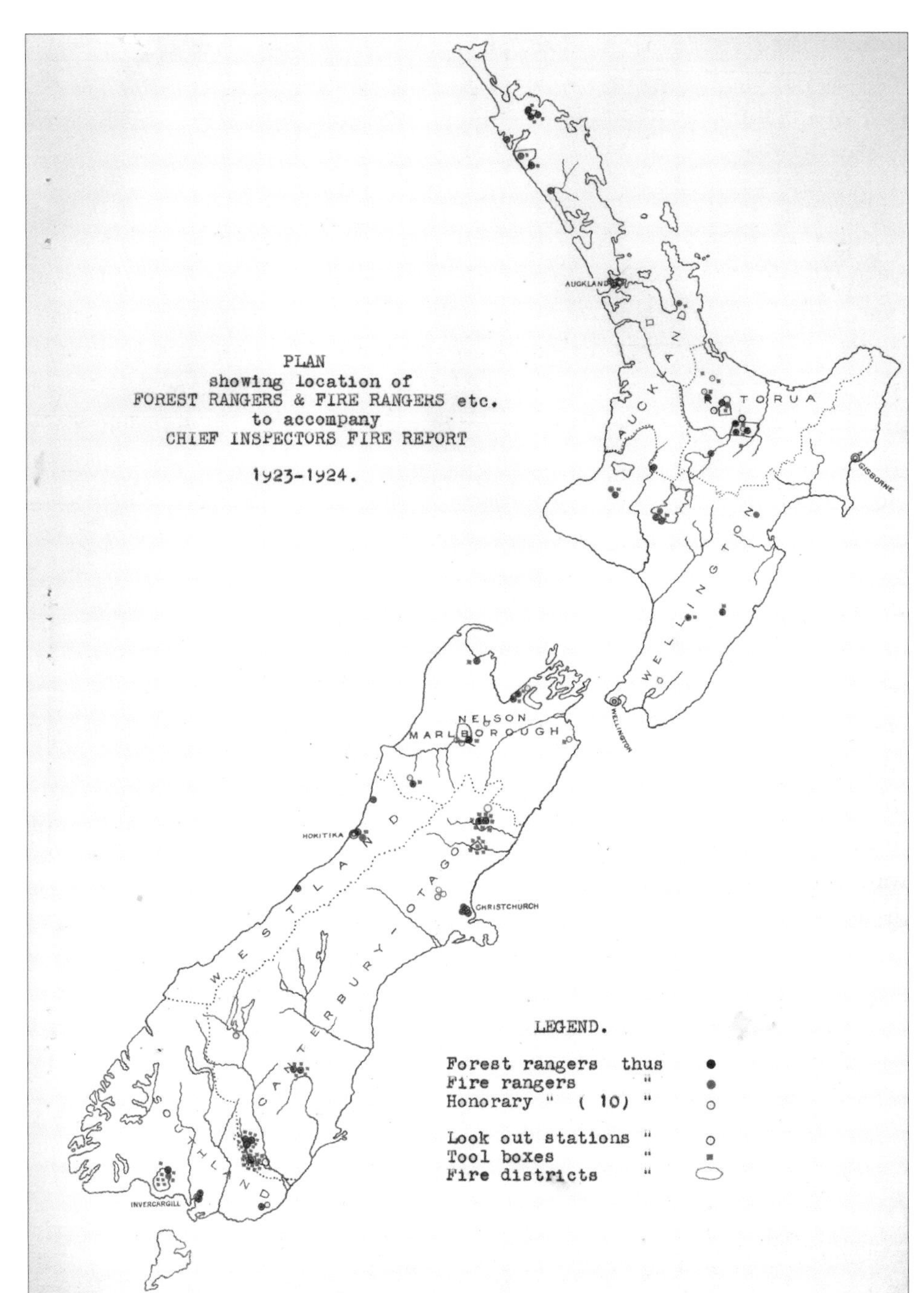

'Showing New Zealand's fire districts, distribution of tool boxes, rangers etc 1923–24.'

ANZ, F1, 12/0, Pt. 1, Annual fire report & fire protection

and, as at least one ranger realised early on, the Service received assistance with its fire-prevention and protection work without incurring salary costs.[44]

The co-operative patrolmen were employed by sawmillers to see that fires ignited by sparks from any hauler or locomotive were put out promptly, and to check that ashes were dumped in the places prepared for them.[45] The idea, again demonstrating the creativity of Forest Service staff, seems to have originated with West Coast conservator A. D. McGavock who, concerned about the number of fires on worked-over bush on the West Coast, mooted it with millers in late 1922. Their positive response led him to forward it to Wellington, where Phillips Turner promoted it.[46]

The fire district provisions were in place by the 1922–23 fire season if not earlier (some fire district plans appear to have pre-dated the enabling legislation[47]). By 1925 districts were judged to be so successful in protecting forests[48] that the provisions (following the Victorian equivalent almost word for word[49]) were extended to allow local bodies, afforestation companies and private owners or occupiers of land of over 81 hectares to apply to have that land included in a fire district. Where there was multiple ownership, unanimity was not required, facilitating the take-up of the provision. By 1927 fire districts had been proclaimed over 607,028 hectares: 29 districts were on state-owned land and two local bodies had used the provisions.[50]

The private fire districts worked in much the same way as the state ones except that the responsibilities and expenses were borne by the owners, not the state. The applicant nominated and paid someone to act as an honorary ranger, who was responsible to the applicant and was in charge of fire prevention and control. He issued permits, conditions and instructions for control. He assessed the adequacy of the fire plans. These had to specify the locations of ridges, firebreaks, fire-tool depots and lookout stations, as well as of telephone lines, water supplies and other means of fire prevention – and if the owner did not draw up a plan, the Forest Service did. Landowners made a deposit of £5 5s to

Fire district notice.
'The proclamation of fire districts is an important link in forest-fire prevention.'
AJHR, 1928, C.–3

'Motor-tractor supplants horses and locomotives and minimizes fire danger.' Although they presented some dangers from sparking, motor engines were a real improvement on locomotives. No illustrations of spark-arresters appear to have survived from this period.

AJHR, 1925, C.–3, facing p. 18

cover inspection costs. Any surplus was to be returned but applicants bore any further costs. Once approved, the application went before the local Land Board. Once the board approved the plan, the fire district was gazetted (all owners or occupiers no longer had to be notified in writing[51]) and the provisions pertaining to state forests on lighting fires, permits and precautions came into force.[52]

The 1925 amendments to the act reflected experience. The fire-ban area on boundaries of state forests was extended from the 'quite inadequate' 45.7 metres to 401 metres.[53] Cattle found trespassing in state forests could be sold or destroyed – their role in opening up the bush and making it vulnerable to fire, as well as in directly destroying it, was recognised. That year, too, following extensive trials to determine the most effective device, the Spark-arrest Regulations made approved nullifiers (spark-arresters) compulsory in all state forests and fire districts from 1 October to 30 April. Thirty-two nullifiers were installed over the 1924–25 season[54] and until the 1940s conservators were asked to supply the numbers of arresters in their districts annually – though, as we shall see, the problem was long unresolved. In 1927, after B. B. Wood of William Wood & Co. Ltd had approached Ellis about 'the constant danger' of fire in flax swamps and the very considerable investments needed to grow suitable flax,[55] Ellis began the process of amending the legislation to enable flax-growing areas to be designated as fire districts. (These provisions, initially seen as validating fire district principles, were subsequently considered unnecessary and expensive and were only passed in 1932.[56]) In 1927, too, amendments to the Police Offences Act provided that anyone lighting a fire in bush, timber, scrub or fernland they did not own could be fined up to £10 or imprisoned for up to three months.

The legislative approaches were far from the only means by which Ellis sought to save existing timber and safeguard young forests for future generations. In a phrase remembered 25 years later,[57] he fulminated against the public's indifference and apathy towards fire:

> [A]pparently 'What is everybody's business is nobody's business.' It is absolutely essential that this enormous drain – which may be conservatively estimated to result in a loss of £1,000,000 per year – be checked, and at once. Every individual citizen should concern himself and make his interest felt in the protection of New Zealand forests against fire.[58]

To create that concern Ellis called into play his considerable propaganda skills to start a huge, multi-faceted public education programme. The key, Ellis believed, was making information available: '[I]t is right that the people should have the opportunity … to be given an accountancy [*sic*] of stewardship of the 7,553,690-acre [3.06 million hectare] forest estate'.[59] His articles in relevant journals promoted his ideas and approaches,[60] and the work of his rangers and fire patrolmen in promoting fire districts and explaining their fire prevention methods helped introduce settlers – and industry after 1926 – to good practice.

To facilitate the conservators' work with Maori, the Forests Act was translated into Maori and 500 copies were printed.[61] Slogans for public posting began to be developed, with Ellis typically seeking input from his staff. Foresters responded by flooding head office with their examples, their careful drafting, colouring and shading demonstrating their originator's care and interest. Although Ellis wanted a focus on co-operation rather than punishment, he also wanted to establish that offenders would be 'prosecuted with the utmost rigour of the law'.[62] In the end efforts such as the first focused and alliterated example below gave way to signs under the Royal Arms and citing 'The Forests Act 1921–22 and regulations thereunder', as illustrated in the second example.

NOTICE
TO
BUSHMEN
Timber saved is timber made
or
PROTECT YOUR FORESTS
Wilful waste brings woeful want
You want work
Your Children want Homes

BUSH FIRES
Every person who, without lawful authority, lights a fire
in, or allows a fire to spread into this State Forest, is liable
to 3 months IMPRISONMENT or a fine of
£50.
A REWARD not exceeding £50 will be paid for information leading
to the conviction of any person guilty of the above offence.[63]

These signs, printed initially on calico and in Maori where appropriate, were distributed annually to all conservancies for posting in the most frequented parts of state forests, (although they were not always displayed).[64] In further publicity drives, information sheets on fire districts, which clearly stated the legal provisions and the penalties, were sent to owners of land included in the districts. Consistent with Ellis's view that 'more good can be done by endeavouring to enlist the assistance [of those at whom the signs were directed] in preventing fires, than by directing attention to the penalties provided by law for offences',[65] an explanatory note sought to promote assistance rather than antagonism:

> This provision is being made to safeguard valuable State property, and it is not intended to harass or restrict you in the reasonable use of fire in land-clearing operations.
>
> Your willing co-operation in the observance of the restrictions imposed under the Act aforementioned will be much appreciated. Remember that, after all, you are only being asked to assist in protecting your own property – the State Forests.[66]

In 1924 the department began to use new and increasingly sophisticated methods of reaching the public. In one hardly exciting example, a calico pictorial fire poster, its red background suggestive of fire, urged its viewers to 'Prevent Forest Fire – It Pays'. The same year, and reflecting the growing exchanges of information and ideas with the outside world, other posters (their rhymes 'preferable to the substitutes attempted in New Zealand') were copied from *American Forests & Forest Life*.[67] In 1926, as it had done in Victoria, the British Imperial Oil Company offered to print a fire leaflet, with the Forest Service to be responsible for its text and distribution. Leaflets were despatched promptly to hotels and tourist spots in Hanmer Springs, Rotorua and other popular forest areas.[68]

Attempts to capture the public's hearts and minds and gain their co-operation did not end there. Films such as *Hearts Aflame* from Harold Titus's novel *Timber* were made available for wide showing. Thousands of posters urged the populace to 'Plant a tree' or advertised trees for sale; guides on planting and forestry, picture advice cards and calendar folders were printed to popularise Forest Service policy.[69] Trees were sold at low prices. Schools, an obvious target, were visited; articles were published in the *School Journal* and the *Education Gazette*, carefully designed to fit the teachers' level of experience; the 'Forestry in Schools campaign' helped establish tree nurseries in state schools. By 1927 'elementary forestry' (sowing, caring for, studying and protecting native flora and fauna) had 'a definite place' in many school curricula, and forestry gained further prominence with school competitions sponsored by timber companies like Ellis and Burnand.[70] (The firm was established in 1903. By 1953 its bush railways, timber yards, mills, factories and 'engineering feats' made it a significant presence in the central North Island.[71])

A ranger posts a fire notice in Westland, c. 1923.

Photo: A. Hansson. ANZ, AAQA 6506, Box 32, folder 12/16, 945.2, AH254

In perhaps the most radical move – which may, more than any other, have

Farm shelterbelts, Canterbury province.

AJHR, 1921, C.-3, pp. 8–9

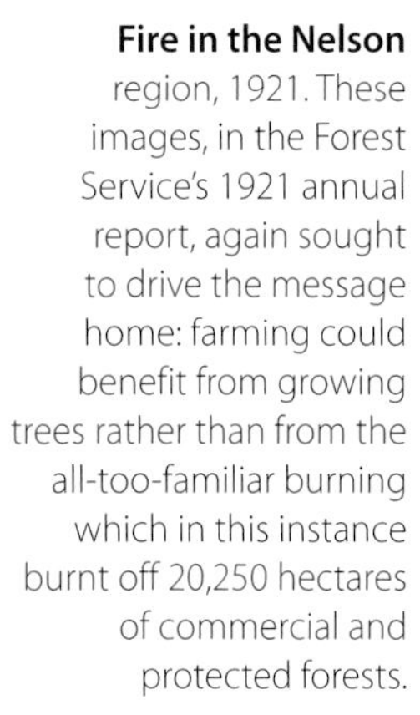

Fire in the Nelson region, 1921. These images, in the Forest Service's 1921 annual report, again sought to drive the message home: farming could benefit from growing trees rather than from the all-too-familiar burning which in this instance burnt off 20,250 hectares of commercial and protected forests.

AJHR, 1921, C.-3, pp. 8–9

benefited the lives of New Zealanders through to the present day – the indigenous state forests were opened up for recreation. In the French Vosges and Jura, Ellis had seen how tourism and recreation assisted the communities' forest conservation and management.[72] Considering this idea in New Zealand, in 1924 he wrote, 'It is only as the community values the forests that the community will demand the proper use and conservation for the forests', and began to add camping places and fireplaces in the forests, extending the opportunities for general public use.[73] Just as some climate change scientists today advocate tourism to Antarctica to create a community of interest in it and its future, so Ellis believed that to give the public access to forest, with accompanying aesthetic experiences and recreational opportunities was to give them a stake in the forests and therefore in fire prevention.

Arnold Hansson standing at a camp in Waipoua Forest, c. 1922. Hansson was a firmly principled Norwegian who had graduated with honours in forestry from Yale. A. D. McGavock, Ellis's successor, dismissed him because of conflict with superiors, but in 1929 Hansson was fresh from some 18 months' work setting up the Westland conservancy and acquiring a reputation as an extravagant foreigner because he drove the Service's Model T car on almost all of Westland's roads.

ANZ, AAQA 6506, Box 27, 901.1, AH 11

The way in which the policies were implemented was perhaps as important as the policies themselves. Ellis worked to maximise co-operation in all sorts of ways. He presented fire districts as a means of relieving sawmillers and landowners of serious liability for fires. He argued that the provisions, far from treating settlers unfairly, had some settlers outside fire districts complying voluntarily.[74] To balance settler and forestry needs and to get land boards on side, he involved the boards in approving fire districts.[75] He had a valuable ally in his chief inspector, Arnold Hansson, who could provide the practical detail needed to implement Ellis's big ideas, translating them into everyday terms and practice.[76]

Hansson and those early rangers and foresters undertook a complex job and they went about it in a way that was then quite novel. They made up a

> *corps d'elite* [on whom fell the unenviable task] … of introducing and administering the new policy at its sharpest end – in the forest and sawmill. Travelling by horseback or bicycle these crusaders rarely knew whether the day's yield would be life-long friendship, or a period of convalescence resulting from violence offered by a sawmiller exasperated beyond control by 'unwarranted government interference.' It says much for the character [of those early rangers] that the State Forest Service … quickly became accepted by the industry as a necessary and competent authority.[77]

Gaining that acceptance meant winning public respect. Hansson set out for his men what was needed:

> A timid man can accomplish little and an overbearing or officious officer can do more harm than good. A strong tactful officer will win the respect of the public. An overbearing officer may seemingly get the well will [*sic*] of the public which will … only last as long as he lets the public do as it likes.[78]

Five years later Hansson again coached rangers in presenting fire districts not as measures for harassing farmers in the reasonable use of their land but as a system with built-in self-interest benefiting forest owners, their staff, local residents and the visiting public. No one, Hansson pointed out, wanted to see any form of wealth or beauty, or even shelter trees, go up in smoke. Moreover, a fire district system, which removed serious liability for fires from sawmillers and landowners who complied with the system, was cheaper than high rates of insurance. The fire district principle meant that there were patrolmen willing, without pay, to be on the lookout for fires and to educate the district about fires. Where they were paid, they would also be available for fire-fighting. The local bodies that created fire districts would 'be assured of more support from the ratepayers in their afforestation activities while participation [would] increase shareholder confidence in afforestation companies'.[79] It was a facilitative approach that informed Forest Service dealings with its public until its disestablishment.

With the Forests Act passed, Ellis initiated a survey of the forest-fire situation and had a provisional protection plan drawn up to operate in the high-risk districts of North Auckland, the Wanganui River catchment, Ruahine Ranges, Nelson province and the lower eastern slopes of the Southern Alps. Nationally all conservators were called on to place fire signs for maximum effect, report fires, take steps to prosecute offenders and develop fire protection proposals for their conservancies. Early each fire season each conservancy received fire forms for reporting fires in state forests and adjoining areas. These forms would become more compact over the years, but the sort of information Ellis sought in the early 1920s would continue to inform the later lengthier written reports that head office wanted. Each fire had to be ascribed a name and number; its location had to be accurately described and if possible sketched. Its extent had to be carefully estimated and the nature of the country burned – merchantable or unmerchantable timber, old workings not regenerating, valuable regenerating bush – noted. Its value had to be assessed, along with that of other property destroyed; the cause of the fire established; and the history of the fire described in terms of its progress, action taken, topography, weather conditions and other pertinent topics. Other information required included what was paid out to non-government servants and whether the affected landowners had helped in the fire-fighting. Each fire was a learning experience, too – what, Ellis wanted to know, could rangers recommend from their experience of fighting that particular fire?[80] Staff were also expected, if possible, to find out who had lit the fires and to prosecute, so the public would know that fires 'were not to be trifled with'.[81]

Partial records were worse than none.[82] Ellis demanded that reports be telegraphed through weekly[83] and his irritation at those who failed to do so[84] further indicates how the information was vital to his demonstrating the efficacy of his regime to his political masters and to the general public. His 1921 annual report published the first results of his efforts. Conservancy by conservancy, it detailed what had been lost and its value, the costs of protection and prevention, and how the fires had been started.

The emphasis on preventing fires could thus take many forms: pursuing prosecutions to discourage potential offenders (rangers were constantly reminded that 'convictions are wanted'[85]); enforcing regulations on spark-arresters, ensuring effective firebreaks and keeping vigilant surveillance. With only hand tools and the hand of god (evidenced by wind shifts or rain), it was imperative to prevent fires getting away.

The exact level of fire preparedness and protection measures taken in each conservancy appear to have differed considerably. Giving a sense of this diversity are conservators' fire protection proposals for the 1921–22 fire season, which had to itemise and budget their manpower and protective infrastructure, maintenance, and plans for further development.[86]

In the Rotorua district in 1923 the Whakarewarewa forests had 13 honorary and no co-operative patrolmen. There were 883 square kilometres of native forest and 22.5 square kilometres of plantations. The district had five horses and one motor vehicle. Improvements were listed as three kilometres of trails, 46 kilometres of firebreaks and one whare. The forest had no lookout stations, cabins or tool depots. In the plantation 37 kilometres of phone lines, 43 kilometres of roads, 10.4 kilometres of trails (all essential for easy communication) and 238 kilometres of firebreaks had to be maintained. A fire in the state forest had burned 15.8 hectares; on private land 173 settlers' fires had burned 4410 hectares, destroying £562 worth of timber and regenerating bush (the latter always a fire hazard). Two of these fires were attributed to logging operations, 44 to settlers burning off, and 148 were of unknown causes.[87]

In 1922, only some three years after Canada's northwest built its first

A toolbox housing fire-fighting tools at the Whakarewarewa plantation, early 1920s.

ANZ, AAQA 6506, Box 21, folder 8/25, 432.3, F149

Hut for fire-fighting equipment, Waipoua Forest, undated. The huts had local variations.

ANZ, AAQA 6506, Box 21, folder 8/25, 432.3, M10,164

'Guarding the state forests from fire: one of the lookout stations in the Rotorua region.' The annual report states the stations were on Maungakakaramea and Mt Kakapiko. Maungakakaramea and Rainbow Mountain were names at times used interchangeably for two peaks very close together. Given that a lookout station is today situated on top of Rainbow Mountain, the first station was probably built there.

AJHR, 1922, C.–3, p. 19

stations,[88] the Rotorua district acquired two 'very stoutly built' lookout stations: one on Mt Kakapiko in the Whakarewarewa plantation and the other on the summit of Maungakakaramea, looking out over the planted areas of Kaingaroa and Waiotapu. In each tower a heavy glass window the full length of each side gave an uninterrupted view in all directions. Telephone connections (a distinct advance on messages relayed by horse, foot or bike) to various ranger stations meant outbreaks could be promptly reported and saved time investigating outbreaks too far away from the plantations to be any danger. In the South Island, lookout stations had been installed at Hanmer Springs and Conical Hill exotic plantations, and lookouts, constantly deployed, had field glasses, block maps of the plantation to help pinpoint the fire, and phones. Neither plantation had had any fires that year and settlers had co-operated willingly with protection work.[89]

Other conservancies, generally large and covered with indigenous forest, had meagre resources with which to carry out their tasks. The Wellington district conservator, responsible for some 5.96 million hectares,[90] reported that his district had one whare costing 7s 4d, two intermittent patrolmen, 23 honorary rangers and eight labourers.[91] Westland spent £14 7s on wages, nothing on equipment and £1 10s on travel and other costs. McGavock, then Westland's conservator, had employed 13 labourers for 76 hours, but no patrolmen. One fire, caused by timber operators' locomotive and hauler, had burned 405 hectares of largely valueless cut-over land.[92]

For the following year economics as well as experience saw Hansson instructing each conservancy to develop a definite fire plan for the 1924–25 fire season such as those that Wellington and Westland conservancies had. These relied on stopping fires in the first place and getting settlers to help with protective measures. Do not, he told conservancies, get tied up trying to cover all bases. Rather, a 'systematic fire protective programme' would identify the region's main hazards (the most valuable in monetary terms should be protected or saved 'at any cost';

A Forest Service field party, c. 1923. Far right is Viv Fail, later manager of Tauhara Forest during the 1946 Taupo fires; S. A. C. Darby is second from right; Hansson third from left. Darby joined the new Forest Service in 1919 and was a ranger in the Wellington conservancy until 1926. It was a huge job – half the sawmills in the country were then in his conservancy and farmers were demanding land clearing. Friendly, shrewd and energetic, Darby developed a special relationship with the disorderly millers, was trusted and respected by Maori, and congratulated by his superiors in the Service. The photo illustrates what was expected of the Forest Service men in those early days: bush-bashing and foot-slogging to get to know the resource.

ANZ, AAQA 6506, Box 29, 902.1, AH17

aim for containment for manuka or fernland), give their general location, note whether the areas were within a fire district and set out the methods for preventing or containing fires.[93] Hansson's unsophisticated and practical approach set in train what by the 1980s would become detailed organisational plans setting out the structures, tasks, available and dependable groups and contact numbers, all vital in fighting major burns.

From the 1921–22 fire season the Forest Service annual reports record settler and farmer co-operation and general public support. A closer look at the experiences of Wellington and Southland conservators in those early years of the Forest Service, however, indicates a rather more nuanced situation.

In May 1921 R. Norman Uren, a ranger in Wellington conservancy's Whangamomona county, east of Stratford, received a letter from the conservancy's senior ranger, S. A. C. Darby. (The content of the letter indicates that Ellis may well have discussed the future legislation's framework and philosophy with his conservators.)

> Strange to say, little or no attention is being paid to the ravages of fire. It is, therefore, necessary that we formulate some sort of scheme of control, as well as a means of educating directly people who live on the outlying sections and whose places are contiguous to valuable timber areas.[94]

Darby asked his men to develop action plans showing exactly where fire patrolmen might work over the dangerous summer months; the boundaries of each fire district; and the location of the patrolmen's headquarters, which was to have a phone. The men were to be properly equipped with tools, have the authority to act and means of getting around, and comply with reporting requirements to the senior ranger.

His men were willing, even if hampered by circumstance. Darby's letter was delayed so Uren had only '21 days in which to ride through Taranaki State Forests, a task which had to be pushed through in the depth of winter, when the papa roads were almost impassible [*sic*]'.[95] He could do little but arrange the fire districts, select patrolmen and get information on how far the roads had been pushed into the huge tracts of indigenous state bush set aside for conservation or protection purposes. He also hoped for a more personal orientation: to go over each patrolman's district with him, check the boundaries, introduce him to the neighbouring farmers and so on.

An ambitious plan was drawn up for Ohakune (assessed as low fire risk), Whangamomona (moderate risk) and Dannevirke (heavy fire risk because settlers were burning to increase grazing). The main danger sources were identified: settlers' burning off, a certain incendiarism, thoughtless and indifferent hunters and campers, and lack of spark-arresters. Suggested means of control included using the legislation, patrolmen, education (including 'the personal touch') and, borrowing from practice of the US, enlisting Boy Scouts' assistance and rewarding them with badges and honorific titles. Eighteen patrolmen would be needed for four months. In total, wages for them and for fire-fighters, travelling and horse allowances, and equipment would cost £2300.[96]

The plan had barely been developed before Darby told Uren that, pending legislation and finance, it could at best be tentative, and that Ellis was already urging his staff to halve their budgets; by late September Darby told Uren that he, like Ohakune ranger Norman Dolamore, would have only one patrolman. A phone for the Ohakune headquarters was also turned down, owing to 'financial stringency'.[97]

In the meantime, in late August Uren had made his initial requisition. The few roads constructed were becoming passable, but many were accessible only by horse. Uren had two horses but no pack saddles, which he intended to borrow from the Defence Department as it always had a lot in store. A tent, fly, bucket and two small billies were also needed.

In early October Uren was again in touch with Darby. Of the potential fire districts he had earlier identified he now recommended one: 7932 hectares of provisional state forest in the Waro survey district. It was the best bush he had yet seen, close to a main road, and the settlers in the vicinity had not yet burned off. The fire district should be proclaimed from November on, he thought. He knew of a good man to act as patrolman – H. C. Roberts of Whangamomona was single and educated, and had a phone, a horse and lots of tact as well as a thorough knowledge of the area and its settlers. Uren also knew several 'gentlemen' with a strong interest in the Forest Service who, as honorary rangers, would look after its interests near their farms. Why not appoint them, and further reduce the fire risk without increasing the salary bill? It was a suggestion that Darby heartily approved.[98]

On 24 January 1922 Uren reported that burning off in the district had started three days earlier, and to date 202 hectares had been burned. He had been notified of and present at each burn, and little, if any, standing bush had been damaged. However, expecting another 364 hectares to go in the Tangarakau Gorge, he had someone there keeping in touch with the setters. Roberts had been most useful.

By mid-February, despite continued dry weather and almost continuous burning, not an acre of the provisional bush had been lost. However, 'the most strenuous work and long hard riding trips' had knocked about his two patrolmen's horses, and he had had to provide others.[99] Nor, with the settlers continuing to co-operate with notifying their burning, were those long days of riding over. To add to his woes, settlers were reporting fires – but by letter! Suggest collect telegrams, Darby instructed.

By the end of March the fire season was effectively over, and on 10 May 1922 the results of the season were finally collated for the whole of the Wellington region. In all, £350 had been spent: three patrolmen had been employed for 94 days and one casual labourer for one day. On state forest land four fires had destroyed a negligible 10 shillings' worth of regenerating bush; on private land 23 fires had destroyed 775 hectares of timberland – a loss worth £1400 ($130,000). Three fires were attributed to travellers, 20 to land clearing and, on state land, one was caused by roadmen.

In June 1922 Uren set out his plans for the next fire season. He wanted the Waro survey district and blocks in the inland Taranaki survey districts gazetted as one fire district. This area would include timber stands with 'the greatest Scenic value in any of Taranaki'. A. W. Bonner would be the patrolman, dedicated to that district only. From his house on Tangarakau Valley Road he could keep in touch with the whole area by making day trips. Roberts would cover Whangamomona county to Bonner's boundary and the Clifton county to Ureti; while Mr Smith of Hurleyville would keep an eye on State Forest 15 and the reserves near his home. Wages would be as last year: 14s per day and 3s for the horse. Uren recommended too that someone go over the whole district in July to arrange burning days. He and Roberts had done that the previous year, had got the whole district tabulated and ensured someone representing the Forest Service was present at all burns – a process that, in 1922, Ellis recommended for other conservancies.[100] Budgeting was essential – only £355 was

Duncan Macpherson. We meet him later as Wellington conservator of forests in Palmerston North, still battling trains that were not using spark-arresters. He retired from the Forest Service in 1938 when he took charge of Parks and Reserves in Christchurch. He died in 1954.

ANZ, AAQA 6506, Box 28, folder 10/42, 901.1, M10,530

available for the entire region – but if he could spend £25 building a small depot with a stretcher, a fireplace, fire-fighting tools, buying chaff for the horses and getting a locker for provisions, he would have something that would support the Service for many years.[101]

On 22 January 1923 a 26-year-old was fined £2 2s (almost $200) for lighting a fire, the first conviction for the Wellington conservancy under the new act. In December's preparations for the 1923–24 fire season, depots were established and patrolmen were active in the most dangerous areas. Although settlers were lighting potentially destructive fires, no damage was done at time of reporting.

Ranger Uren's energy, his preparedness to put forward ideas and his superior's willingness to listen to them, and the elements of devolved authority were part of the environment that Ellis fostered. However, we need only look at the Southland conservancy to appreciate how widely experiences differed.

Southland conservator Duncan Macpherson wrote at length to Ellis in July 1921 on the challenges facing Southland's state forests, where millers were active. Milling regions could not be treated like afforestation areas, Macpherson thought. Firebreaks, useful for stopping grass, fern or tussock fires so they could be beaten out, were impractical in an environment where flying sparks could ignite trees 100 to 200 metres away. Two early proposals (one of which was to minimise expenses by selecting men 'who are not frightened to do a lot of hard work in the way of cycling'[102]) were costed out and submitted. One cost £338 all up; the other, more ambitious but dealing with the ever-present danger of sparks from fire engines, cost £774. As if anticipating Hansson's later advice not to try to cover all bases, in September he developed a more budget-conscious version. One fire patrol would be employed for four months at Riverton to travel around specified state forests and educate on the fire danger, particularly in relation to engine sparks; one of the rangers could attend to the rest. He ticked off Stewart Island as not needing attention because its general conditions did not favour fire, and the Invercargill area because a milling company owned all the timber. But at the head of Lake Wakatipu, G. Reid, the manager of the Glenorchy Scheelite Mining Company, who ran a small mill at Paradise, might keep a general eye on the fire situation, including fire lighting and timber poaching, for £3 a month. Although Ellis approved the plan and left the decision on the patrolman 'entirely in your hands', Macpherson wanted further reassurance of the devolved authority. Ellis restated his position: 'I have to state that provided you keep within your allotments and the authorised rates of pay there is absolutely no necessity for you to refer to me for authority to engage the fire patrol men.'[103]

The Southland statistics for fires for the wet, cold 1921–22 summer and autumn show no fires in or adjacent to state forests.[104] Macpherson proposed the same arrangement for the 1922–23 fire season, costing them out at £100 and noting that he was not planning any improvements or maintenance. The results were similarly pleasing. His prevention, protection and detection measures cost £53 more than anticipated, but there was only one fire from a burn-off and only over two hectares of valueless private cut-over land.[105]

In the following fire season the whole country was affected by drought, particularly from December to February. In Southland, as elsewhere, the number of fires soared – particularly on private lands. But as earlier generations of conservators had found, enforcement posed its own challenges. In 1923 Macpherson was sure musterers working in the area at the time were responsible for a fire. But without reliable evidence he could only write to the manager warning of future liability and, in keeping with Ellis's regime, asking for help with suppression work.

In 1924 Macpherson began tackling more aggressively what Ellis described as 'perhaps the greatest enemy to the forest ... the locomotive unequipped with spark-arresters'. The problem was not new, nor limited to Southland. Ellis was already concerned about the 'negligible' liability New Zealand Railways faced for the fires on its tracks.[106] In 1921 engineer Alex Entrican had urged conservators to encourage millers to use spark-arresters for machinery as an important preventative measure.[107] There is no record of how many companies complied, but in April 1924 Macpherson reported that locomotives belonging to James More & Sons (logging in a state forest under licence) were responsible for four fires that month. Three were in smouldering sawdust and had gone unnoticed; one could not be put out without water, 'for which some time was spent in getting buckets'. This had left the mill a wreck, all engines in the factory irrevocably twisted and only the chimney, boilers and perhaps the engine salvageable. Moreover, More & Sons' trains and 'the Otautau Timber Co's loco cause fires on practically every day on which a strong North-west wind is blowing.

Taringamutu Totara Co.'s sawmill, c. 1910. This mill was in the central North Island, at what is now Taringamotu, close to Taumarunui. But the set-up – the drying logs, the mill's proximity to the line – illustrates the milling practices Macpherson was dealing with.

NL, ½-055330-F

Cinders about an inch in diameter are ejected when barking up a hill under pressure.' Spark-arresters, which caused the engines to lose power, were taken off as soon as the engines were out of sight. Further precautions were essential. Macpherson thought the best method of dealing with the problem was to have a jigger with a kerosene bucket and a hand-spray pump following the train. 'If this is insisted on it is probable that the miller may discover that some existing spark arresters are not so bad after all'[108] – but to date the company had not taken kindly to either his suggestions or his attempts to enforce the legislation.

The following day Macpherson wrote to More & Sons' manager, describing the circumstances of the fires, pointing out that the wind was causing them to spread and that his employees were doing nothing to extinguish a small fire near a cottage. Their locomotives appeared to be responsible for fires in the area. He recognised that the much-advertised spark-arresters appeared to be of little practical use, and thought having a jigger equipped with a hand pump and water following the locomotive was an effective, practical response. He asked to be informed of what More & Sons planned to do.

It was a courteous letter, recognising the problems the company faced and written without officialese. But at the time Ellis was accusing millers of price-fixing, and More & Sons' manager was in no mood to co-operate. The company was doing all it could, he replied tersely, and no government timber had been lost through any fault of the company's. Spark-arresters had been fitted at the beginning of the fire season and were removed only at its end, the day before Macpherson's visit. He wholeheartedly rejected Macpherson's suggestion. His employees were always on the lookout for fire, and '[w]e do not think from our experience that there is any necessity to have a fire brigade following the locomotive'.[109]

That correspondence ended. In August Macpherson started concentrating on the next fire season. He identified the different threats faced by the different areas of the conservancy, drew up new fire districts and nominated patrolmen to seek co-operation from mill-owners, timber workers and farmers, and to encourage early burning around the danger areas. He also drew up the budget for the next fire season.

There were two fires out of season. One Macpherson thought had been started by a youth shooting pigeons, but this was denied in the face of clear evidence against him. In October small fires in grass roots and in smouldering peat, which were almost impossible to extinguish, burned a cottage, stacks of wood and 14 hectares of regenerating bush. This time More & Sons was the culprit, but Macpherson considered it would be very difficult substantiate a claim to the loss of 14 hectares of regenerating land in a court of law.[110] In addition, in this instance More's men had done a lot of the fire-fighting. Macpherson thought that the threat of action, the loss of cottage and grazing rights (already disputed), and the cost of the men's time would be sufficient deterrent.

Then in January 1925, after damage to 101 hectares of cut-over bush, More & Sons was again in the spotlight. This time Macpherson took firm action. On 23 January patrolman S. H. Herron recorded, in writing, that a Mr E. Trail said not only that the company's engine was the cause of his fire but also that each time the locomotive passed his sawmill he or one of his men had to follow it to extinguish the fires. A few days later Macpherson informed the company that he was holding it responsible. It had a period of grace in which to fit spark-arresters on both locomotives and haulers; in the meantime it was to maintain fire-fighting equipment, including spray-pumps and buckets, on locomotives and haulers alike. If it failed to comply its operations would be suspended. He would be in touch or would visit shortly. A few days later he cancelled More & Sons' grazing licence, which had expired some months earlier, and asked it to remove its stock.

But the issue, Macpherson knew, was bigger than More & Sons. His attempts to educate the milling community were not going well, and on 3 February he took a stronger line. He circularised all millers, informing them

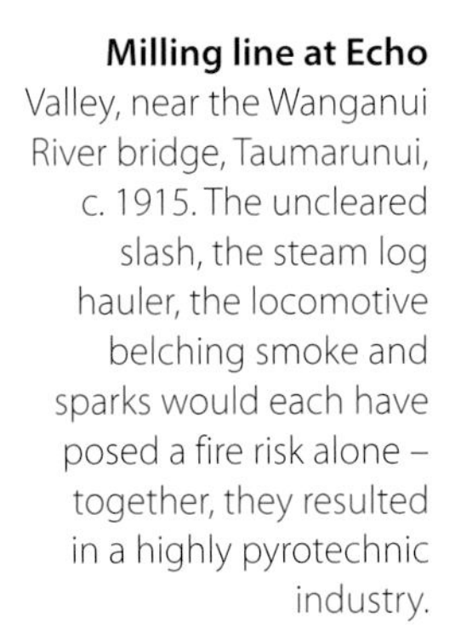

Milling line at Echo Valley, near the Wanganui River bridge, Taumarunui, c. 1915. The uncleared slash, the steam log hauler, the locomotive belching smoke and sparks would each have posed a fire risk alone – together, they resulted in a highly pyrotechnic industry.

NL, 1910–20, PAColl-5521-16

that the Forest Service was holding millers responsible for fires caused by their operations and for damage done to both worked and standing bush. He went on, effectively, to develop a fire plan for them. Locomotives and haulers were to be fitted with a satisfactory make of spark-arrester or one approved by the Commissioner of State Forests. He didn't expect immediate compliance but, until the arresters were fitted, patrols with satisfactory fire-fighting equipment were to follow locomotives. All fallings or potentially flammable dry wood were to be cleared to a 45-metre radius from haulers. Full buckets of water and a sufficient reserve were to be kept handy, plus a spray-pump for use near the hauler in the heat of the most dangerous days. Drivers were to be made responsible for maintaining these, as well as for extinguishing all small fires or for seeking help. If the ranger or fire patrol found they were not fulfilling these responsibilities, drivers or the auxiliary fire patrols should be sacked, 'in [millers'] own interests as well as in ours'. In conclusion, the circular stressed co-operation, not mere enforcement. Rangers, fire patrols and tool depots were available in emergencies, Macpherson pointed out, and most operations had been 'wonderfully clear' of fire over the previous four years. However, if recent fires continued he would have to close operations in hazardous conditions. He also appealed to the operators' self-interest. Calculating costs such as wages, investments, profit, haulage and gear, he concluded that for every acre of green bush burned, £60–120 'goes up in smoke … That is the position. Will you do your best to help? I think you will.'[111]

A firebreak being maintained with a Caterpillar grader and Super Mogul grader in Kaingaroa State Forest, Rotorua, 1927. The grader's success in removing topsoil down to the unweathered pumice and thereby eliminating the weed problem led to plans (probably stymied by the Depression) to use such equipment more often.

Photographer: L. Cockayne.

ANZ, AAQA 6560, Box 21, folder 8/20, 432, H1719

The letter to More & Sons had some effect. The company continued to disclaim liability for the fires but argued that it was working on minimising the danger from sparks. It made various claims: it had fitted spark-arresters that Railways used, it would fit a better type if it could find one, and it was urging its employees to do all they could in case of fire. The rest of the letter was taken up with the cancellation of the company's grazing licence, which More considered unreasonable and a 'petty revenge'.[112]

The millers' overall response was less satisfactory. Macpherson found that they challenged his authority to make fire-prevention demands (a letter in which he sought head office confirmation that he could stop the engines suggests he too had momentarily lost his nerve). Macpherson also considered that educating the men was an ineffective, short-term solution. He doubted that the millers took him seriously, as they went on operating. The only provision the millers did not object to was that drivers not co-operating be sacked. In the meantime, the case against More & Sons had run into complications. Although his fire patrol could counter More's denials of responsibility, Mr E. Trail was refusing to testify.

Ellis backed Macpherson to the hilt. In late February 1925 Inspector Hansson visited. The outcome was not very satisfactory. It was raining, and numerous small fires, apparently caused by locomotive sparks, were smouldering. A ranger had been working on them all day. More's spark-arresters (an expensive type) on its two locomotives were proving far from satisfactory – one could not ride on the logs behind the engine 'without getting one's coat and hat burnt in numerous places'. However, Southland & Otago Co-operative Timber Co. Ltd was trialling a Cheney Spark Nullifier, which, if successful, was likely to be adopted. But at the moment, Macpherson reported, the millers thought them too good to be true, and were trying to shelve the arrester question for the moment. Action against More & Sons for the October fires was dropped because it was too difficult to prove a claim in court for the destruction of regenerating bush.[113]

The Forest Service focused primarily on preventing forest fires. They educated the public to change social attitudes; they prosecuted to discourage potential offenders ('convictions are wanted', rangers were reminded[114]); they endeavoured to enforce the new regulations around spark-arresters; they ensured effective firebreaks and vigilantly patrolled them. Prevention, even more than today, was crucial. This was hardly surprising. Fire-fighting tools and equipment were improving and beginning to be mechanised but suppression methods were still very limited. No wonder a ranger's handbook stressed the imperative of getting to a fire promptly, and with enough assistance to get it under control quickly – if possible before the heat of the day. Some general principles were laid down: plan the attack after inspecting the fire and keep posted of progress on all sides, take advantage of lulls in wind, and attack small fires at their head and large ones on their perimeters. With adequate water supplies unlikely, shovelling earth was the most efficient way of putting out a blaze – generally a man at the end of a shovel could control at least 46 metres of a line in the time it would take to get water to control 5 metres. Two or three shovelfuls were also ideal for throwing against burning logs or cooling down a blaze. Branches, sacks or shovels could be used as flails to extinguish grass fires or small fires established on a firebreak. Once fires were under control their edges had to be dug in; burning logs moved from hot beds, their embers scraped off and covered with earth; and any flammable material had to be moved to a safe distance.[115] (These seemingly basic methods continue to apply to practice today, though with a deeper knowledge of fire behaviour and efficient deployment of fire-fighters, as well as a greater ability to use water and chemical treatments effectively, fire-fighters can now take a more scientific approach to fire suppression and extinction – and tackle considerably bigger fires.)

The Forest Service's interests in preventing fire were not unique. Private forests were being planted over this period, particularly in the bush-sick country of the central North Island where plantation owners did not need to compete with farmers. In the mid-1920s New Zealand Perpetual Forests Ltd (later New Zealand Forest Products), planting in the Taupo–Tokoroa area, was keen to assure existing and future bondholders of the fire-prevention activities it used to minimise fire risk. In 1927 the company claimed it had planted more breaks than the bonds subscribed, and was ploughing them and keeping them clear of weeds. Further, a ranger's cottage was built on each block, telephone connections were installed between the cottages and an 18-metre lookout tower was built at Maraetai, where a man with a phone watched 'when there is any possibility of fire'. Patrols were established and maintained over the Christmas period, and some 250 permanent employees working in the forest were available in emergencies. The firm, according to its 1926 report on progress, also worked to enhance fire protection by developing 'friendly terms with many neighbours, whom it would be reasonable to suppose are sufficiently wide awake to see that the neighbouring forests are and will be a considerable advantage to them'.[116] Was this gloss, born from a concern to allay fears of fire and profit loss? Bondholder reports, commercially focused, tended to minimise any risk and, while the firebreaks and the tower were real enough, the company literature mentions neither training nor tool depots. From the late 1920s history shows a company inadequately resourced to deal with the many and persistent fires on its property and falling back on Forest Service protection (for which it was charged and for which it paid).[117]

During his time as director Ellis continued to laud the fire prevention and control system he had instigated – the fire district system, organised patrols and public support had 'satisfactorily solved' the problem of fires and their control,[118] and losses in the state forests were almost negligible.[119] Was there any substance to these claims?

In 1955, almost 30 years after Ellis retired, Alex Entrican, then Director of the New Zealand Forest Service, paid fond tribute not only to Ellis's 'colourful personality' and optimism but also to his considerable administrative foresight in implementing the government's indigenous forestry policy.[120] That administrative ability saw the development and implementation of a fire policy that achieved immediate results. Between 1895 and 1920 an average of 40,000 hectares of bush, regenerating growth, scrub and grasslands had been lost to fire each year.[121] Over the first five years of the implementation of the provisions of the Forests Act such loss was dramatically lessened. Weather, as always, continued to influence the outcomes of any particular year, but between 1922 and 1927 an average of 10,400 hectares were lost per annum, from an average of 239 fires. Indigenous virgin forest amounted to 14 per cent of the total forest loss; the rest was cut-over forest and scrubland.[122] The total loss was a quarter of what had been burned previously and, more tellingly, only 500 hectares were from state lands.

But if Ellis's protections were working for the state forests, some 49,267 hectares of forest and scrub on private lands had been lost.[123] '[I]nadequate supervision and indifferent protection', particularly during land clearance operations, which left the fire-blackened hills and stumps all too reminiscent of World War I battlefields, were major culprits.[124] Willing settler co-operation in protecting plantations and voluntary adherence to fire-district regimes outside fire districts went only so far.[125] The repeated careless burning and the difficulties encountered in bringing successful prosecutions illustrate that the mindset of the settler right to burn still thrived. Issues over the use of spark-arresters by the government's own Railways continued to vex the Forest Service in the next two decades. These demonstrate the continued remarkable lack of concern about fire and its effects. Ellis had begun to attack that attitude but the battle was by no means won.

Chapter Four

Consolidation – and change: 1928–47

It is easy to characterise the Forest Service's fire-fighting policies and practices in the two decades after Ellis's departure as a continuation of the Ellis years. The areas in exotic forest, both state and private, increased. As in the Ellis years and despite all efforts, fires seemed an inevitable component of the dry years. Unless locals acted, fires outside fire districts that were not directly endangering property or lives were left to burn out. Within fire districts the Forest Service played a critical role that went beyond fighting fires with heroism and hoses, shovels and sacks. Increasingly sophisticated fire plans, backed by sophisticated forestry engineering to facilitate quick responses, were designed to make such effort effective. The preventative measures developed under Ellis continued as the Service promulgated more fire districts, and further huge effort went into educating an ever more mobile public.

Some individuals and government agencies took up the challenge enthusiastically. In 1941 Canterbury county councils' fire prevention scheme suggests that the ideas the Forest Service promoted were being put into practice in the form of public education and practical approaches to fire prevention (including using spark-arresters, keeping all water races full, making sure roadsides were clear of flammable scrub and grass, fire detection and notification, and effective penalties and fire bans).[1] But such a detailed scheme was atypical. In the 1930s we begin to see a divergence, in mood rather than substance, from the exuberance and hope of the Ellis years. The Forest Service's comments on its ability to prevent fires and on the public's willingness to change its attitudes to burning were more guarded; increasingly officials perceived the need for a tighter regulatory framework and tighter fire-control practices that they recognised were at odds with the general population's perceived god-given right to burn. Officials' concerns were borne out in 1946 in particular when the devastating Taupo fires, seen then as a national calamity, demonstrated only too well the limitations of the Forests Act and the existing fire-control and prevention system.

The investment in exotic forestry and its potential return over the two decades after Ellis's resignation provides a context for the Forest Service's concerns about fires. The planting that Ellis had instigated continued apace: in 1930 the 102,385 hectares of softwood plantations constituted the 'largest area of state-owned and state-planted forests in any single state in the Empire'.[2] By 1944, in the central North Island alone, 4047 hectares were planted at Whakarewarewa, with an additional 121,497 hectares in the Waiotapu and Kaingaroa plantations, where cattle sickness, later found to be a result of cobalt deficiency, had prevented pastoral farming.[3]

Moreover, with the Forest Service increasingly recognising the value of the indigenous forest for timber, rather than just for protection forest, reservations of that resource grew apace. [4] By 1948, of the total landmass of New Zealand, 3.73 million hectares (13.9 per cent) were reserved as state forest. Of that area, 2.53 million hectares (almost 68 per cent) was indigenous forest, permanently reserved to protect the land and to protect the bush itself against fire and wasteful timber extraction. (The Forest Service, like many today, favoured what would now be called sustainable logging over 'lock-up-use-not' management.[5])

Between 1927 and 1947 the estimated value of forestry production to the economy doubled from £3.6m to £6.6m.[6] Private afforestation contributed to this return, spurred on by entrepreneurial opportunities offered by the perception of a timber famine, by the chance to piggyback on the Forest Service's experience and work, and by some impressive profits from older, more established plantations.[7] Over 1930–31 a total of 89,435 hectares of private exotic forest had been planted, 6715 hectares in that year alone.[8] However, by then the rate of planting in what was characterised as 'the get-rich-quick afforestation boom of the 1924–28 period'[9] was beginning to slow.[10] (Timber companies' profits between the two world wars were affected by the vagaries of domestic demand, imports from particularly the US and Canada, the Depression, industry restructuring and the trees' immaturity.)

Fire remained a constant threat. Both Alexander Douglas McGavock, who succeeded Phillips Turner, and McGavock's successor, Alex Entrican, constantly reiterated the message that Ellis had so passionately espoused. Fire was 'the outstanding forest abuse of the country',[11] 'the outstanding menace to the Dominion's forests'.[12] Had the sources of that threat changed? The Forest Service (like its commercial counterparts) did burn to clear land, with the occasional breakaway causing relatively minor damage.[13] During the Depression perceptions that planting gangs at Golden Downs were lighting fires to get the higher wages that fire-fighting attracted led to an immediate clamp-down – anyone found with cigarettes or matches beyond the camp was to be instantly dismissed and planting gangs were to be given time in lieu, not wages, for any fire-fighting.[14] But the outstanding threat to both state and private forests and land came from outside. The old problems of the match farmer and sparks from logging machinery and Railways persisted. Gas-producers, developed to circumvent World War II petrol shortages, were attached to running boards or bracketed onto the backs of some cars and sparked badly as they inefficiently burned carbonised coal, a variety of wood and other fuels;[15] and an increasingly mobile public exacerbated the threats to the country's forests, gaining in value as they aged. The damage even limited fires could inflict was considerable – in 1931 the loss of 22 kauri trees to fire meant the loss of over 68 kilometres of millable timber;[16] in Canterbury £2500 was lost when fire burned 61 hectares of young plantation trees.[17] In wet years the accumulation of unburnt slash simply added to the fire hazard.

The extent of fire damage during the McGavock years is difficult to establish. McGavock was a deeply private man who, until his last years, and then only in the right company, talked little. His 'task and duty' he construed as 'understand[ing] and carry[ing] out the instructions and the policy of his Department and [keeping] completely silent about it'.[18] Perhaps reflecting his belief that information on fires in plantations was 'psychologically' bad for the general public,[19] Forest Service annual reports during his directorship stopped detailing the incidence and extent of fires that Ellis had published. These were

A settler's fire, Pohokura, Taranaki, April 1931. The smoke and the evidence of earlier fires on the hills in the forefront shows fire (with all its attendant hazards) still being used to effect major clearances. However, from the 1930s, with the immediately useful agricultural land largely cleared, this practice was diminishing.

ANZ, AAQA, Box 21, 8.30, 435, no. 6725

Alexander Douglas McGavock. His long association with forests and forest administration began in 1891 when, as a young cadet, he joined the Department of Lands in Invercargill. At that time the department was highly involved in active and intense forest administration and Crown forests were more financially important than settlement. He was conservator of forests in Hokitika in 1921, and Director of Forestry from 1931 to 1938. He died in 1958.

ANZ: AAQA, 6506, Box 28, 11/2, 901.2, no. 15673

reinstated when Entrican became director. From 1940 to 1945 fire destroyed annually an average of 5800 hectares of mixed forest cover (bracken, fern and scrub, second-growth and regenerating timber). The state lost an average of 1100 hectares from an average of 30 fires annually,[20] so the majority of these fires were outside state forests. It was a considerable improvement from Ellis's day.

Three particularly bad fire years at the end of McGavock's time demonstrated the fine line between control and disaster. In the 1937–38 summer extraordinarily hot and dry conditions in some parts of the country saw humidity fall to 38 per cent (40 per cent was internationally recognised as dangerous) and temperatures rise to 29.4°C, creating tinder-dry conditions. On the North Island Central Plateau from 29 November to 2 December, 38 fires in state forests and eight elsewhere

'Fire – the worst enemy of the forest.' An undated reproduction of the picture of the beginning of the 12,000-hectare burn seen in Chapter One, used to drive the message home in 1929. In 1939 spring conditions were also extremely hazardous. However, the Forest Service did note that the only fire in a state forest that year was restricted to 'a bare 64 acres in a solid forest of 22,000 acres … in itself adequate testimony to the efficiency and speed of the alarm system and assembly methods'.

ANZ, AAQA, 6506 Box 21, 8/29, 435, no. H.67

broke out virtually simultaneously but apparently independently. Together they burned 1139 hectares – 728 hectares of milling bush and 410 hectares of scrub, cut-over bush and waste. Over 40 million board feet of mature indigenous timber was damaged, and little was salvageable. The state lost about 17 million board feet over 364 hectares. The monetary loss to private companies was not published, but preliminary – and confidential – reports indicated 'the greatest losses of commercial timber suffered since 1918, the figures certainly exceeding 20 million sup[erior]. F[ee]t. of standing timber'.[21]

There were bad fires, too, in the Far North during the 1941–42 summer and subsequently across the Wairakei–Taupo district in 1943–44. New Zealand's first aerial surveillance painted a picture of the Taupo district fires that from 24 December 1943 until 3 January 1944 burned over 15,540 hectares. Although army and Forest Service efforts limited the destruction of exotic timber to 28 hectares (the rest was scrub and fern), the threat of fire was brought home. Taupo township and the Wairakei Hotel were both endangered; patients from a nearby asylum had to be evacuated. The 'stark, blackened boles' to which the landscape had been reduced were an ironic reflection on the area's billing as a popular scenic resort.[22]

Those fires were mere precursors of what was to come. During a drought that particularly affected the top and central North Island came the disastrous fires of 1946. Sixty-two separate fires within state forests swept over 6609 hectares, 50 per cent of which was indigenous forest (although almost all of the 28 million board feet damaged in the Mangapehi and Oruanui forests was judged salvageable). Only 65 hectares of exotic forest were lost from a fire at Tairua, outside the Taupo area; no monetary value was ever placed on the worked-over forest, tussock and scrub that made up the rest.

The private forestry and sawmilling companies lost huge sums. Twelve sawmills and three box factories (not all in the plateau area) were destroyed. Outside state forests, 311 fires swept over 574,000 hectares: 13,223 hectares of privately owned exotics mostly in the Taupo area, 4423 hectares of indigenous bush where 525,000 board feet were destroyed, and a further 214,686 hectares of cut-over exotic forest, tussock and scrub. Although Entrican attempted to address the issue of the private companies' vast amount of scorched exotic timber,[23] it was basically unsalvageable. The forests were largely young trees, with low yield; bushmen, who were anyway needed for more profitable work, were in short supply; and there were 'very real' difficulties in persuading anyone to work on such an 'uncongenial and dirty type of logging'.[24] Moreover, towns and settlements had been threatened, roads (briefly) closed, and the army and fire brigades from all over the North Island had rushed to Taupo. Capitalising on the sense of a national emergency, Entrican advanced an issue he had been

Alex Entrican, Director of Forestry 1939–61, in 1956. His determination to professionalise his service was reflected in a variety of measures ranging from replacing '[t]he higgledly [sic] piggledy manner in which motor vehicles are usually parked in front of Forest Service Offices and Stations' with organised parking places and 'proper observation of good parking practice', to importing the latest meteorological weather stations, and inviting free and frank advice from his conservators.

ANZ, AAQA, 6506, Box 28, 902.1, no. M2006

thinking about for some time: the need to construe fire protection in terms of protecting rural areas as a whole, rather than forests specifically.

Like his mentor Ellis, Entrican (whom Ellis recruited as a young engineer in forest products in 1921[25]) was university educated. A workaholic, he was passionate about his department to the extent that in later years he would be dubbed Mr Forest Service.[26] He worked to professionalise the organisation. The staff developed business and personal relationships with overseas counterparts and travelled internationally, and his employees saw him as providing the Forest Service with a clear direction. He was variously remembered as a 'gruff, no-nonsense man' who was 'by no means automatically pleased';[27] and though his wrath could terminate or stall a promising career,[28] there were occasions when he apologised to people whose actions he had misconstrued.

Entrican's background, his approach, his energy and breadth of reference in addressing fire protection issues stood in stark contrast to that of McGavock, who had risen through the ranks, first as chief clerk and receiver of land revenue, then, in 1920, as the West Coast conservator in Hokitika. Though '"apparently … blind to the value of professional training" on which Ellis and Entrican put such store',[29] McGavock was highly versed in the country's land and timber legislation, and encouraged his conservators to pursue legal avenues in preventing or discouraging fires. Both men were constrained in what they could do, one by the Depression, the other (initially) by World War II and its aftermath (the rough scraps of paper on which all departments' business was conducted until years after indicate graphically the austerity the war imposed). But the relatively constrained approach that McGavock brought to fire protection contrasts with the energy, commitment and new initiatives with which Entrican addressed the subject.

By the 1940s, where blame for the fires could be established (although often this was impossible) it was sheeted back to three main culprits: New Zealand

Railways' locomotives (especially on land outside state forests), settler carelessness, and burning off. Other causes, though minor, reflected the changing times. Relatively minor conflagrations were sparked by two plane crashes in or close to state forests in 1940 and 1942 respectively, and by Verey lights fired as a signal to aircraft near a Canterbury state forest in 1944. In 1943 and 1944 New Zealand and US troop-training exercises in state forests were restricted because of concerns about fire.[30]

The Forest Service had a sufficiently large staff to deal with most of the fires within exotic forests and fire districts. Private companies, unprepared and lacking resources, called on the Forest Service staff and work gangs when their forests or forest boundaries were threatened. Locals, often including those responsible for the fires, helped (or were impressed); a volunteer brigade might even be on hand. But with all its limitations of coverage, the Forest Service provided, by default, the country's only organised rural fire response.

Undoubtedly fire-fighting brought moments of adrenalin rush. In 1930, emerging victorious from a 14-hour battle with a gorse fire, ranger Braudigam and his men adjourned to 'a good supper given to us by Mr & Mrs Muir of the Wairoa tearooms, to both of whom we owe our thanks, we left and got home at 2.45 a.m.'. Eight years later a Rotorua paper reported excitedly:

> Whipped and spurred by the relentless gale, the fire fiend came raging over the scrub and manuka-covered hills while frenzied calls from look-out to look-out went ringing into the State Forest headquarters. From there urgent calls went out to every available man, some 40 miles away, to go with all speed to the scene of the danger.[31]

But however exciting the fight, dealing with even small burns over the spring and summer months required frustrating, demanding work. Men had to fight or stand by for hours. Where fires scorched or destroyed regenerating and merchantable timber they had to investigate and report on them and, if possible, find and prosecute the culprits. The outbreaks in the Auckland conservancy over the 1937–38 fire season illustrate the time, money and often unavailing effort involved.[32]

On 11 October 1937, with the season barely under way, a B. May of Whangamata lost control of a fire on his land. Impressed labourers and Forest Service staff fought for a total of 29 hours to get it under control. On 29 October a fire, probably from sparks from a bush hauler, started in State Forest 97, where Ellis and Burnand was milling. There was no prosecution – no milling timber had been damaged, no one had seen the fire start, and the ranger reported that Ellis and Burnand was doing all it could in terms of patrolling tramlines and instructing staff in safe practice. On 7 January 1938, when a small burn outside the fire district was threatening the Riverhead plantation, locals were called in to help put out nearby logs and stumps, which could otherwise smoulder for days. On 15 January A. Pettman, who very probably knew of the fire district provisions, burned close to Waipoua Forest without a permit. A day later a fire lit by Frank Gill, burning without a permit, got out of control and threatened a plantation. It cost the Forest Service £3 15s 9d in wages to put it out. On 17 January Henry Moore, who had been refused a fire permit, disregarded the fire ban and lit two fires. On 18 January two fires in different sections of State Forest 97 burned cut-over, valueless timber; no men were working in the area and the milling company did not appear to be at fault. On 14 March the Forest Service foreman reported three fires on private land lit by owners who were all burning without permits: D. Dear said he did not know his earlier permit had expired; P. Pederson claimed he did not know he needed one; and C. Bracey, although told he had to have a permit, had made no move to get one. A day later fires on sections owned by Burton Gorrie, A. E. Williams and Henry Moore were all lit without permits.

Some of the fires were more than nuisance value. On 11 December Maurice Jensen, after long negotiations about developing substantial firebreaks to qualify for a fire permit, burned bush specifically excluded from permitted areas. Although four men sent to patrol the area assessed it as safe, it flared up again – and witnesses suggested Jensen himself was the culprit. It took two days and wages for 22 men amounting to £28 19s before the fire was out. Strong winds over the area meant the smouldering debris was watched for a week, at a further cost of £93 3s 10d. In spite of some rain, large logs were still burning at the time of reporting. In March at Whangamata W. Stables died of burns sustained in fighting a fire that had got away. (His history of burning without a permit perhaps explains conservator R. D. Campbell's somewhat terse comment on Stable's age and inappropriate involvement with burning operations.) Nationally the Forest Service spent significant sums on fire-fighting – £805 in 1935 ($75,900 today), £415 in both 1936 and 1937 and £805 in 1938 (McGavock pointed out that spending more to lose less suggested effective protection). During 1939–40 fire-fighting in state forests cost £448; in 1941, dealing with fires that burned over 1521 hectares inside state forests cost the Forest Service £531 9s, with an additional £198 12s spent on fighting fires over 1550 hectares outside state forests.[33]

Those direct costs did not include the time-consuming work of reporting fires. Until 1945–46, when a new reporting system required only fires within 16 kilometres of a state forest to be reported,[34] all lookouts, patrolmen and those staff at headquarters had to record all fires, in and out of fire districts and irrespective of whether or not they threatened forests. Notes jotted in an abbreviated diary form were the basis for subsequent more detailed, sometimes lengthy, reports on the fire's general characteristics (the colour and form of the smoke and flames), the weather conditions and, as far as practical, any findings of a later examination of the burnt area – what had been damaged, where (indicated on meticulously drawn maps that accompanied most reports) and how badly. Most importantly, these reports set out any evidence that might lead to successful prosecutions – and prosecutions were sought whenever possible.[35] Generally the reporting officer would recommend whether to prosecute, the conservator endorsed both report and recommendation, and the paperwork was then forwarded to the Director of Forestry. Finally, a broader picture was developed by locating all fires on the maps at national and forest headquarters.

Considerable Forest Service resources went into bringing prosecutions. By the late 1930s the Service was recouping some costs from culpable operators – despite the difficulties inherent in doing so. In the Auckland conservancy, for instance, Henry Moore pleaded guilty and was fined £2 12s, Gorrie was fined £3 10s with 15s costs, Dear was fined £2 0s 9d for costs of suppression, Bracey had to pay £2 10s and court costs of 10s. Gill, who had lit two fires, had to pay £10, costs of £1 13s and £2 16s 11d towards costs of suppression. The fines were a fraction of the £50 legislated for illegal burning, and the Service had to battle with what now seems an almost wilful misunderstanding of the rangers' work. Forget the legislation – farmers needed to burn, the defence lawyers argued, and readily pleaded extenuating circumstances. Gill was represented as a poor, single man unable to make a living on the farm, who was working as a truck driver for £4 16s a week. His farm, his only asset, was heavily mortgaged.[36] Moore himself wrote twice to Wellington, appealing to the Minister of Lands in righteous indignation:

> The offence I am summoned for was lighting a fire a few days after rain … there was absolutely no danger to the forests … Now, Sir I have a wife and nine children to keep, and am on a very poor and heavily mortgaged farm, I am expected by the Corporation to farm it, but cannot farm it if I cannot clean it up.[37]

The manager of the State Advances Corporation took up Moore's cause, suggesting that ranger Biggs be asked 'to let up a bit in the execution of his duty'.[38]

Jensen's case (which was finally dismissed), and others, reflect the difficulties the Forest Service faced in achieving successful prosecutions. Rangers had to amass a sufficient weight of evidence to preclude magistrates simply choosing to believe one witness over another. Witnesses were not always credible or consistent. Particularly where the apparent perpetrator denied charges, admissible evidence to link suspects to crimes was often difficult to find – even compelling circumstantial evidence was not enough. The courts themselves could frustrate efforts by imposing trifling fines or convicting without charge, or dismissing claims for damage for unauthorised fires or even the charges themselves.[39] Courts also often accepted defence lawyers' pleas of extenuating circumstances (and indeed Forest Service staff sometimes cited them too).[40] In 1946 Entrican set out what his staff had to do to get evidence that the courts might consider robust. They were to get written or verbal reports from anyone who had seen the fire start; at least two officers had to attest to verbal evidence; the men had to look for evidence that precluded all other suspected causes, such as the absence of anyone else in the vicinity, or the direction of the burn; if spark-arresters were at fault, expert evidence had to be cited to prove the arresters were unsafe and inefficient.[41]

The constraints of this period limited the Forest Service's ability to capitalise on the increasing range of fire-fighting equipment that was becoming available. However, new methods of attack were slowly adopted and increasingly sophisticated conservancy fire plans focused on the efficient use of men and equipment. Firebreaks, lookouts, communications developments and (perhaps inspired by US and Canadian publications[42]) fire-prediction instruments were all about facilitating prompt suppression; the amount of land protected under fire districts grew steadily; and from 1938 massive effort went again into public education.

Suppressing forest and bush fires to minimise the area burned was (and is still) a rural fire-fighter's primary objective. In general terms men worked to establish a fire-line, thus minimising potential dangers. Work then began on separating, cooling and smothering. Hand tools – axes, hooks, mattocks, shovels, rakes and hoes – were essential for cutting a base line for back-burning, separating out burning fuel from unburnt at a fire's perimeter, and removing surface litter to reduce fuel. Back-burning from natural barriers or firebreaks also helped isolate fires.

The tools of the trade were still simple. Sacks, flails, shovels, even green branches were used to smother fires; hand-torches, flaming branches or raked-out burning leaves were employed to burn out debris and vegetation that might otherwise further fuel the fire. Buckets and knapsack sprays were valuable in small fires or for mopping up. Technological advances, largely postwar, provided lighter and standardised tools as well as better means of transporting water and men and of applying water to fires – motorised pumps meant faster delivery. From the early 1930s a developing science also provided a better means of determining the sorts of hoses, nozzles and pressures needed to use water to maximum efficiency.[43]

Fire engine, Hanmer plantation, 1927. Although this carried (some) water, a pump and a hose, and hand tools could be shoved on to the tray, there were no seats for fire-fighters, and the vehicle's off-road ability would have been highly constrained. It was a far cry from what was advertised for the Gwynne's engine.

ANZ, AAQA, Box 21, 8/25, 432.3, no. 3335

Portable fire engine, Karioi Plantation, 1931. Note the bigger pumping engine, and the specialised tool storage space at the rear.

ANZ, AAQA, 6506, Box 21, 8/25, 432.3, no. 6607

In the late 1920s and early 1930s new mechanised equipment was coming on the market, and new sales techniques saw promotional leaflets and information flooding into the Forest Service head office, where little action was taken. In 1927 Gwynne's 'Baby' fire engine, especially recommended for small country towns and rural districts, could handily be used as a trailer, dragged by a car, or used as an independent self-propelled unit. A small, lightweight fire pump,

Pacific motor fire pump, side view, 1927. A number of conservancies had these pumps, which 'under certain conditions [were] most efficient'. Evinrude pumps were mostly preferred from the mid-1930s – until they were replaced by the generally unloved Paramount pump. In this photograph the man on the right is carrying the hose. How this method would have worked with a heavy, wet hose is anybody's guess.

ANZ, AAQA, 6506, Box 21, 8/25, 432.3, no. 3303

Pacific fire pump, back view, 1927.

ANZ, AAQA, 6506, Box 21, 8/25, 432.3, no. 3304

which could be operated by 'unskilled' labour, was also available.[44] 'There are no assets in ashes – No dividends in debris' claimed the manufacturer of the two-stroke portable 'All-Canadian-Made WAJAX FOREST FIRE PUMP', which apparently embodied experience, design, workmanship and honesty.[45] (Later Wajax models would become the standard Forest Service portable pump.) New Zealand's Colonial Motor Company, which assembled and distributed fire engines, sent, unsolicited, its specifications for the Ford trailer pump and equipment it was supplying to the Ashburton County Council.[46]

By the end of the 1930s the conservancies had varying numbers of portable mechanised pumps – generally Evinrudes that delivered 5460 litres per hour. Knapsack pumps, which held about 16 litres of water (the Indian was generally judged better than the Holland or the Vermoral), were also acquired as supply and import restrictions permitted. Water tanks crated for immediate loading onto trucks, Hauck burners, chemical fire extinguishers for buildings, and binoculars of variable quality and water bottles (two items frequently borrowed from the Defence Department) were held variously across the conservancies. Often staff had to improvise, such as by using buckets made out of five-gallon oil drums. However, until after the 1946 Taupo fires, shovels, spades, slashers, fern-hooks, forks and sometimes fire-beaters were the main fighting tools. Few questioned whether these tools would be adequate when the trees had grown bigger.[47] In the huge central North Island forests hand tools were adequate for putting out the inevitable 'small fires' occasioned

A spray pump in action, Kaingaroa State Forest, 1945. John Barber, a trainee in 1950, recalls that the pumps were 'a four-gallon capacity container, strapped to the back with a wand. This was pressurised either by a lever action valve at the base of the container or trombone action on the wand, a concept still very common today in garden sprayers. It was an early strike or mop-up weapon. They always leaked, were reasonably effective [but] needed constant refilling.'

ANZ, AAQA, 6506, Box 21, 8/24, 432.3, no. H1813

Dennis trailer fire pump in action in 1945.

ANZ: AAQA, 6506, Box 21, 8/25, 432.3, no. H1688

by careless burning off, trespass and motorists. Indeed, many foresters there thought water was unnecessary, based on the particular conditions of the area: friable pumice soil, a scant water supply and the expense of ensuring a static supply in tanks or reservoirs. Instead the Forest Service relied on 'early detection and ... quick operation of manual methods for fire-extinction'.[48]

But despite the primacy of such methods, it was not until 1947 that tools

A trailer fire pump at the ready, probably Rotorua conservancy, early 1940s.

ANZ, AAQA, 6506, Box 21, 8/26, 432.5, L4431

were allocated across conservancies on a rational basis. Conservators were then instructed that at all times light or supervisory vehicles were to carry one or two Indian pumps (secured against damage), two long-handled shovels, one axe and one slasher. Workmen's transport was to have sufficient hand-tools for all the men, plus three or four Indian pumps. At small stations everyone was to have a water bottle; 50 per cent of the men were to have Indian pumps, long-handled shovels, slashers and axes; and 25 per cent grubbers and fire-rakes. Other gear included first aid equipment, emergency rations (tea, sugar, condensed milk, thermettes), files and axe-stones, electric and pumice torches, and diesel fuel containers.[49] That must have been something of a wish-list. A year later postwar shortages saw the Auckland conservancy short of 114 water bottles (not even army ones were available) and 80 Indian pumps (a poor exchange rate precluded their purchase).[50]

The vehicle situation was little better. Although by the 1940s each district had one truck always loaded with fire tools and ready to go, the poor quality (and low number) of those available jeopardised the Forest Service's responses. In 1938 the vehicles covering the vast forest of Kaingaroa were one Ford 70cwt V8 lorry, one 70cwt Morris leader truck, an 80cwt Comber diesel truck, and three half-ton trucks. Whakarewarewa plantation had one 30–50cwt Bedford truck and one half-ton van; Waiotapu had a Ford van and a half-ton V8 truck. Ranger L. H. Bailey at Whakarewarewa had to use his own car to get the

'The Quad fire engine demonstrating its powerful jets at the Kaingaroa State Forest, Rotorua, October 1945. The newspapers were full of photos of these engines. Unorthodox the vehicle might have looked, but media and officials marvelled at its tough construction; manoeuvrability; speed over rough country; and ability to traverse wire fences, bulldoze through trees or negotiate one in three grades. Its track-grip tyres, made to withstand machine gunfire, were virtually unpuncturable.

ANZ, AAQA, 6506, Box 21, Folder 8/24, 432.3, no. M8863

gang to a fire;[51] reimbursement for petrol costs was permitted a year later.[52] At Eyrewell Forest in Canterbury and at Ohakune patrols got around on bicycles. In 1944 the Canterbury conservancy had 23 vehicles, but wartime shortages of equipment and men had reduced the state of the trucks and tractors to a lamentable condition. Only four trucks were in some sort of working order, and two of those were heavy on petrol and needing increasing maintenance. The rest were described as 'life finished – does 9 mpg', 'unfit for use', 'should be written off and sold' and so on. Tractors and graders were shuffled between conservancies. One of the six tractors at Eyrewell was over 19 years old; only the 15-year-old second-hand one from Rotorua was any use; the rest were 'valueless', 'unsuitable' or 'nearly worn out'.[53] With Kaingaroa's little Austin graders incapable of maintaining the 842 kilometres of formed roads and 336 kilometres of firebreaks, the breaks were rapidly deteriorating.[54]

Specialised trucks for rural fire-fighting purposes had been on the market since 1939. But with enclosed cabs, a wide wheel base, multiple forward gears, two-speed rear axles, and firmly secured equipment such as 27 headlamps, chemical fire extinguishers, pumps, hose and hand-tools, they were expensive; neither McGavock nor Entrican moved beyond the enquiry stage.[55] Ahead of the 1944 fire season, and after discussions with urban Fire Service fire inspector R. Girling-Butcher,[56] the Forest Service bought 17 Ford Desert Mule chassis from the army, which no longer needed them for the war effort. One was

The Quad engine demonstrating the flexible spray equipment.

ANZ, AAQA, 6506, Box 21, Folder 8/24, 432.3, no. 1804

modified and tested. It had fire-fighting equipment specially designed for the chassis: an 1818-litre tank, a high-pressure geared pump for replenishing water supplies, 305 metres of 1.3-centimetre canvas hose with plain and fog nozzles, and nine metres of suction hose. Two permanent standpipes, which could be operated in a complete circle, shot powerful jets of water from both the back and the front of the moving vehicle. It carried a crew of eight, as well as a complement of shovels, axes, spades, ropes, picks, eight Indian knapsack pumps, first aid equipment and food.[57]

By mid-1946 all 17 Desert Mule vehicles, or Quads, as they came to be called, had been modified and distributed. The following year, after the experience of the Taupo fires, they were further modified. In addition, 28 specialised water tankers were being constructed. These could travel over rough country, pick up water from any source, supply it to engines or pump it directly onto fires. A further nine Desert Mule chassis were bought as hose-layers and four were fitted with old pump units from the Emergency Precaution Scheme (developed to deal with civilian needs in air raids or invasion). Similar equipment was on order for the other vehicles.[58] By 1948 nine more engines had been purchased and all conservancies had been issued with detailed lists of a considerable range of hoses, branches, nozzles, fire-fighting tools and equipment to be carried on all Quad engines, tankers and trailer pumps. They had also received clear instructions on maintenance, and the first equipment officer, whose primary responsibility was equipment maintenance for the whole country, was appointed.[59]

A tanker in use at a fire course in September 1955.

ANZ, AAQA, 6506, Box 21, 8/26, 432.3, M1557

The new vehicles gave the Forest Service an unprecedented ability to get teams to fires and to pour water on them. But maintaining fire readiness was difficult. Equipment was not standardised within regions or across conservancies so some of it was under-used,[60] and response times may have been delayed. Training courses on using the new mechanised equipment had to be set up and staff rostered on.[61] Within conservancies protocols had to be instituted so that 'someone' (shades of Ellis) was responsible for maintaining the equipment properly, keeping accessories in place and batteries charged – or making sure spares were available. Many staff were likely to have had relatively little hands-on experience of machinery,[62] so conservators had to pass on minor operating details, such as ensuring that no one continued to run the pump units on their daily check if they were not actually pumping water! To make matters worse, vehicles were unreliable. For a brief period in 1946, for instance, Quad engines and fire tankers could not be used as general-purpose trucks or water-carriers because '[p]lant that must be kept at 100% condition throughout the fire season should [not] be exposed to breakdown by being constantly used for a purpose outside its original duties'.[63] There were turf wars, too. A debate over whether the sirens needed to notify men working in exotic forests to prepare for fire-fighting and assemble rapidly could be mounted on non-law enforcement vehicles went on for some years.

Curiously, the first practical fire-fighting courses were only offered in 1946, and not until 1947 is there any mention of training in the 'one-lick' technique of fire-fighting or using a fire trail to back-burn. By then identified as permitting faster forward movement than the step-up approach (in which men worked along a line, moving forward as a body), the main advantage of the new technique was that less training was necessary. In back-burning, for instance (the same principles applied to putting out fires), each man or group was given one of a variety of tools, and each moved along the fire trail, partially felling whatever was on the trail with one stroke and then moving forward so that by the time about 50 men had passed by, a well-defined trail had been constructed. Axemen generally went first, slashers followed after them, grubbers cleared vines and opened up the trail, fire-rakers cleared more by pulling away slash, then a second set of grubbers cleared broken sods and earth.

With the track now available for back-burning, men came with torches made of 19mm piping. Filled with diesel, they were plugged at one end and at the other had a wick made from frayed rope. The men, back-firing as they moved along the trail, were followed by others with fire-rakes and long-handled shovels, controlling the fire edge. Men with knapsack pumps patrolled continually to put out smouldering logs, stumps or any burning vegetation on the forest floor. Light trucks with portable radio or runners acted as communication links between the fire team and headquarters.[64] This method, used by Australian bush brigades since at least 1937, was adopted in New Zealand surprisingly late.[65] But once it was taken up, Forest Service staff, in the aftermath of the Taupo fires, lost no time in spreading the word. Tramping clubs held evenings on preventing and fighting forest fires, and the Service showed a film of the method and demonstrated it to local body and forestry staff.[66] Safety precautions were contained in a simple rule: 'If it gets too hot on your face, get out.'[67]

By the early 1940s, conservancies' fire plans had broadened in scope. The 1936 Hanmer fire had demonstrated that station buildings had to be adequately protected, and in 1939 the question of how families and settlements, particularly within the Rotorua–Taupo area, were best protected was first raised (and was not fully resolved in 1944[68]). Also at issue was the safety of the driving public if fire broke out in those great central North Island forests.[69]

The plans were more sophisticated than those developed 20 years earlier. The typically detailed Golden Downs forest fire plan for the 1942–43 fire season linked specific people to specific tasks. (The innovation of linking job classifications to specific tasks, so that staff knew their role in any fire-fighting situation, would not come until the 1980s.) In the event of fire, ranger J. R. Rodger was to move the fire truck to the fire tanks, then go immediately to the locality to organise the fire-fighting gang. There designated foremen were to be in charge and ensure the men in their own gang had the appropriate tools

Forestry workers being trained in the one-lick method. Roy Knight, a woodsman in the 1950s, remembers, 'You had to be in voice distance with the next person down the line, because if a spark came over your line you'd yell, "I need help!" Then one either side would come and give you a hand, then go back to their normal place. It was a boring job at times, believe me, but sometimes it was quite a successful job and you'd look back the next day and think, Hell, we've done a good job on that.' The original Forest Service rakes, used for some decades, were made out of old agricultural mowing knives that were angled and had handles attached.

Southland News, 13 October 1950

(hand tools were still critical). If there was no Forest Service officer present, Rodger had to decide how to attack. In open stands with no canopy the men were to converge with beaters to the head of the fire from the flanks and rear, cut a fire-line ahead of the fire or, as a last option, systematically back-burn. In older stands with a canopy the men were to converge on the head of the fire from the rear, cut the burning canopy so it fell and could be beaten out, and then rake the ground. Other options were to fell the trees inwards, damping down the rear edges, or to systematically back-burn. Rodger's administrative duties, as outlined in the plan, included receiving fire reports during weekends, ensuring the completion of routine fire prevention duties and being alert to the potential for fire.

The delineation of other staff responsibilities was similarly precise. D. L. Bryant, for instance, was generally responsible for daily meteorological readings and recording fire lookout reports. In a fire he was in charge of communications: reporting the fire, getting trucks and vehicles to the scene, impressing outside labour. Relevant telephone numbers, ranging from those of local sawmills to Wing Commander Manhire's at Tahunanui, were listed.

Blacksmith W. Hannen, responsible for the overall assembly of the Myers pump, had to check that bits of equipment such as hose couplings, spray pumps and blow-lamps were functioning, and that water tanks, Myers fittings and all other fire-fighting equipment was on fire trucks every Friday afternoon. W. A. Robertson, after being alerted by the fire alarm, was to take over the fire truck from Rodgers, fill the water tank and report to the storeman-clerk when ready to leave. The forest lookouts, forbidden to smoke or use wax matches (which were still available into the 1960s), had to report immediately any fire in any exotic or indigenous forest, fire district or adjoining properties. Lookouts had to know the names of all topographical features so they could give a precise locality. In a fire they were to keep watch and could not leave their posts without headquarters' approval. Other individually named staff members were to ensure that tool depots were properly equipped. Different men had responsibilities for different areas: ensuring that public notices and fire signs were put up in designated districts; that first aid kits were complete and emergency rations sufficient; that telephone lines were well maintained and portable phones regularly checked; and that water storage tanks were kept full. Finally, whoever was the most junior staff member was to be a runner and to act as a backup to Bryant, Rodger or the foremen.[70]

To achieve the rapid response that was essential to checking incipient fires when the primary fire-extinction methods were shovelling and beating, the Forest Service also relied on forest engineering, surveillance and, particularly from the 1940s, meteorological instruments that identified times of high fire hazard.

Tractors and graders gradually replaced shovels and picks in constructing firebreaks. Hauck burners were used from 1926 to burn off the vegetation. After the success of the grader in Kaingaroa in cutting down to unweathered pumice in 1927 and eliminating cultivating operations, the Forest Service planned to use such machinery more extensively,[71] and in 1940 was purchasing special machinery for removing stumps from firebreaks.[72] In 1920 Ellis had first signalled his intentions to open up the forests with roads, packtrails and telephones to provide ease of access and mobility to patrols and fire-fighting units. Although in the early years of the war, vehicle restrictions hindered road-making, by 1943 some 1175 kilometres of roads and firebreaks had been constructed in Kaingaroa alone.[73] By then, too, modern machinery was to some extent making up for manpower shortages by keeping roads and bridges in good repair. The issue of adequate water supplies for fire-fighting in the big exotic forests at Hanmer, Balmoral and Kaingaroa, first formally raised in 1940, only began to be fully addressed after the 1946 fires when six concrete reservoirs, each containing almost 55,000 litres, were constructed at Balmoral;[74] this solution was later applied to other forests.

Ranger J. Rodger, by then ranger in charge at Eyrewell Forest, tapping a telephone wire to send a fire warning to radio- and telephone-equipped headquarters and start the general alarm. The metallic circuits (two wires, two sets of insulators and insulator pins), as well as being more expensive to construct, could not be strung over trees or along fenceposts like the old single-wire, non-metallic circuits. However, because the return path was by wire rather than through the earth, there was less interference.

Free Lance, 24 January 1951

By 1941, 595 kilometres of new telephone lines, with small phone boxes for reporting fires or incidents, had been built along main roads in exotic forests and metallic circuits were replacing earth circuits which, though cheaper, were more prone to interference. By 1946 the lines extended 742 kilometres, of which 163 kilometres were metallic circuits. At Kaingaroa the new lines, connected to better-equipped exchanges, helped prevent delays in reaching the operator and minimised interruptions. In 1940 the Forest Service persuaded all owners in the vicinity of the big state exotic forests to agree to the use of their phones in emergencies.[75] In 1947 even closer co-operation with local telephone exchanges was developed: when a member of the public reported a fire through their local exchange, it would pass the message on to the appropriate Forest Service officer or alternative staff member, both of whose contact details were on hand.[76] It was a considerable advance on the pigeons suggested in 1921.[77] Radio-telephone was installed at both Kaingaroa and Rotorua in 1941 and the new switchboard was attended 24 hours a day. (On one occasion a phone did not work, 'probably' because of a thunderstorm and 'not improbably' due to the operator's inexperience, reflecting the problem of quickly upskilling staff.[78])

The new equipment worked well during and after the 1946 emergencies. New equipment such as radio lines between key offices, mobile radio stations at key locations, portable lightweight radio for steep terrain and equipment for laying field telephone cables achieved almost complete North Island coverage.[79]

Since 1922 lookouts in towers had assisted detection. By the 1940s, well placed high above the exotic forests, lookout men used direction-finders and increasingly accurate maps showing essential topographical features – firebreaks, internal and external access tracks and roads, houses, buildings, fire towers, telephone lines and water supplies[80] – to give cross-bearing on any fire sighted. After they notified headquarters with this information, rapid action

The Eyrewell lookout tower was built 30 metres (109 steps) high to give adequate surveillance across the flat Canterbury Plains.

Free Lance, 24 January 1951

A lookout station, 1939. More usually, lookouts were sited on high points. This stolid building, built to withstand mountain-top gales, demonstrates the more general pattern of lookout construction at the time.

AJHR, 1939, C.–3

followed. The lookout equipment was simple, so skilled and expensive staff were not needed. Trustworthiness and alertness were critical – and not assumed. Lookout men had to ring through every two hours from daybreak to nightfall. The ranger rang as a drill several times a day and occasionally tested them by releasing smoke and then recording reporting times; constant practice got them to a standard where they could refer to any point in their area without consulting their map.[81] Communications were enhanced in 1941 by radio installations in lookouts, along with wind chargers. In 1942 alone, lookouts spotted and reported 3591 fires,[82] but the system was not foolproof: fires might be behind ridges, fog might obscure vision. More significant constraints on the operational effectiveness were the years, if not decades, of delays in construction because of wartime shortages of materials and builders[83] (only after the Taupo fires were new lookout stations constructed at Rotoehu, Eyrewell and Ashley forests[84]), inaccurate maps and the wartime lack of good binoculars.[85]

In the mid-1920s in Canada and the US, aircraft were adding a new dimension to both surveillance and the transportation of men and equipment. In 1929 the *Evening Post* enthusiastically reported aerial surveillance in Ontario; the *New Zealand Herald* followed up with reports that in Australia aircraft from both private firms and the air force were used for forest fire detection.[86] By contrast, in New Zealand a private firm's offer for a similar service in 1930 was summarily rejected,[87] and by 1934 Washington State had concluded that aircraft were valuable for detection work only as a supplement to lookouts.[88] However, as planes began to prove their worth in the war they began to be used in fire protection – first in 1941 in the Eyrewell Forest. There they advised fire-fighters on outbreaks and lines of fire,[89] providing an invaluable overview denied to those on the ground.

Aerial surveillance in the 1944 Taupo fires again demonstrated planes' potential contribution to fire-spotting. In an almost prescient move the Air Department (set up in 1937 to administer both service and civil aviation) co-operated with the Forest Service in the 1944–45 fire season and planes patrolled some 2833 hectares over the Central Plateau. New types of planes and air-to-ground radio were also trialled. During the 1946 Taupo fires planes from the air force and the New Zealand Aerial Mapping Service flew over Auckland and Wellington conservancies and over the Urewera Range, Coromandel and Tauranga, radioing information to headquarters at Taupo or direct to ground crews so outbreaks could be fought with maximum efficiency.[90] After the 1946 fires, in efforts to improve detection and suppression, aerial patrols from Rotorua covered over 2.02 million hectares from the main-trunk line to Whirinaki River and from the Bay of Plenty to Tokaanu, and other pilots were encouraged to report.[91] By 1948 the air force provided permanent aerial patrols in the Rotorua conservancy and could extend them to other areas at short notice.[92]

After lengthy enquiries of North American forestry officials,[93] Entrican saw to the installation of ever-more sophisticated meteorological instruments in both exotic and indigenous forests as part of the surveillance system that, from 1940, he made a high priority. That year work began on erecting 20 new weather stations within fenced enclosures. Each louvred shelter held a psychrometer with its wet and dry thermometers and a small hand-worked pump for dragging air across the bulbs – rather more sophisticated instruments than those introduced in 1926.[94] Inside the shelter, on the left, was a moisture-indicator scale for weighing fire sticks. (The wood in these sticks closely represented the conditions in which scrub, fern and dry timber ignite. Because they got wet in rain, dried out slowly under calm conditions and more quickly with wind, the cumulative effect of variations in atmospheric humidity could be constantly assessed.)

In the centre a hydrograph recorded the wet and dry thermometer readings. A rain gauge and fire sticks on wires just above the ground were placed outside. These instruments provided an objective assessment of fire hazard conditions: lack of rain, low relative humidity, warm temperatures, and high wind (which increases the rate of evaporation and fire-spread) recorded by Beaufort scale estimates were supplemented by anemometer readings. From this information and the state of the fire sticks it was possible to assess fuel flammability. More equipment was added to the shelters in successive years. Inexpensive cup-type anemometers that instantly read wind speed were bought in 1940, and careful investigations were conducted to identify the best wood for the fire sticks, to best represent different sorts of vegetation.[95]

Theoretically readings of the relative humidity taken three times daily (and more frequently in fire-hazard weather) alerted staff to impending dangerous conditions. Particularly over hazard periods, reports were telegrammed through to head office at set times. These were assessed in the light of Meteorological Station data, and warnings immediately broadcast where necessary. Although Entrican was chary of blanket proclamations where the legislation could not be enforced,[96] the 1940 regulations empowered him in extreme conditions to radio instructions for all equipment in forests that was likely to give off sparks to cease operating immediately. He used this power in 1946 to ban fires and the use of any steam engine whatsoever in the Bay of Plenty, Rotorua and Taupo counties.

But, as the Service soon discovered, the new equipment could not contribute to fire safety unless the staff were interested, able to recognise risk and act appropriately.[97] After the 1946 fires a fire-hazard meter was developed to assist staff to recognise risk.[98] By 1947 the meter had been installed in 25 stations,

'Louvred shelter with fire-control instruments for predicting dangerous fire conditions.'

AJHR, 1939, C–3

giving good national coverage, and there were plans to extend a radio network over the whole country.[99]

With the director and conservators sure that creating fire districts served to contain burning and protect and preserve forest, the provisions were extended to cover areas of peat and flax in 1932 and then to sand-dunes in 1941. Over the 1930s and 1940s several new fire districts were steadily approved. In 1932, 42 districts covered 818,852 hectares. Ten years later a total 61 districts covered nearly 1.4 million hectares. Thirty-nine of these were in state forests (1.1 million hectares), eight were private (149,183 hectares), and 14 (165,664 hectares) were administered by government agencies and local bodies. By 1946 another 10 districts, again mostly state-administered, added a further 404,686 hectares – fire districts now covered a total of 1.96 million hectares.[100] Provisions for burning within these fire districts were increasingly tightened. Not even smoko and billy-boiling were sacrosanct. If foremen thought the risk was too great, gangs were not allowed to boil billies; smokos were to be held only on firebreaks; all cigarettes had to be properly stubbed out before men went back to work; and matches, lighters and cigarettes were to be left in coat pockets, not taken into the forests.[101]

Provisions for burning off in summer months also became stricter. In January 1943, when the Service was instructed to maintain adequate fire protection above all else, burning off during the fire season was prohibited unless the conservator had assessed the fire conditions as safe and the provisions to prevent fire-spread as adequate. Any instructions to remove protection measures for fire had to be signed by the director himself.[102]

McGavock, himself a keen outdoors man,[103] noted almost every year the steady increase of trampers and outdoor activity that helped inculcate 'in the minds of the younger generation [a] sound forest "sense"' that would safeguard the 'forest heritage';[104] tramping clubs began erecting huts in state forests and were increasingly willing to turn out voluntarily to battle forest fires. (Ellis would have approved.) Perhaps because of the Depression and its aftermath, fire education programmes aimed at the general public lapsed in the early 1930s but in 1938 they began again on an enormous scale, with new initiatives whose outreach was intended to encompass an ever more mobile and urban population.[105]

The examples of US anti-fire propaganda – leaflets, cartoons, comic strips, articles, photos, radio ads and billboards – sent to Entrican after the US entered World War II suggest that New Zealand's initiatives were not novel. But the New Zealand programme did not link to obvious wartime enemies in the way that the US campaign did, with its images of a Japanese soldier holding a burning match to his face against a forest fire background, and films such as the anti-Japanese *Wood for War*.[106]

Nor did New Zealand's anti-fire efforts rise to the heights of, say, the US Department of Agriculture's hard-hitting *Hell and High Timber*, the story of George Spelvin who declares the forests 'our priceless heritage' one moment and flicks a cigarette out of his car window the next. Within minutes a fire is started that takes two weeks to quell. Two towns are annihilated; the National Guard is called out; animals are frenzied with terror. Men come 'off the fire-lines day after day, half-naked … with blackened faces, and eyebrows burnt off and shirts hanging in shreds from their blistered shoulders', land is ruined, water poisoned, fish killed, and tourist and recreational potential lost forever. 'Throw that cigarette anywhere you want,' the story ends. 'It doesn't matter where you flip it now. There isn't anything left out here to burn. There won't be for another hundred years.'[107]

But within the New Zealand context the effort was significant. A series of 14 stickers with anti-fire slogans was printed for government use during December 1938.[108] The slogans were intended to attract public attention and co-operation with messages such as, 'When leaving make sure your billy-fire is really out', or 'After bush fires come floods, soil erosion, monetary loss'. However we might read them now, they captured imaginations then. The Public Trust alone wanted 30,260 stickers. The AA, the North and South Island Motor

'Yours in Trust', an American Forestry Association poster. New Zealand publicity never rose to such heights.

ANZ, F1, 12/0/3B, Fire prevention policies and signs 1938–45

Unions, New Zealand Forest Products and the Dominion Federated Sawmillers' Association – big industry players – got behind the scheme; private individuals wanted the stickers; schools and one motorcamp proprietor extolled their educational value.[109] The following year, when the campaign focused on the dangers of spring burning – 'September fires can spread like March fires. Begin

The subject matter of the stickers ranged widely, covering the contribution trees could make to everything from aesthetics to industry.

ANZ, F1, 12/0, Pt. 3, Fire protection general 1929–38

your forest fire precautions NOW'[110] – 210,000 stickers were distributed to government departments, relevant unions, companies, associations and individuals. By 1947 members of the public were suggesting slogans they had seen overseas: 'One tree will make a million matches. One match will destroy a million trees' was a particular favourite. Stickers went out to an ever-widening group including stock agents, oil companies, the Junior Chamber of Commerce and the Tararua Tramping Club. All were ready to assist.[111]

The campaign was not run on stickers alone. In 1937, inspired by Tasmanian example,[112] the Director-General of the Department of Post and Telegraph offered to run slogans in postmarking machines – if the Forest Service would bear the cost of manufacturing the slogan plates. It did, and would do so again. Private industry also played the corporate citizen. Beehive safety matches had a slogan printed on each side of their wrappings. Some targeted smokers: 'Knock your pipe out carefully'; others the general public: 'Don't destroy live trees'.[113] Tobacco firms were approached for similar contributions.[114] Bookmarks, first printed in 1939 with a picture of burning bush on one side and instructions on fire safety and the need to be careful with matches and campfires on the other, continued to be produced in ever more attractive formats. Print runs rose from 50,000 in 1939 to 200,000 in 1945 and 1946.

In 1941 posters appealing for public co-operation over the fire season were displayed in about 50 selected post offices. Over the next two years they were distributed through Forest and Bird and displayed in schools, railway stations and post offices. (Railways, typically as we shall see, considered that as it was already displaying posters on the various phases of the war effort, room for Forest Service posters would be limited.[115])

Calico fire signs were reprinted in 1938 (paid for by Internal Affairs) for posting on the roadside, initially in the flashpoint Rotorua–Taupo district.[116] A year later 50 vitreous enamel ones were displayed at the public access points to the area where it was anticipated they would have the most impact. (Wartime austerity and pressure on printers limiting the print run.)[117] In 1945 Forest Service officers turned their attention to developing standardised signs with short, snappy slogans. Why not, asked one, copy the continental practice of placing signs a few chains apart, each bearing one word that combined to create a warning?[118]

Not until 1947 were colour combinations determined. Chosen for instant recognition by children and adults alike, road and forest signs had green lettering on a white background, and fire signs had the red lettering on a white background that is still used today.[119] Erected amid concern that road signs might distract motorists (something of an issue at the time) they were taken down at the end of each fire season so that they could have a fresh impact the following year – and to lessen the chances that they would be used for target practice.[120] Roadside arrows that named each lookout were erected on main highways adjoining Whakarewarewa, Waiotapu and Kaingaroa forests to create 'a good psychological effect on road users'.[121]

The press contributed with a barrage of editorials and comment over the fire season. The *Rotorua Morning Post*, for instance, on 10 October 1938 castigated New Zealanders for their 'indifference' to fire and 'absolute stupidity that says little for our national pride and consciousness'. It reiterated the importance of individual responsibility and highlighted conditions imposed on fire districts or flammable environments.

The Forest Service did not rely on the written word alone. In 1937 AA patrolmen were first appointed as honorary rangers to form a chain of protective patrols along highways.[122] Their work was primarily preventative: patrolling in their working districts and spotting and suppressing small fires before they got out of control.[123] It also provided opportunities for educating campers and tourists.[124] Head office pressed the conservators to make the relationship with the AA staff work. Although initially the men on the ground had mixed reports on their effectiveness,[125] by 1941 educational value was seen in the Canterbury AA rangers' work, which involved assisting in the Canterbury local rural fire prevention scheme and reminding picnickers to put out their fires.[126] Fire warnings first broadcast nationwide on radio in 1937–38 (but then suspended because of security fears during the war) alerted the public to fire danger, and radio programmes, sometimes featuring the Commissioner of Forests, emphasised the appalling annual cost and destruction of fires and sought public co-operation – 'the most important single factor in the prevention of forest fire'.[127]

But public attitudes do not change overnight. Despite the education programmes, individuals and timber and milling companies continued to act with scant regard for future profits, and too often there was not enough evidence for successful prosecution. In 1939 a Mr Rutter admitted to lighting fires in and around Tongariro National Park 'to get through fern and scrub to clearings where he had to work but he has not lit any fires since being warned by our Inspector'.[128] In February 1941 a grass fire on a private plantation at Oroua Downs (north of Himatangi) destroyed 10,000 17-year-old *Pinus radiata*, valued at £5000. There were no firebreaks, and lupins and other vegetation grew close to the boundary.[129] Just over a year later, when all timber stocks were vital to the war effort, the Owhango box factory went up in smoke, losing 10,000 board feet of its stock of 200,000–300,000 feet of white pine. A spark from a smoke stack had started the fire and the factory had no spark-arrester, no fire-fighting equipment, and no extinguisher or hoses – 'not even an outside tap'.[130]

By 1939 the Taupo Town Board was representing to the government its concerns with the fires that annually burned scrub in and about the township. In January 1944, as patrons were being evacuated from the Wairakei Hotel because of fire, another fire was started by sparks from a local mill hauler. Many mill staff were still away on holiday and the company kept what men it had in the mill, leaving only two to deal with fires; this one continued to flare up for over two weeks. Another fire had a good hold in fern on a face close to the Forest Products site. The Taupo Totara Timber Company, an indigenous timber milling company and one of the 'big' players in the industry,[131] continued to work its locomotive without a spark-arrester even as fires flared along its line.[132]

Perhaps even more extraordinary were the circumstances surrounding some of the 49 fires that burned in the Taupo district during the devastating 1946 season. Timber companies' machinery and employee carelessness were again implicated, but members of the public also contributed. Many of the fires broke out too far away from easy access points, suggesting they were not accidental.[133] Local farmers, wilfully breaching the regulations or pleading ignorance, continued to burn off; garden fires got out of control; hunters' fires lit to singe the hair off pigs were left burning; kids played with matches and motorists flicked cigarettes out of car windows.[134]

Work on changing public attitudes continued as a critical component of fire prevention, particularly after the Taupo fires. During 1946 and 1947 conservators showed Forest Service slides of the fires to relevant bodies such as catchment boards to emphasise the 'need for general fire protection';[135] and the commissioner's radio talks on the summer fire danger became even more frequent.[136] A massive £5000 campaign was planned for the 1947–48 season (Cabinet signed off on most of the expenditure). With designs to encourage permanent display, neat and attractive posters were to be developed for all country hotels and boarding houses. Advertisements were to screen in all country resorts and selected rural theatres. Leaflets, posters, radio talks, games and a 16mm film, *Fire Danger*, were prepared for school use. Scouts and Boys Brigades were to have leaflets to paste into their handbooks; Railways service cars were to have neat, framed signs alongside the Railways sign – but not, it was specifically noted, among the other advertisements; and the National Film Unit was approached about including a forest fire prevention item in its weekly news film. The material resulting from this campaign was high quality, with organisations in Queensland and South Africa requesting permission to use a leaflet and various posters.[137]

In 1947 the Commissioner of Forests painted a somewhat gloomy picture of '*hundreds* of small fires [breaking] out in scrub, bush and grassland each summer', in spite of the legislative and practical measures taken to prevent them. The result was 'a huge bill of destruction, erosion and land deterioration', far greater than most New Zealanders realised.[138] The Taupo fires focused everyone's attention but for some time it had been clear that the Forests Act 1921–22, however innovative at the time, was inadequate for dealing with the scale of the problem.

In 1939 Entrican, still only acting director and showing an initiative that the more rules-bound McGavock had not evinced, asked all conservators for their views on the existing statutory powers and on some administrative and personnel questions. In his first annual report, citing sawmillers' voluntary co-operation in complete 'blackouts', he raised the possibility of a ban on all fires during periods of abnormally high fire hazard.[139] Reform was in the air.

The move was doubtless prompted by two factors in particular. First, using a more sophisticated conception of forestry than previously developed, Entrican blended elements of earlier arguments with some progressive ideas to stress the complex inter-relationship of exotic and indigenous forestry that was vital to what he termed 'New Zealand's land use problem'. Forestry, he argued, was 'not alone the planting of trees; nor the production of timber; nor the provision of high-country grazing; nor the protection of water-sheds; nor the preservation

A small area of beech killed by fire in Burwood, Southland. This image by John Johns, the Forest Service's talented photographer, graphically records the impact of fire on the environment.

ANZ, AAQA, 6506, Box 21, folder 8/30, 435, M1914

of wild life; nor the perpetuation of historic, aesthetic, scientific and primeval values; nor the development of recreational uses'.[140] The Forest Service's role was to keep its own non-agricultural land in maximum productivity by avoiding erosion and flooding, thus safeguarding the country's ability to produce. Controlling fire, 'the paramount agent of destruction', was the challenge.[141] The issues were not new (McGavock was among those who raised them during the 1930s); nor were they framed in the ecological terms that Rachel Carson's *Silent Spring* achieved in 1961. But in drawing on the notion of interdependence, Entrican was ahead of his time in arguing for a social obligation to protect vital community interests. This would have legislative implications for tackling the problem of fires.

While initial plans to amend the constraining Forests Act had to be shelved,[142] Entrican moved rapidly to implement regulations to address the second factor that prompted his spirit for reform: the longstanding problem of fire caused by sparks from operating machinery (including trains) in the forests. In 1939 the Cheney spark-arrester was still being used only 'to some extent in New Zealand'.[143] The problem lay not only with loggers and millers: the act did not bind the Crown. Difficulties with the government-owned New Zealand Railways had increased from the late 1920s, particularly on the North Island Central Plateau. Duncan Macpherson, transferred from Southland as conservator, informed Phillips Turner, 'I could follow the course of the railway as far east as Waiouru, by the fires being lit by the sparks from locomotives, some 20 fires being visible at the same time.' The Garrett engines represented the greatest danger – they produced a terrific draught that hugely increased the risk of sparks when the engines were 'pulling up a stiff grade at full throttle'. Moreover, Macpherson argued, community relations were threatened because sawmillers saw that the regulations did not apply to Railways.[144]

These reports would form part of what became an on-going saga. It began in August 1929, when outside the official danger season Macpherson reported that a locomotive had started a fire in the Karioi plantation. The sparks that some grades of coal threw out constituted a serious menace, he noted, and suggested that head office liaise with Railways.[145] Railways' general manager replied only to Phillips Turner's second letter. If the Forest Service wanted to plough more firebreaks on the Railways reserve he had no objection. Railways' spark-arresters had been inspected and were in good condition; in addition, hard coal would be substituted for the more spark-prone soft coal before the dry weather set in. In the end, and to avoid a stalemate, the Forest Service ploughed the firebreaks and Railways agreed to bear one-third of the cost.

By then Macpherson, perhaps smarting from his Southland experience, had the bit between his teeth. Over the next year letters flew from his pen as he reported further fires, the risk to mature forest valued at £2.5 million and the surfacemen's carelessness when burning off as they cleared and repaired the tracks. In November 1930 Phillips Turner, feeling that additional clout was needed, took the unusual step of asking his minister to take up the issue with the Minister of Railways.

There, for the moment, and despite Macpherson's further promptings, the matter rested. Macpherson sprang into print again in 1937 with reports of three fires caused by Railways, 'the worst offenders in this respect'. In both the US and Canada, trains had to use diesel, not coal, he noted, and recommended the same practice for New Zealand.[146] McGavock again prevaricated. Two years later, Railways were still the main source of outbreaks.

The Forest Service took precautions at a local level. In the Ohakune district one man patrolled the lines during the fire season, and field staff from the district office patrolled the western side of the district to Erua. Erua plantation staff watched for locomotive fires, and a district patrolman stationed at Erua patrolled the tussock country to the Spiral at Raurimu and as far as the Chateau, a particular danger area.[147] A few months later Entrican, who as far back as 1921 had encouraged the general use of spark-arresters,[148] took up the matter with G. Mackley, Railways' general manager.

A somewhat intransigent Mackley would not agree to patrols following trains over the fire season. From his perspective the exotic forests had been planted long after the railways had gone through, and the Forest Service therefore should 'carry their own risk and provide their own safeguards'. However, Entrican did wring some concessions: Railways would ensure spark-arresters were in good condition, issue instructions to dump ashes and flammable materials safely in specified locations, extinguish fires at the incipient stage and report them and even, if the Forest Service specifically requested it during times of extreme hazard, implement Railways patrols. Moreover, the Forest Service was authorised to call on all Railways staff for help in serious situations, although their assistance would be 'subject, of course, to obligations ... to their own service'.[149]

In January and again in September 1940, after the usual rash of fires around Raurimu,[150] Entrican wrote to Mackley again, 'deem[ing] it a favour if you could see your way clear to issue special instructions to locomotive and other crews' working in the area to exercise particular care in disposing of clinker and ashes.[151] In the 10 months from January to the end of October, fires caused by Railways' locomotives had cost the Forest Service £16 5s in wages and salaries. Railways' concessions had not substantially changed the situation: it had supplied express engines with half soft and half hard coal; goods engines had a 3:1 ratio of hard to soft; and all engines were fitted with a Waikato (though not the recommended Cheney model) spark-arrester.

These measures had not substantially changed the situation. After one night train had gone past, an officer reported 'many standing dead logs were on fire with streams of sparks being blown from afar ... and some logs actually blazing'. There was snag-burning, and an outbreak in the forest. 'The fires were reaching their height in the evening with many areas brilliantly afire in the Ohakune and Rangataua flats ... South of Ohakune the locomotive ... was broadcasting sparks until the Karioi forest vicinity when it ceased, apparently through bringing into use of an arrester' – which forest officer Reid found 'most gratifying'.[152]

In 1940, after input from both industry and conservators,[153] the Forest (Fire-prevention) Regulations 1940 came into force. They superseded the 1925 regulations, which had specified that from 1 October to 30 April and unless exemptions had been granted, no external-combustion engine was to be used in a state forest or fire district unless equipped with an approved spark-arrester. The new provisions aimed to eliminate all careless and uncontrolled burning near all forests[154] – protection on the boundaries, as ever, was critical. The provisions covered all fire districts, including privately administered ones. Based on conservators' advice that the rural populations displayed an extraordinary proclivity for burning at any time other than the dead of winter, the fire hazard season was extended from 1 August to 30 April the following year.[155] Over that period no steam external- or internal-combustion engine of any sort or gas-producer, apart from ordinary farm machinery, was to be used in fire districts or state forests without a Forest Service-approved spark-arrester. To be approved it had to stop sparks or flame from entering funnels or exhausts and prevent live coals or fire escaping from ash-pans or fire-boxes, and the operator had to make sure that any ashes dumped would not cause fires. In a significant change from the past, a uniform ban was imposed during the seven-month fire ban season – no burning matches, lighted cigarettes or cigars, pipe-ashes or ashes from gas-producers, or any other burning or smouldering material were to be left without being totally extinguished.

The onus of responsibility for fire safety and suppression, in a number of ways, was pushed back onto the individual operator or permit-holder in an effort to impose a mutual obligation for protection. Permits to burn did not protect permit-holders from claims for damage and were not to be used if wind or any other conditions could spread the fires. No longer did the Forest Service have to request help: if fires were endangering a state forest or parts of a fire district, all owners or occupiers of adjoining land had to immediately report the fire to the nearest fire officer and do all they could to extinguish or control the burn. Moreover, for the first time, the regulations specified that anyone in a state forest or fire district, whatever they were doing, had to immediately act to suppress a fire and report it. In a provision that specifically caught up mill and logging operators, right-holders in a state forest were similarly responsible for any fire within half a kilometre of their boundaries, and on regularly used routes up to half a kilometre outside the state forests. In addition, right-holders had to provide and maintain fire-fighting equipment in readily accessible places. Finally, if weather conditions created an extreme fire hazard, the Director of Forestry could order the suspension of all logging, sawmilling and other operations.

Today we can see that New Zealand was finally catching up with the 1911 British Columbian fire-protection provisions or Victoria's 1928 Forests Act, on which Entrican modelled his regulations.[156] At the time Entrican was conscious of the tightrope on which he was treading. The disastrous fires in Victoria, Australia, had shown that it was essential to have powers to deal with fire hazards; impose obligations on factory hands who, knowing of fires, made no move to extinguish them; and deter people from leaving fires burning, particularly on the Rotorua–Taupo road. At the same time Entrican was well aware that the farming community was concerned its clearing activities would be unfairly curtailed and that any conditions seen to be unreasonable would be unworkable.[157]

With the regulations passed, Entrican pressed his conservators to administer the new fire laws and regulations strictly. South Island conservators' responses indicate that their millers and farmers were co-operating to some degree, while government departments, the AA and honorary rangers were all doing their bit.[158] On the other hand, Railways remained largely intransigent across the country.[159] In 1946, six years after the new regulations, 200 of the recorded 311 'spot' fires outside state forests were caused by locomotives on the main-trunk line. Promptly suppressed, they did little damage but the level of effort involved in patrolling and suppressing incipient fires went down in the collective memory of Forest Service staff.[160] Jack Barber, who joined the Service in 1948, remembers a 'probably apocryphal' King Country story of Railways men throwing clinker out to see how fast the Forest Service men would respond.[161] Even when the Forest and Rural Fires Act 1977 eventually bound the Crown, the problem was not fully resolved.[162]

Only better rain, Entrican wrote to Ellis in wry despair in 1940, would provide a real solution to the fire problem. The new North American fire-hazard equipment had only served to heighten the sense of alarm. Now the Forest Service knew that every two or three years native forests would be in extreme danger and that risks in the exotic forests were increasing. The 1940 regulations, cramped by the authorising legislation, were 'not as foolproof as we would have liked' and he hoped to amend them. Make the onus of proof fall on the defendant, he advocated; oblige those with land adjoining forests to keep it clear of flammable material; and require forest owners to maintain sufficient fire-fighting personnel.[163] And tackle the problem of gas-producers. They posed a far greater risk than could be solved by a decision about where ashes from their fireboxes might be dumped. One Forest Service staff member reported seeing a car equipped with a gas-producer from which live embers

Pakihi vegetation resulting from consistent burning (burn to burn) in an area originally worked for silver pine, 1954.

ANZ, AAQA, 6506, Box 21, folder 8/30, 435, M1146

were possibly escaping; worse still, flames were leaping from its firebox and, as the car drove steadily on, fires sprang up in its wake.[164]

In two successive years Entrican tightened the legislation further to address these issues. In 1941 an amendment to the Forests Act provided that men within an eight-kilometre radius of a fire district or state forest could be called in to help control or extinguish fires, mandated the minimum of equipment required by those working in state forests, and established recreational and camping areas. Also included were the first provisions dealing with the problem of traffic in areas where fire-fighting was going on.[165] The following year the regulations were expanded to strengthen fire control and prevention. Logging operators had to have 'efficient' spark-arresters as well as a means of preventing live coals or fire escaping. Patrols with specified equipment had to follow steam engines within 30 minutes of their passage and extinguish any fire. Tramway owners had to remove flammable material within a state forest or fire district or in adjoining areas; if they did not, a Forest Service officer could. Provisions concerning gas-producers were tightened to prevent those used on cars from causing fires, and the minister could prohibit their use on roads or in adjoining state forests (although Entrican had wanted to go further and ban all gas-producers[166]). Permit-holders, too prone to thinking fires were 'safe' only to have them flare up again, not only had to help with fire-fighting: they were not to leave the area unless the fire was safely out. Finally, no one was to enter a state forest or private fire district without a valid permit, or without the permission of the owner or occupier; and permit-holders, like everyone else, had to observe their permit's conditions.[167] By 1943 Entrican had started work on a nationwide campaign to remove the threat of the small farm-burning operations that in North America were destroying trees, fences and lives.[168]

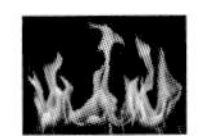

There was, however, one class of land for which no agency or organisation was responsible: the thick scrub-covered land that covered 80 per cent of the Rotorua conservancy. In 1938 the Taupo County Board, concerned at the way in which the bad summer fires had been allowed to 'rage, for weeks at a time' in and around Taupo, wrote to the Commissioner of Forests wanting better and more co-ordinated protection. The government recognised the problem, but finding a solution was problematic. The town did not have the 1000 people that the Fire Brigades Act 1926 required before a district could be declared a fire district and have a brigade formed, and it was unable to finance a brigade itself.[169] The fire district provisions under the Forests Act could not be used as the only afforestation companies that could apply to form a private district had already done so. Nor did the provisions confer free fire protection: the Forest

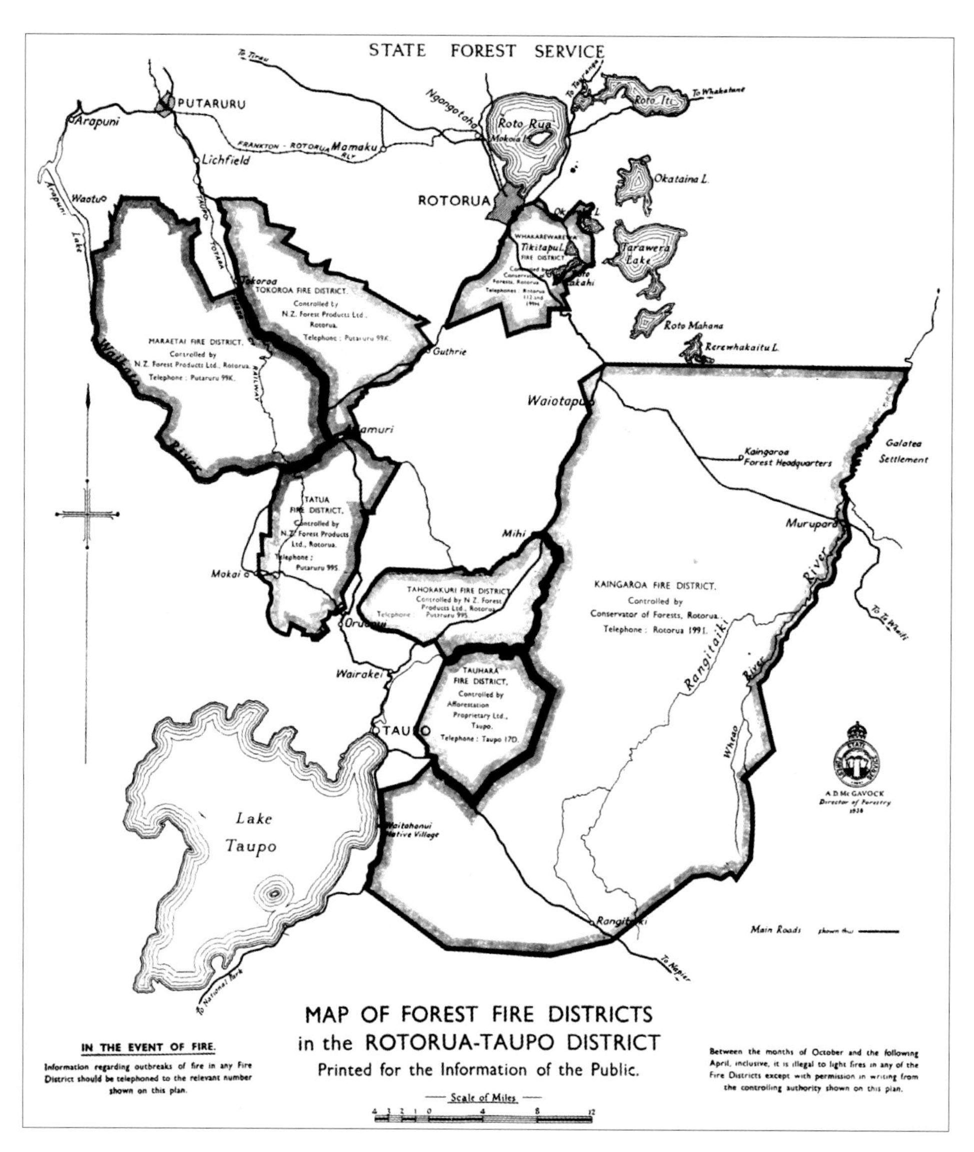

Map of forest fire districts, Rotorua–Taupo area. Although the area was studded with private forest fire districts, a review as early as 1939 showed that 'private-forest owners continued to rely too much on "fire district law"', while their practical measures to protect their own forests were 'insufficient'. The Forest Service wanted appropriate powers to compel them to carry out essential fire-prevention measures.

AJHR, 1939, C.–3, plate 16, facing p. 16

Service could not, even under 'the most elastic interpretation of the current statutes', justify the expenditure.[170]

A temporary, and hardly satisfactory, expedient that would allow the council to make its own bylaws was eventually found under the Electric-Power Boards Act 1925. As a consequence, in early 1941 the Taupo Road Board sent notices to all ratepayers banning all unpermitted fires over the summer months and requiring householders to keep their sections clear of flammable rubbish.[171]

But the scrub, fern and self-sown pines surrounding all the forested areas, Taupo township and many of the scenic reserves continued to pose a 'major hazard'. As the land was not in any fire district there was no control over who lit what, and Maori especially (whom many saw as 'notoriously careless' with

Gas-producers fitted onto cars and trucks. 'Use N.Z. Coal – Save Petrol, Save Money' urged the caption. The equipment was developed to circumvent the severe wartime petrol shortage. In New Zealand most gas-producers burnt carbonised coal. They did not provide the same power as petrol, starting could be tricky, and their use demanded careful driving and engine maintenance.

ATL, Eph-b-Petrol-1940-01-front

fire) did a lot of hunting. Two-thirds of the fires in the district were in that wasteland and, although the Forest Service probably had sufficient staff to deal with outbreaks in its own forests, the commercial plantations also relied heavily on them. With the Forest Service stretched and the commercial plantations unwilling, no one was prepared 'even to investigate' fires in those wastelands, 'either to determine the cause, or the extent and potential hazard to property, unless the fire is manifestly an immediate threat to the township or some of the forestry areas' – by which stage it was too late to prevent serious damage. Eventually, and with some Forest Service reservations,[172] it was agreed that, as a temporary expedient, the government would station a Quad at Taupo, a local fire brigade would be formed, and a committee with representatives from forestry interests and the Taupo Road Board would control and finance seasonal fire-fighting in the problematic areas.[173] It was, of course, on exactly that wasteland that the 1946 Taupo fires began.

In the summer of 1945–46 New Zealand – especially the Auckland, Waikato, Hawke's Bay, Bay of Plenty and Rotorua districts – was plunged into 'the worst drought since 1918'.[174] Hot, dry conditions, and from mid-January exceptionally low relative humidity, forced the main hydro stations to restrict their output in February and the public were asked to conserve electricity.[175] By the second week of February fire reports were taking up many column inches in newspapers. In Northland, where the failure of root and hay crops had already created a serious shortage of rough feed, an 'unprecedented series of bush and farmland fires' destroyed thousands of metres of millable timber and threatened scenic reserves. Small settlements would have been lost but for local effort.[176]

On 11 February, at Wairoa, just west of Tauranga, volunteers and the Tauranga Fire Brigade battled for eight hours to save a pine plantation. A fire in the Te Whaiti district (on the Rotorua–Waikaremoana road) was brought under control. '[S]cores of fires' were racing across Hawke's Bay pasture and timber country, and hundreds of men and women battled to the point of exhaustion – back-burning, chopping and sawing firebreaks. High winds constantly defeated their efforts.[177] Wellington's urban brigades were stretched as gorse fires sprang up in the suburbs.

In the Waikato, fires sweeping over vast areas of peat and swamp were almost out of control. Men and women were fighting a fire in the Motumaoho Swamp that had been alight since before Christmas; there were fires in the Te Mawhai district and between Ohaupo and Ngahinapouri, and between Lake Road and Paterangi. A wind change might drive fires around Ngahinapouri into the high country; farmers were out at night checking for outbreaks. At Motumaoho, peat and grass fires were burning furiously. Some ill-equipped volunteers from Hamilton and Cambridge arrived readily, rescued a herd of 60 cows from the danger zone and localised one outbreak. On 13 February seven grass fires on the outskirts of Hamilton had the brigade out; the swamp fires in the Te Awamutu district had freshened, endangering autumn and winter feed; in the Te Araroa

Pictures of the 'widespread damage to farms, bush and buildings' interrupted the usual Weekly News summer coverage of yachts, horses, swimming and beauty contests. 'The menace of peat fires in the Waikato' shows farmers digging a trench to try to save a millet crop.

Weekly News, 20 February 1946, p. 25

and Cape Runaway area fires had destroyed thousands of hectares of feed, miles of fencing and a Maori meeting house and dining hall valued at £240.

Two days later the *Dominion* reported numerous fires in the Auckland district, and that in the South Island eight hectares of Selwyn Plantation Board forests near Burnham had been lost. But conditions there were easing. On 20 February settlers, borough council workers and volunteers battled to stem the spread of outbreaks around Omanawa (at the edge of the Kaimai Range) and Judea (just outside Tauranga). By then the Hawke's Bay fires were only smouldering but fires scattered across the Bay of Plenty area, some in gullies where experienced bushmen never expected to see them, created such a hazard that Entrican banned the use of all steam-powered logging equipment. In North Auckland, too, fire conditions were again hazardous.

The fire situation considerably taxed small urban local brigades. In the month to 22 February, for instance, the Hamilton Fire Brigade had received 70 calls (up from 48 the month before). With engines often passing each other in the street on the way to fresh outbreaks, the brigade was so stretched it had to have its machines serviced while on the job. It was struggling to organise staff to respond to calls as well as maintain an adequate response force for Hamilton city; for that reason it had been unable to respond to the request for a trailer unit for the Taupo emergency.

Nowhere were the conditions worse than in the Taupo–Rotorua district. Down to the roots of the bracken and grass, the wasteland was tinder dry. The sporadic fires, which had tended to be simply a nuisance, did not die out as usual. Towards the end of January 'a continuous fire or series of fires ... between the Tauhara and Tahorakuri plantations and Lake Taupo ... wakened up daily with the heat of the sun and threatened from point to point as the wind veered'.[178] But on Saturday 8 February the situation moved well beyond control. A cigarette butt among the bracken and fern on what is now State Highway One was possibly responsible for the 'blazing holocaust'.[179] Sweeping through thousands of hectares of forest, it became established in the Australian-owned Afforestation Proprietary Company's substantial block of over 10,000 hectares at Rotokawa (east of Taupo on the Rotorua–Taupo road). Some 7000 30-year-old pines were destroyed. The fire then moved into another of the company's areas, burning 4500 hectares of 20-year-old pines.[180]

By noon on Sunday, as new fires were constantly called in, some 6070 hectares of Forest Products land were reduced to charred stumps. By late afternoon a freshening wind was pushing a front of fire 64–80 kilometres wide towards the Bay of Plenty. On Monday 11 February, while a further outbreak 18 kilometres down the Napier–Taupo road destroyed an estimated 12,140 hectares of mature timber, the chief worry was the 20,234-hectare block of native and exotic forest between Oruanui (where one correspondent claimed a fire had burned for days even before the end of January) and Atiamuri. There freshening winds fanned the flames in the native bush 'in to a fresh lease of life'.[181]

The fires ran through fern and scrub and, whenever they reached the forest, howling winds propelled them forward in hops of close to half a kilometre. Within a couple of days millions of feet of pines were 'a dead loss', reported the *Waikato Times*, apparently unaware of the pun, and much more was likely to go. 'Men who have worked their lives [in the forests] state that some of the losses cannot be made good within several lifetimes.' In the Oruanui forest some of the best stands of totara and matai in the country, representing 3 million metres of millable timber vital to the country's postwar reconstruction efforts, had been reduced to charred stumps, the *Waikato Times* claimed.[182]

Towns, too, were endangered. At one stage on Sunday 10 February the Wairakei and Spa hotels and Taupo township had only a firebreak and a stretch of scrub between them and the fires; a fire on a spur above Lake Okataina also demanded attention. By Monday evening some 640 civilians were involved. The whole of the Orakei Korako valley had been burned out. With flames travelling almost five kilometres in 20 minutes, Atiamuri residents, some of whom lost everything, were hastily evacuated to Putaruru; a day later Putaruru itself was on alert. Taupo, covered with a heavy pall of smoke and with a night sky lit by the glow of fires, was momentarily safe, though closely watched incipient fires in the scrub spelt constant danger.

The authorities' response was immediate. In the early days of the fire a call went out for one of the Emergency Precaution Scheme trailer pumps stationed at Rotorua to protect the scenic and forest areas. But the attempts to stop the fire crossing the river failed as fires straddled the Taupo–Rotorua road. Reinforcements were called in.[183] The Forest Service impressed all able-bodied men in Taupo, Putaruru and Rotorua. Soldiers and sailors also arrived from Waiouru on Saturday evening. Fire engines, some setting off within minutes of the request, came from Wellington, Rotorua and Auckland – and these were vital as the four Forest Service engines had to be reserved to protect Kaingaroa Forest. Volunteers were there in their hundreds – bushmen from Putaruru, local men in white shirts and women in summer floral dresses beat at the flames.[184] Impressed men had to be fitted out with army-issue denims, boots and fire-fighting tools. Truck owners and others threw 'their resources into the fight' and battled for over 24 hours, snatching sleep only when reinforcements arrived. 'I saw them,' said Wellington MP Charles Bowden in one of many tributes,

> grimy with smoke and sweat, singed and scorched by the intense heat, eyes bloodshot from exposure and lack of sleep and almost dropping in

> their tracks from weariness and yet battling on against the devouring monster. It was their courage and willpower that kept them going.[185]

'Fire-fighters passing through a pine plantation on their way to burning trees near Atiamuri.'

Weekly News, 20 February 1946, pp. 24–25

On 12 February Taupo was again threatened, and hopes that the Atiamuri block might be under control were dashed when, late in the day, the fires once again flared. Flames shot 30 metres into the air; the intensely black manuka smoke mingled with white smoke rose some 1524 metres. Nothing he had seen in Italy compared to 'the show at Taupo', said a recently demobbed Roger Mirams who was filming for the National Film Unit.[186] The fires that had been on the borders of the Kaingaroa Forest were dying there but moving steadily westwards. A big new fire, in the Hauhungaroa Range, had also flared in what Commissioner Skinner, stationed at the fire-fighting headquarters at the Spa Hotel, described as the best stand of native timber the Dominion had left. The bush fire at Oruanui was still uncontrolled, despite the efforts of 70

'The scene of desolation after the fire had swept through the bush at Tatua, north of Taupo.'

Weekly News, 20 February 1946, pp. 24–25

fire-fighters, but three large bulldozers had managed to construct an effective firebreak around the Oruanui village and mill.

In the Rotokawa forestry area 100 men had successfully back-burned over a large area, extending the safety margin considerably. By now the plane used for fire-spotting had arrived and Public Works men from the Karapiro dam project swelled the fighting force, which now totalled 222 soldiers, 300 navy personnel and Public Works men, and 200 impressed civilians; later, soldiers not yet demobbed would be brought in. Local bakeries were able to slacken their pace slightly when the armed forces established their own kitchens in the domain. The fire brigades from other regions brought their own rations, blankets and overalls (as well as fire-fighting equipment) and the Public Works men brought in food from their own camps.

By 15 February the Atiamuri fire was almost under control and the lack of wind created a small respite. But many men were still battling the numerous scattered fires and were patrolling the entire district to prevent further outbreaks. For the army, unused to dealing with bush fires, the situation was frustrating. 'There are times when one feels that all is finished with, and that the fires have gone out,' but then they would break 'out in all directions' making it an 'almost impossible' situation to deal with …'[187] On 19 and 20 February several outbreaks in and around Taupo township again, briefly, stretched the fire-fighters. Telegrams had already gone out for specialised medical equipment; now the army urgently needed heavy vehicles, slashers and shovels.[188] Authorities considered that positions should hold provided the wind did not rise, but the 200 Maori living in Nukuhau Pa were evacuated, Taupo residents planned their escape, '[e]very available man and piece of equipment' were marshalled, extra engines were rushed in from Wanganui, Palmerston North and Te Aroha, and water lorries and piles of hose were assembled. Control was achieved at a price. Sapper Donald James Moire was hospitalised with bad burns, a Forest Service officer and Public Works man both needed treatment for burns and thousands of metres of hose were damaged when a Wellington engine was surrounded by fire.

Nor was the Rotorua fire situation stable. On 19 February a new fire crossed the Okere River and spread around the lake. Although the army and air force were called out, it was still burning a day later. By 22 February an almost continuous chain of fires burned in the Mamaku Ranges between Ngatira and Okere, and a couple of days later fresh reinforcements of air force fire-fighters arrived from Auckland. A fresh outburst of fire near Mokai village also demanded further firebreaks and back-burning; the Wanganui brigade was again called out to protect the Forest Products plantation. A welcome rain at Rotorua did little to extinguish fires and near Tokaanu another fire demanded intensive work.

On 27 February the *Dominion* reported that most outbreaks were under control, though water continued to be pumped onto all accessible fires and, with humidity rapidly falling, any wind could make the position serious again. Then on 5 March it rained heavily. The crisis was over.

Fighting the fires had been a massive logistical undertaking. Food and rations for those fighting as well as those on standby had to be organised immediately. The local women in the War Services Auxiliary made tea, coffee, sandwiches and scones for the impressed men until the army and Public Works men took over the following day, and the local YMCA distributed free Patriotic Issue cigarettes and worked to secure further supplies.[189] A somewhat makeshift fire control headquarters was established in the Spa Hotel, but no one was ever 'quite sure whether the next telephone call would come from parties of fire-fighters or Mrs. Jones in Room 45 who had not been supplied with a new towel'. The headquarters was then moved to the stage of a public hall, confined to this area because the rest of the hall was in use. The result was that 'weary men who had fought the fires for days and nights on end without respite' might telephone HQ only to hear background noise – the strains of "For he's a jolly good fellow" and the popping of corks'.[190]

The navy and later the army manned the radio at a fire headquarters at Atiamuri. In the King Country, army personnel with radio facilities at various towns provided additional links in surveillance, reporting fires identified from ranger stations for forwarding to Wellington, Auckland and Palmerston North. The Taupo area itself had to deal with the problem of burnt-out poles delaying phone communications. At the height of the emergency over 2500 men were

A naval party setting up radio communications in a pine forest near Atiamuri. The accompanying article, sub-headed 'Material for vast bonfire in Taupo region', commented on the fire hazard that had been 'accentuated by the uncontrolled spread of broom, gorse, blackberry and self-sown pines. In this valuable scenic area in recent years camp sites have become overgrown, and the lake front for miles has become an impenetrable jungle of blackberry.'

Weekly News, 27 February 1946, p. 20

employed:[191] at any one time a highly diverse group of local residents, Public Works men from the hydro-camps on the Waikato River, the armed forces, fire brigade men (volunteers and permanent), the impressed men and the Forest Service had to be equipped and deployed.[192] It seems remarkable that only one person was rather badly burned; likewise, although several fire brigade appliances were badly damaged and over 1.5 kilometres of hose was destroyed, the level of loss could well have been higher.[193]

Although the private companies contributed funds – sometimes under duress[194] – the fires cost the government over £51,500 ($107 million) in overtime for Forest Service staff, for contractors and individuals brought in to stand by, patrol and dampen down, and for the armed forces' labour, vehicle hire, cartage, fares and lost tools. Nor did that sum take into account a longstanding feud, not resolved until February 1947, over the wages of timber workers and returned soldiers whom Entrican had verbally impressed.[195]

Vegetation fires continued to burn both in and outside state forests. Some saw these as the inevitable result of settlers burning off.[196] However, by now there was a pattern: fires in state forests were fewer and less destructive than those in private forests. During the 1946 emergency only 65 hectares of exotic forest were lost at Tairua; nothing was destroyed in the Kaingaroa forest, although the fires reached its boundary; and the Forest Service logging operations in Te Whaiti forest were entirely undamaged, though fire had swept over hectares of adjacent cut-over areas of both Maori and privately owned land. The Forest Service's protection methods were effective – the longstanding problem was to get private operators to adopt them. Given the legislation, regulations and the educational initiatives in place, the behaviour of some individuals during major fires makes it difficult to avoid concluding that changing the population's mindset was an uphill battle that would take longer than the 30 years over which the Forest Service had been working.

Entrican's report to Parliament a few months after the fires emphasised the minimal losses the Forest Service had suffered because of its own fire-fighting efficiencies,[197] but not all were convinced. The *Gisborne Herald* deplored the lack of precautionary measures. Acting Prime Minister Walter Nash, the paper claimed, had only reiterated 'the old warnings against personal carelessness'. But with afforestation on its present scale, firebreaks, although never the full answer, should have been developed earlier, and a comprehensive air surveillance scheme should have been organised to spot incipient fires. 'New Zealanders in a querulous mood – a pardonable attitude in the circumstances – may well ask what the department intends doing by way of exemplifying the truth that an ounce of prevention is worth a liberal poundage of cure.'[198]

Girling-Butcher's assessment of the brigades' efforts raised critical issues about control. The urban Fire Service, he considered, had been effective in its primary function of protecting built-up areas, and its heavy equipment had been invaluable, even if the emergency had rendered its assistance 'totally inadequate'. Without directive control, for instance, Johnsonville's hose-laying Fargo engine and about 1.5 kilometres of landline had been taken by a permanent Wellington crew and retained for about two weeks, leaving Johnsonville residents somewhat ill-protected.[199] Moreover, without any adequate system of control and communications, particularly in the initial days, Fire Service units had not been used effectively, and the Forest Service's reluctance to relinquish direct operational control had deprived the Fire Service of opportunities to support forestry teams.[200] Lack of adequate equipment further mitigated against the brigades playing the necessary support role, especially when it came to transferring water to Forest Service mobile and manual pumps. For that to be done satisfactorily, brigadesmen needed pumps of over 900 litres per minute and reserve stocks of overland and delivery hose. Critically, too, not only should they be working the hoses, Girling-Butcher reckoned, they should be directing the operation because they had a higher level of training than did the Forest Service men, who were trained for the smaller mobile units.[201] MP Charles Bowden was the first to call for a royal commission, not to enquire how the fires started – a matter 'more or less of academic interest' – but to make sure no such fires happened again.[202]

Meanwhile officials were already working to improve protection, particularly around Taupo. At the end of 1946 Girling-Butcher's perseverance had established a structure for voluntary mutual assistance involving the township, the Forest Service and Forest Products. (However, although this structure, which would become the Central Fire Authority Committee, was seen to work well, the Afforestation Proprietary Company never joined, and various terms of the agreement were often questioned. In 1953, with developments in the fire practices of the township and the major companies, the committee was deemed no longer necessary.)[203]

Within the Forest Service, whose heavy truck and equipment shortage highlighted its own vulnerability, Entrican began liaising with the army and Post Office in mid-1946 to ensure that vehicles and wireless sets would be available in the Rotorua–Taupo district for the next fire season; in the following years such arrangements were expanded nationwide and continued into the mid-1950s.[204] Taking a broader perspective, forester Courtney Biggs, who had played a major co-ordination role throughout the emergency and who would become the Forest Service's first fire control officer, set out the considerable number of issues the fires had raised. The most important ones related to developing a centralised control and co-ordination system, establishing a supreme fire district fire authority (had this been in place 'the fires would never have reached dangerous proportions'), having a system for co-ordinating all government departments, ensuring private forests became part of a Forest Service fire plan under a unitary authority, and creating an 'essential' link with the national Fire Service.[205]

By June 1946 perceived deficiencies were being rectified. The development of a fire-hazard meter, new lookout stations, better radio communications, the construction of 3637-litre water tankers and hose layers (and the permanent air patrol in place by 1947) have already been noted. New education efforts were spurred on by the knowledge that, in the middle of the emergency, a sawmill operated a slab fire within a few metres of one of the huge exotic forests and fire-fighters and patrolmen had ignored it. The regulations were displayed in easy-to-read poster form in sawmills for visitors and workers, and leaflets explaining workers' obligations under the regulations imposed on them were put in their pay packets.[206]

The emergency had dramatically highlighted the shortcomings in the Forests Act and in the Forest Service's ability to organise control and prevention. Entrican, reporting to his political masters, stressed the need for the power to declare fire districts in a crisis; pending legislative change, a huge Kaingaroa–Rotorua fire district, covering 542,280 hectares, was promptly gazetted to forestall any future emergency. Forest Service officers or local authorities had to be able to proclaim total fire bans at times of high fire hazard, and the fire district concept needed to be extended to other rural areas in what Entrican termed 'fire emergency districts'.[207] Moreover, if forest and rural fires were to be dealt with on a national basis – and that was now an inescapable conclusion – legislation was needed to maximise co-operation between the Forest Service and the Fire Service, itself undergoing a long-postponed major reorganisation.[208] On 18 November 1947, after two conferences to ensure the bill's smooth passage, the Forest and Rural Fires Act was enacted.

That legislation established a structure in which urban fire brigades could play a defined (but circumscribed) role in rural fire-fighting. However, the transport, equipment, training for other than static fires, and vital adequate financial backing for that role were not available to the occasional fire brigade whom we last left doing its best in the 1908 fires. It is now time to survey briefly New Zealand's fire brigade history, to watch sparsely equipped, territorially limited groups emerge from poorly funded bodies of men into somewhat better equipped organisations – but organisations whose role in rural fire-fighting was still constrained.

Chapter Five

'With hearts ahigh and courage aglow': Fire brigade history to 1977

The little settlements that increasingly dotted the New Zealand landscape were a constant fire risk. Wooden structures, roofed with canvas, wooden shingles or thatch (forbidden in New Plymouth), and surrounded by wooden fences, often stood either alone or alongside others on roads bordered by grass that was yellow and dry in the summer. Unkempt gardens and hedges, upset candles, oil lamps combined with the lace curtains and drapes so loved by the Victorians, open fires, chimney sparks and New Zealand's ever-present wind added to the danger. The limited capacity of wells and tanks did not guarantee a secure water supply for fire-fighting, and streams were not always usefully positioned.

The Nuhaka Brigade was formed in 1960. Before then

> [t]he locals were without any protection whatsoever. There were occasions when we residents would get a phone call at night ... The night Kupu Winiana's home at Waipuia just over the dry creek bed [caught fire], some of the family ... lived higher up on the side of the hill. They had a well and a tank and that was our only chance of doing anything. So a bucket brigade had to walk up hill for about 100 yards in the dark, with only the flames from the house to show the way ... Then there were a few cottages around past the Manutai Marae neat the Tahaenui Cemetery ... Well, that was not so bad, because the river nearly always blocked, and water was fairly easy to get. That was one thing with living as it was then. Every household had tubs etc, that were absolutely essential. No piping of water in those areas ... Then there was Paul Mitchell's home ... There was no chance of a save at that fire at all. For in going up the paddock it was as light as day.[1]

Apart from the detail of the phone call, the ingredients of that recollection

– willing locals being roused, the bucket brigade, the fortuitous supply of water that made the difference between a save and a loss and the all-too-real possibility of fires well out of control before help arrived – could have placed it in the previous century. Fire brigades like Nuhaka's were formed in both the 19th and 20th centuries largely because locals took the initiative to move beyond buckets and organise formal protection for their towns and homes. Yet, essential as their role was, small-town brigades in particular, and the fire inspectors appointed to advise and assist them, battled for decades in the face of significant fire loss, public apathy and carelessness. Only in 1906, after decades of effort, was the first specific fire brigade legislation passed; not until World War II equipment was declared surplus did many brigades begin to get adequate equipment; and the fêtes, parades and other fundraising activities that were essential into the 1970s for small (often rural) brigades indicate the precarious financial environment in which many operated.

What did this situation mean for rural fire-fighting? In the early days, lacking specialised training and sophisticated equipment, brigades' contribution was largely limited to providing manpower – purportedly disciplined – to the general effort. By the end of World War II, better equipment and greater mobility allowed more specialised input; from the end of the 1970s further reorganisation and technological improvements enabled brigades to provide a first-strike force in quelling vegetation fires.

Many strands in the New Plymouth brigade's early history characterise the experience of brigades across the country. By 1855, and fortunately for the township, the army brigade had a hand-operated pump machine to work alongside civilian bucket brigades. This allayed concerns with fire control. But in 1864, as the army was gradually withdrawn, local anxieties started to be voiced. Two years later an elaborately structured volunteer brigade was formed. Membership was limited to 100; the 90 volunteers paid sixpence a month to a members' relief fund, and honorary non-active membership was a guinea a year. Insurance companies subscribed £200 ($21,000 today) to purchase an efficient Merryweather hand-operated pump, and another was bought with funds raised from public subscription.

But good equipment was difficult to obtain, the publicly funded engine did not work well, and in 1869 the insurance companies, only too aware of the brigade's inefficiencies and the loss of a lot of property, planned to withdraw their machine to Wanganui. The proposal roused considerable local ire and had to be referred to a judge, who ruled the pump be handed over in July 1870. The brigade disbanded. Although it was re-formed under new officers and with 28 active members in 1873, the town council refused to give any assistance and the insurance companies failed to even acknowledge a letter asking for a new fire engine. The brigade again disbanded, and was in recess for most of 1877.

Then in 1878 the borough council agreed to employ 15 firemen on a salary of £2 10s per annum, and the New Zealand Insurance Company helped buy new equipment. Ten years later the brigade, its bell cracked, its uniforms in bad repair and its two reels struggling to bear the weight of wet hoses, was making invidious comparisons: while the New Plymouth council was refusing to raise its annual grant of £65 to £90, the Wanganui council had given £200 to its brigade and insurance companies had donated another £150. The New Plymouth council finally agreed to lift its grant to £70.[2]

The issues inherent in that story – the roles of government, the insurance companies (which had been an important force in the development of English brigades) and local bodies; the brigades' precarious hold on existence; and the availability and nature of brigade equipment – we explore more widely below.

Central government, which considered that city councils or boards and insurance companies should support brigades, provided almost no assistance to those early brigades. Auckland's City Board Act 1863 facilitated the brigade's formation by exempting members from most of their militia duties. The

Wellington's Fire Brigade, with members standing on what appears to be a horse-drawn steam engine en route to a fire, sometime between 1900 and 1909.

ATL, ½-14840-F

Wellington Central Volunteer Fire Brigade got infrequent payments for engine upkeep in return for protecting government buildings over the years; eventually the £10 the government contributed in 1868 became an annual payment and increased slightly.[3]

It took decades for the government's stance to change. The Municipal Corporations Act 1876, passed after the provinces had been abolished, made central government responsible for local government administration. Here was an opportunity to impose a uniform approach to how local bodies dealt with urban fire risk. But instead, the provisions round that risk were essentially permissive. Councils were not required to establish brigades. Moreover, council-appointed fire officers did not have to be brigade members – a provision that might be seen as undermining the brigades, which wished to elect their own officers, and the volunteer contribution and spirit behind them.[4]

United Fire Brigades' Association certificate of membership, 1903. From the association's inception, such certificates, along with the games, badges and honours lists and boards, were very important in developing brigade members' esprit de corps – an essential ingredient for any successful volunteer brigade.

ATL, Eph-D-Fire-1903-01

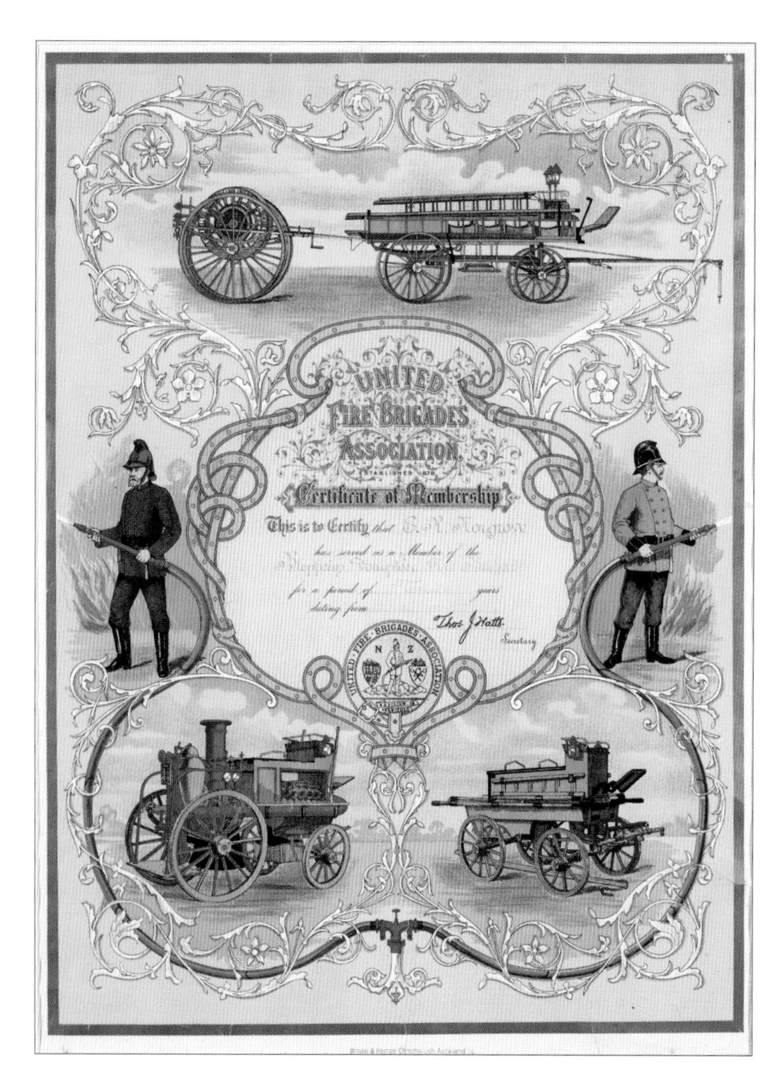

In 1878 the United Fire Brigades' Association was formed to promote the strengths of the brigades, alert citizens to the need for fire awareness and fire prevention and develop links with government that would lead to legislation governing fire-fighting and provide an adequate funding base.[5] The association was singularly unsuccessful. It did manage, almost a decade later, to attract what became an annual grant, and later again some assistance for its members to attend games and conferences, but it failed to get any legislation into the House, let alone passed.

The municipal legislation had a predictable effect on insurance companies and municipal bodies. Insurance companies, both overseas and in New Zealand, were unfettered by any requirements for set contributions and behaved seemingly arbitrarily. In 1864 Melbourne's highly efficient fire brigade, which had fought fires so successfully that it removed the incentives to buy insurance, fell victim to its own success. Faced with falling premiums, the insurance offices, which had largely funded the brigade, found it 'more conducive to profit to break up the brigade than maintain it in an efficient state'.[6] In Dunedin, Lyttelton and New Plymouth, insurance companies provided equipment, and they wanted it well used. As in New Plymouth, one company threatened to remove the Lyttelton Volunteer Brigade's appliance, which in this instance had a salutary effect on brigade formation.[7] But there was no certainty. In 1878 the Port Chalmers Volunteer Brigade failed to receive its £60 due from an insurance company.[8] In 1882 insurance firms were contributing generously to the Christchurch and Invercargill fire brigades, and very generously to Auckland's, but refused to contribute to Wellington's. Such actions may have reflected the fire history of the various towns – in towns with poor fire records the expenditure was unlikely to have a good enough return. But there were too many insurance officers chasing limited funds and failing to vet adequately the clients they took on, and self-arson by clients seeking insurance payouts was too common.[9]

Occasionally local councils offered support, generally by making small donations towards new gear. Geraldine's town board, wary of the fire danger in Canterbury's high winds and appreciating possible insurance benefits, recognised the need to establish a volunteer brigade as soon as possible. In 1888 it offered to construct a water-race and form a brigade under its supervision if the fire insurance company would grant a concession of 25 per cent off its premiums. A few months later it decided to strike a council rate of a shilling to provide funds for a fire engine and appliances, identify a suitable site for an engine shed and advertise for a fire engine. (Typically, and in a pattern that repeated itself over the decades, the engine was a hand-me-down – purchased in this instance, along with a hose reel, from Gore for £40.) In 1889 the brigade was formed.[10]

Without support from the local council or businesses, maintaining a brigade was hard. Not every employer would allow his employees to attend a fire when the bell went, and volunteer enthusiasm ebbed away.[11] Palmerston North's brigade, first formed in 1877, was disbanded and the money handed back to subscribers because it could secure neither insurance company nor borough council support. There was a brief resurgence in 1881, but by the end of 1885, with no funds for training materials or fire-fighting, the brigade was declared defunct. (In this instance the story had a happy ending – the brigade was promptly resurrected with the council's financial backing.)[12]

Any brigade that lacked council input had to improvise with its fire-fighting equipment. Clyde's first volunteer brigade (formed in 1868) made a reel from a cheese crate and a buggy axle and wheels, to which it fixed a length of rope for hauling. Its hose was leather riveted together; its bell tower was constructed from an old boring plant that came from a neighbouring county council.[13] The five citizens who formed Oamaru's first volunteer brigade in 1879 had a hand-pump and a hose reel; the creek and individual tanks provided the water.[14] In the Manawatu in 1886, even as the *Feilding Star* pronounced on the 'disastrous [bush] fires ... not only in every part of the colony but at our very doors', the council merely resolved to pay half the salary of a night watchman if the insurance companies would pay the other half. It also agreed that, in the event of fire, the Manchester Rifles' bugler would sound the alarm so that anyone who wanted to help could gather in the town square. For some time the only equipment the town had was a hand-operated pump bought by councillor Alexander Ferguson.[15]

A picture builds up of many small towns gambling on not having a major fire. If enthusiasts wished, they could form a volunteer brigade, financed by whatever public funds were available, but bucket brigades could do little more than extinguish minor fires and try to prevent larger fires from spreading. This failure to back and fund a proper brigade reflects a certain faith in do-it-yourself. Just as newspapers informed their readers on how to burn the bush, so they published advice on fire-fighting tactics and fire management. Settlers were told the value of planting poplars, willow alder or periwinkle around their homes, and were told how to construct 'very effective beater[s] ... from a strip of stout hide ... fastened on a handle of wire' – a far more durable tool than a branch.

Settlers had also arrived with their own knowledge of the value of draping wet branches over shingle roofs and wooden walls. They were prepared to fight to protect their belongings[16] and were remarkably effective. Property may have been lost, but few lives were (and this is true not just of 19th-century colonial New Zealand but of other countries and other centuries[17]). Settlers must have at times pondered on how useful brigades could be without public water supplies and on streets impassable in winter. Would Feilding's hand-operated pump mounted on a wheeled platform for trundling to the fire really make a difference? Such cumbersome equipment had to be got quickly to the scattered homes that made up many early settlements. Once there, it needed willing manpower; a proper steam fire engine relied on horses and a volunteer engineer. None of these conditions could be assumed to exist.

Moreover, the outlay required for quality fire equipment would have entrenched settlers' doubts about the value of a fire brigade. Rollo Arnold estimates that for effective fire protection a small town needed at least a well-mounted bell (the oft-used church bells were frequently not central enough or loud enough, and could be heard as calls to worship), a fire pump, ladders, a transport vehicle, and housing for the equipment, amounting to at least several hundred pounds. Towns of more than 2000 needed a steam fire engine and

New Zealand's first Merryweather self-propelled 'Fire King' steam fire engine, photographed outside the Central Fire Station in Wanganui, between 1919 and 1921. The plaque on the left of the engine reads: 'Merryweather London first grand prize patent steam fire engine'. The Wanganui brigade bought the engine in about 1903, just before the makers moved into motor fire pumps. The machine was heavy, moved slowly, and it took four to seven minutes to build the requisite pressure, which could only be maintained by stopping periodically. Yet the engines – with a coke- or oil-fired boiler, 454.6 litres of water, which was sufficient for about half an hour's running, and a fuel capacity for four hours of operation – dominated the market and were exported the world over until 1922. In 1921 the Ohakune Fire Board bought this engine, part of a pattern of second-hand purchases by no means unusual in New Zealand.

ATL, 1/1-009810-F

horses – in 1889 a new engine imported directly from England cost about £275 ($50,000). Even voluntary brigades were expensive to outfit properly and to run. Understandably, small towns' disgruntled citizenry who wanted passable roads and footpaths, street lighting and permanent water supplies – the very amenities that would add to the efficacy of any brigade – loudly voiced priorities other than fire brigades.[18]

By the end of the 1880s the main towns had 'reasonably efficient' part-volunteer and part-professional brigades. Smaller towns were not so well placed.[19] However, in 1906 members of the then 104 brigades in the United Fire Brigades' Association of New Zealand finally had 'the pleasure of seeing placed upon the statute book the long-considered and long hoped for legislation – the Fire Brigades Act'.[20] But the act set up fire boards without providing funding, leaving them 'without sixpence to their name', and leading to the conclusion that the system was 'unworkable'.[21] It was promptly replaced by the Fire Brigades Act 1907 (which was amended by the 1908 act, itself also promptly amended).

The 1907 act established fire districts for Dunedin, Christchurch, Wellington and Auckland. The Governor could also declare other fire districts if the population of the area numbered more than 2000 and where a ratepayers' poll showed that at least 15 per cent approved the move. For those areas not declared fire districts, the provisions of the Municipal Corporations Act 1876 applied until a new act was passed in 1920. Each fire district had a seven-member board comprising nominees from the Governor, the contributory local authority and insurance companies doing business in the district.

There were three critical sets of provisions. First, and in contrast to the earlier act, there was some funding – £200 for each of the four main centres and an extra £200 for Wellington for protecting Parliament and government buildings. The other boards were funded for one-tenth of their (ministerially approved) estimates of annual expenditure, up to £50. A related provision, forward thinking for the time, required insurance companies and local authorities (through levies or general rates) to pay half each of the remaining estimated expenditure.

Second, the act established how boards should operate, covering matters such as brigade appointments, fire brigade by-laws, establishment of a permanent force, resignations, terms of office, offences, borrowing money, leasing or purchasing equipment and providing fire escapes. A host of machinery provisions dealt with how money and property were to be disbursed if the area ceased to be a fire district. The third critical provision related to a government-appointed salaried inspector and deputy inspector who were to inspect all brigades controlled by fire boards at least annually, give advice and assistance, and report to the minister.

Thomas T. Hugo, inspector from 1909 to 1930–31, was the first of the two committed and intelligent men who held the position. His successor, Roy Girling-Butcher, was more of a 'big picture man'. He held the position until 1950 when he became chair of the new Fire Service Council, a body arising from a reorganisation of the Fire Service that he had largely instituted.

Thomas Hugo, New Zealand's first Inspector of Fire Brigades. He held the job from 1909 until his retirement in 1931. Coming from a Devon family of seafarers, he began his picturesque and varied career with an adventurous time at sea in the early 1870s. He arrived in New Zealand as a ship's captain in 1890 before spending some time as a surveyor in the Australian interior. He returned to Wellington and took charge of its fire brigade in 1899. Although he was something of a martinet, he achieved a remarkable increase in the brigade's efficiency and his considerable ability was recognised with his appointment as inspector. When he died in 1933, aged 73, his large circle of friends remembered him as a congenial companion and as a popular figure on Wellington bowling's green. A fire engine bore his coffin and led the funeral cortege through the town to the Karori cemetery.

Hugo's reports began in 1909 and were furnished annually until his retirement. They form an invaluable record of his visits to the brigades under fire boards' control (an achievement in itself, given transport at the time), along with his assessments of their equipment, their fire drills and their instruction. Hugo also advised local bodies on anything from water-supply schemes to the purchase of machinery, he tested imported machinery, he supervised the manufacture of locally made plant and appliances. He regularly deplored the country's fire loss, in which it led the world.[22]

So how effective was the 1907 act? When it came into force only 21 out of a possible 46 fire boards were established (in 1908 the districts in Waimate and Alexandra were disestablished because they did not meet the population requirements).[23] Yet even the partial protection offered in those districts was timely. By 1909 New Zealand's fire loss was 'so excessive in proportion to its population' that it represented a serious drain on the community's wealth.[24] Ever-rising figures did not allow complacency. In 1917 insurance companies paid out some £5000 ($578,000); in 1922 the total was £765,310.[25] By 1929 the average per capita loss rose to an all-time high of 22s 6d – a staggering figure compared with the English loss of 3s 9d per head in 1925.[26] Hugo was clear on how this longstanding loss should be addressed. Fresh from overhauling the Wellington Fire Brigade, he wanted brigades to be better equipped, insurance companies to stop over-insuring, and public education to be implemented.

The causes of loss through fire damage, although easy to identify, proved hard to address, and Hugo's restatement of issues he identified in the first couple of years into the job form a refrain that lasted into the first half of the 1930s. As in the rural fire scene, public apathy and carelessness dogged the urban one. In 1908 a majority of councillors in Wellington and Napier voted against authorising the ratepayer poll needed to establish fire district status.[27] Preventable fires were not prevented. Efficient locally made automatic fire alarms and automated sprinklers, available from at least 1925, were not installed in many of the larger shops and factories. To Hugo's frustration, the construction industry could not be persuaded that it made economic sense to build with heavier frames. Nor would it line with lath, plaster or any of a range of patented, reasonably priced and at least partly fire-resistant materials that were coming on the market by 1928. Instead, builders opted for cheap half-inch match-board lining covered by scrim 'and the whole then disguised by further coverings of pasted-on wall paper which when dry, is nothing more or less than a quick-fire train throughout the whole building'.[28] Also problematic were the ways insurance companies vetted for risk, their preparedness to accept owners' valuations without question, consequent over-insurance and, as Girling-Butcher generously phrased it, 'the temptation towards incendiarism'.[29]

As with rural fires, preventing fires is better than fighting them. As early as 1924 Hugo estimated that at least 50 per cent of fires were caused by carelessness that could be prevented. He advocated education programmes such as those in Canada and the US, where by 1924 there were public education programmes and school-level fire-prevention programmes, and special college-level programmes were being developed to educate fire officers on preventing and extinguishing fires. By 1933 North American fire-prevention weeks included presentations via public lectures, broadcasts and the press on the inadequacies of normal insurance policies and the consequent need for care.[30] Yet, although the Forest Service was by this time running its education programmes with some success, the New Zealand government was prepared to do no more than dabble in fire education. Only three fire-prevention education efforts were run – in 1932, 1935 and 1936. The broadcasts, school visits, general publicity, demonstrations, leaflets and circulars on how fire alarms worked were deemed effective but it would be a long time before such initiatives reappeared.

Children with fire engine and firemen, Government House, Wellington, 1957. Such public events (as in this photo when the public could visit Government House) gave fire brigades opportunities to push their education message. Small brigades used the occasions to hand around the hat as well.

ATL, EP/1957/4338/F

Ronald Girling-Butcher. Born in 1887 and the son of George Girling-Butcher of the Mines Department, he joined the staff of what was then the Colonial Laboratory (later the Dominion Laboratory) under Dr J. S. Maclaurin in 1901. Eight years later he became Inspector of Explosives, in charge of the Explosives Branch of the Department of Internal Affairs – a position that would assume more responsibilities when the Explosives Act was extended to cover the safety of cinemas (leading to a close association with the film industry). In the mid-1920s he travelled overseas to gain expertise in controlling and storing dangerous materials. On Maclaurin's retirement in 1930, Girling-Butcher became chief inspector, but left that job in 1931 to succeed Hugo as Inspector of Fire Brigades. Thereafter he chaired the Fire Service Council until his death in 1952.

Installation of automatic fire alarms and sprinklers was compulsory in a number of US and Canadian states, and in Melbourne and Sydney, but Hugo's calls for such a legislative change in New Zealand in 1929 and 1930 went unanswered. A building construction bill, intended to update the outdated municipal by-laws, was delayed in 1933, then dropped altogether a year later because the government intended instead to develop a set of model by-laws. As Girling-Butcher pointed out in 1934, the issues Hugo had canvassed over the preceding 26 years had still not been addressed. Indeed, two years later Girling-Butcher's own assessments of measures needed to address fire loss traversed familiar ground.[31]

But Hugo's years of cajoling, advising and reprimanding did make a significant difference to brigade efficiencies. The issues he had to address differed markedly between the main centres and the small towns. In 1915, for instance, the Christchurch brigade was despatching its large motor pump, which needed seven men to operate it, to large stores, factories and mills outside the town boundaries, leaving the city and its large department stores, full of flammable materials, unprotected. Unless the brigade recruited three more men, Hugo recommended, it should not travel to fires outside the town boundary.

The smaller centres faced problems of an altogether different order. Dannevirke, for example, in 1915 needed a way of getting men and equipment to a fire, a modern alarm system for summoning the force, and automatic alarms in its large warehouses. Hamilton's brigade was below authorised strength and attendance was unsatisfactory, as was the mains water-supply pressure and volume. The gear and appliances were very dirty and the equipment was deficient. Palmerston North's mains did not ensure the whole town was properly defended, and Petone's horse that kept balking (not atypically) needed replacing with a fire engine. In Tauranga the brigade was not up to strength (a longstanding problem) and the men needed more drilling and better uniforms and equipment.[32]

Although after Hugo's first visit many of the brigades lifted their performance, particularly with getting to practice, others did not. Advice on equipment and its maintenance, means of transport for men and equipment, training and response times was ignored from one year to the next. For instance, in Hamilton, two years after his initial inspection in 1915,

> various matters came under my notice that point to considerable carelessness in the care of the appliances, and a total disregard of my teaching and advice … the hand-pump and hose still lying on the floor of the station, and the hose, which should have lasted at least ten years, now useless; no hand-lamp available, yet one comparatively new lying on the ground with the glass broken … dragging and scraping boots along the new hose when wet to flatten it – and this last is particularly bad, seeing that at my previous visit I had given some advice as to the care of hose, particularly condemning such action.[33]

If that were not enough, the following year the brigade exercised poor judgement in agreeing to despatch 11 men to Horotiu freezing works, about 12.5 kilometres away and outside the town boundary. 'Even with the brigade at its authorized strength, that is altogether too many men to send such a distance away, and under the present circumstances, absolutely unthinkable,' Hugo fulminated.[34]

In 1916 the Dannevirke brigade had started commandeering a car to get its men to the fire in better shape than when they had had to drag the reel. But by 1918 the brigade was failing to carry out fire drills or care for the hose, and other equipment was in poor condition. Rotorua had not bought the recommended reel and Woolston still lacked an adequate means of summoning the men to a fire, all the more necessary as the town lacked a water supply and the board relied on a chemical engine and other first aid appliances. At Balclutha the foreman and three of the 15 brigade members were working at the freezing works nearly five kilometres out of town, so it was essential the brigade be maintained at full strength because as many hands as possible were needed for the 'very exhausting process' of dragging 'the reel, hose and other equipment to the fire'.[35] A 'very serious' wartime manpower shortage, first identified in 1916, further compromised brigade performance. Dargaville needed more

The Petone fire station and the Shand Mason manual hand-pump, 1892. This pump was drawn by horses (or men); like the Merryweather, it could deliver about 456 litres a minute. It was called 'the killer' because the men had 'had it' after a few pumps at the long pump handles. Debris in the water supply easily slowed its progress.

ATL, APG-006301/2/G

men as they had to manhandle the equipment to the fires. Gisborne had four boys in its muster in an attempt to get the requisite numbers. Unsurprisingly, Hugo wanted all able-bodied men and youths who were too old or too young for armed service to help with fire protection as their patriotic duty. He also recognised that recruitment, which drained the towns of suitable men for the brigades, impacted on more than brigade membership:

> A matter which I consider it necessary to comment upon is the management of costly brigade motor machines. After acquiring a motor appliance generally some local chauffeur is appointed driver and 'motor expert' to the brigade, and having considerable spare time on his hands in most cases turns his attention to electrical inventions … with the result … the machine … suffers in consequence … I do not refer to any of the principal centres, where the professional Superintendents assume entire responsibility for the practical workings of their brigades, and do not allow any 'tinkering' – the word is used advisedly – with their motor engines.[36]

Hugo did not limit his criticism to amateur mechanics. In 1916 the United Fire Brigades' Association, the professional body that provided invaluable support, games, training and the resulting esprit de corps, had recommended adopting a particular type of hydrant, 'but the pattern referred to is the most

Firemen with the Station Street Brigade handcart and hose reel, Upper Hutt, 1922. Paeroa was not alone in having second-hand, out-of-date equipment that would have been little use in rural fire-fighting.

ATL, 1/2-022392/F

faulty in design and defective in construction of any standpipe [Hugo had] ever seen in use for practical fire-brigade work'.[37] However, that is the only recorded instance of disagreement and in 1925 Hugo approved of the continued improvements in the games, and the way new events had been introduced and old ones modified to conform better 'with more modern and practical ideas of fire-fighting'.[38]

The equipment, in many instances, was not a match for those 'modern ideas'. Most of the small or new brigades had to borrow or buy second-hand apparatus. In 1911 Ohakune's volunteer brigade acquired a Shand Mason hand-pump, probably made in the 1880s. Ponies, borrowed for the occasion, dragged it to fires; once there several members manhandled the long handles to deliver an inadequate 455 litres a minute. In 1921 the hand-pump was superseded by the Merryweather fire engine, the first motor-propelled engine in Australasia, bought from the Wanganui brigade. With its iron-shod wheels, it took four to seven minutes to reach the requisite 32 kilograms pressure, and the journey to a fire had to be broken while the pressure was brought back up. Women with prams had no difficulty overtaking as it made its cumbersome way forward.[39] For the first 31 years of their brigade's existence Paeroa's volunteers hauled their hose reel to fires. It was mounted on a single axle sitting between two large wheels and was on loan from Thames. Only in 1926 did they acquire a model-T truck, with a fire engine body added, which had an unprecedented top speed of about 72 kilometres per hour.

By the late 1920s there was considerable inconsistency between towns in the types of equipment brigades owned. Some could afford new gear, a few had no alarms and only hand-drawn equipment, many lacked an adequate supply of smoke-jackets, helmets or masks, and a lot had no general utility vehicle or jumping sheets.[40] But whatever the equipment, the men's dedication and effort were paying off. By 1928 Hugo judged that 'the majority of our brigades [had attained] a high standard of fire-fighting efficiency'.[41] He still brought shortcomings to the government's attention, but fire-fighter turnout, training and attendance were improving and continued to do so.

Town-planning initiatives also contributed to brigade efficiencies. In 1933, all larger towns and cities had street alarm systems that could promptly summon brigades, although this was true of only half of the settlements in fire districts. Water reticulation had long presented problems – although by the 1930s most towns had high-pressure water, the mains were too small.[42] However, in 1930 Christchurch and Auckland, following innovative Australian practice, had brought together engineers, building inspectors, electricity and tramway officials and the superintendent of the fire brigade to discuss such infrastructural problems.

Legislation, too, contributed to the changing picture. The 1908 amendment act, while tightening some provisions to prevent interest groups from unduly influencing decisions over whether to create a fire district, strengthened town boards' power to acquire land for a fire station, as well as their ability to borrow to get land, erect buildings and buy plant.[43] In 1913 the population requirements for towns applying for fire district status was dropped from 2000 to 1000, and fire insurance brokers and agents representing companies not operating in New Zealand became liable to contribute to fire boards' costs. The 1914 amendment act also provided for adjacent fire districts to unite.

But by 1920, as the larger cities expanded and amalgamated with contiguous boroughs, questions were being asked about levels of fire protection and fire boards' ability to buy the expensive equipment they needed as well as to provide the married quarters that would help retain the long-service members who were vital to all brigades.[44]

In 1926 a new (and uncontentious) Fire Brigades Act[45] facilitated the formation of fire boards. Boroughs, city or county councils, and town and road boards of locales with populations over 1000 could apply with the support of only 10 per cent of ratepayers. Government contributions to fire defence increased for larger towns and cities with populations over 6000. Overall, fire boards were operating in an improved environment. Instead of borrowing at high rates for capital expenditure, they could now establish a fund, as well as a sinking fund to repay loans, and could (with the minister's approval) use general rates to buy land, put up buildings and make additions and improvements. They could also provide pension funds for permanent brigade members, a measure that would help retain older, experienced men.[46] In 1932 another amendment act successfully addressed some ongoing operational concerns.[47] However, these changes did not affect those predominantly small towns under municipal authorities that wanted to do things on the cheap.

The number of fire districts gradually increased. In 1930 the original 21 fire districts had grown to 50; by 1933, 60.2 per cent of the population had some degree of protection – just over half were in fire districts and 9.4 per cent had protection offered by municipal boards. Moreover, although New Zealand's fire loss was likely to remain relatively high due to its wooden houses with 'flimsy interior linings' and its widely spread residential districts,[48] the extent of the loss began to fall in 1933. The figures may have reflected an international trend associated with the Depression, but the results were still marked. In that year the loss equalled 11s 5d per capita, well below the rate in other countries with a high fire loss such as Canada and the US.[49] Four years later New Zealand's lowest fire waste ever was recorded. At 6s 8d per capita it was still high when compared with Britain (4s 5d per capita), but better than the 8s in Canada and 8s 5d in the US.[50]

By the mid-1930s the investment in equipment – and the nature and variety of equipment – was improving. Although some small towns still used a hand-held bell to alert the men, most had by then paid the £50–100 needed to install a system routed through the town telephone exchange. Larger towns generally had a street system with an attendant on duty; even in smaller towns men were sleeping over. But too few small-town brigades were throwing canvas sheets to protect against water damage, or using sawdust and towels to prevent water spreading, or employing mops and brooms to mop up; and only a few city and large-town brigades had smoke protection equipment, even though the value of two types had been clearly demonstrated.

The fire engine remained the major problem. By the mid-1930s most

Members of the Silverstream Fire Brigade and their Dennis fire engine at the newly established Silverstream Fire Station, 1930. The fire engine was a clear improvement on what had gone before.

ATL, APG-1793-1/2/G

brigades had only hose tenders, suitable for transporting men and equipment. New high-powered trucks adapted for fire-fighting were available for £600–900. The engines' pumps were invaluable for increasing the flow, especially in a number of secondary towns where reticulation systems had been installed for domestic rather than fire-fighting purposes. These engines had been adapted to local conditions, and their improved braking and acceleration allowed them to be driven safely at 38 kilometres per hour in traffic and considerably faster on the open road. Yet about half the town boards reflected their ratepayers' concern about capital expenditure on what they continued to construe as potential rather than actual risk, and failed to capitalise on these innovations. Small towns in particular were prone to invest in outdated hand-me-downs – by 1936 even 15-year-old engines had high load-lines, inadequate braking systems and poor metal that should have meant they were used as back-ups at best.[51] The considerable variation in expenditure on fire-protection services, even between towns of comparable size and populations, Girling-Butcher attributed to the resistance of insurance representatives on some fire boards (particularly in the South Island) to increase board expenditure to maintain efficient services, and not even trying to fulfil what he thought should be one of their primary functions – setting standards for brigade equipment, training and organisation.[52]

Where municipal authorities – perhaps lacking a fire board's singular focus – ran the brigades, parsimony was particularly pronounced – and their ratepayers' reluctance to make capital expenditure ironically created delays and increased costs further. Girling-Butcher could not help observe the contrast between the local authorities' inefficiencies and the considerable dedication the brigade members brought to the job. Most were volunteers who worked for no or little pay, who sacrificed wages, laboured at fundraising and attended drills; those who belonged to the United Fire Brigades' Association also attended games to good effect. But though those men were generally responsible for whatever levels of efficiency the brigades achieved, systemic issues were clearly not being addressed.[53]

In 1934 Girling-Butcher had first canvassed, at a meeting of the municipal associations, the idea of a single, New Zealand-wide board to control all fire districts. The idea, erroneously associated with a desire to extend government control, was opposed.[54] Three years later he proposed a framework that continued to include volunteer forces (their involvement was vital given that even major cities lacked financial resources), while achieving a compromise between the Australian highly centralised, one-board system and New Zealand's 'extreme decentralisation'.[55] By 1939 the major national organisations associated with fire protection had approved the general principles. Then war intervened.

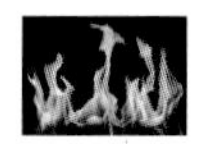

Our narrative does not examine particularly the wartime Emergency Precaution Scheme or the Emergency Fire Service with its auxiliary units, an alternative to territorial service established in the four main centres to assist the existing brigades in fire-fighting.[56] For the small-centre brigades that are our central focus, the important elements are the availability and distribution of equipment, and the impetus the war offered for reorganising the Fire Service.

In 1941 the Colonial Motor Company NZ Ltd, distributor for Ford (which had begun to manufacture fire engines and fire-fighting equipment in New Zealand in the early 1930s), became responsible for producing the country's fire engines and trailer pumps. Exigencies of war and standardisation speeded production, and by the middle of 1945 a considerable supply of trailer pumps and engines as well as ancillary fire equipment was available. Each pump carried 121 metres of hose, plus branches, standpipes, breeches and related equipment. Hose, which could not be manufactured here, had arrived in such huge shipments during the threat of the Japanese invasion that at the end of the war 137,160 metres were surplus. Importantly, too, Cabinet, in agreeing to the purchases, had requested that emergency equipment should, as far as possible, be suitable for postwar fire-protection work.[57]

By the end of the war, aiming to better equip the regular paid fire brigades and free up their older equipment for purchase or donation to other brigades, the government decided to hand £128,000 worth of war-surplus emergency equipment to the regular brigades (those controlled by fire boards and municipal authorities) but retain the title itself. The main centres and ports got the big, heavy equipment and thousands of metres of hose; most of the smaller regular brigades received a Ford standard pump, fully equipped apart from the hose. Almost overnight fire boards in towns such as Balclutha, Cambridge, Dannevirke, Eltham, Feilding, Eastbourne, Matamata, Mataura, Reefton, Stratford, Te Puke, Thames and Waverley, as well as brigades under municipal authorities (such as Arrowtown, Inglewood, Opunake and Roxburgh) had modern equipment. That equipment in turn opened the way for adjacent brigades to provide emergency assistance in out-of-district areas. However, brigades in county areas (Belmont and Stokes Valley just out of Wellington, or Bulls, Clyde, Granity, Hanmer Springs, Reefton and Waikino) continued to miss out.[58]

By 1945 Girling-Butcher was pushing ahead on reorganising the Fire Service. But the drafting of the legislation floundered over two issues. One was Girling-Butcher's proposal for the government and underwriters to fund the 60-odd municipal brigades, a vital measure if the Fire Service was to both protect buildings in built-up areas and transfer water for Forest Service mobile and manual pumps to fight fires in rural areas.[59] The other, longstanding issue, first raised at a United Fire Brigades' Association meeting in 1943–44,[60] concerned how the vitally important volunteers, comprising 80 per cent of the Fire Service, saw their role. The volunteer force had provided the first organised fire-protection service, and was the base from which, almost without exception, the permanent service had grown.[61] Although by World War II in settlements of over 1000 most volunteers were getting a small honorarium to compensate for any loss of wages or damage while on duty, their motives were not primarily pecuniary. Most had joined out of a sense of community or because of a hobby interest. They elected their own members, were involved in local fundraising efforts and were responsible only to public opinion. They did not want to be included in the award with its requirement for compulsory union membership, and they did not want to be defined as part-time firemen.[62] Agreement was finally achieved in August 1947, but the legislation was delayed, then further slowed by the commission of inquiry into the disastrous Ballantynes fire in Christchurch on 18 November that year.[63] Not until 11 October 1949 was the Fire Services Act finally passed.

The new act, attempting to move towards centralisation while still retaining levels of local control, set up a Fire Service Council that would provide essential government direction; and urban fire authorities, with responsibility for fire protection in urban fire districts, which would capitalise on what was perceived as the country's valuable local government structure. The council, funded by equal contributions from the government and insurance companies, comprised representatives from government, the insurance industry and fire organisations. Its major responsibilities included ensuring that urban fire districts complied with the act, co-ordinating units for rural fire-fighting, researching fire prevention and encouraging prevention activities, and establishing training schools and courses. The council was also responsible for inspection and for standardisation of plant and human resources within brigades, and it approved or determined brigades' annual finances.

For their part, urban fire authorities dealt with operational matters at a local level. Existing fire boards in urban areas and those formed in united urban areas had representation from government, insurance companies and local bodies and were responsible for urban fire services. In urban and secondary urban fire districts, local authorities established elected fire committees or fire brigade committees (the latter comprising a ministerial appointee and local body elected members only).

The level of district was determined by the nature of the fire protection available – towns with similar levels of infrastructure were now expected to supply the same degree of protection. With one or two historical exceptions, urban districts had reticulated water and had to supply full services; secondary districts had a water supply and had to supply services to a standard approved by the Fire Service Council, albeit a lower standard than that required in urban districts. As in earlier legislation, there were provisions for fire-fighting services to contiguous areas. Fire districts' funding came from central and local government. Estimated expenditure had to be presented to the council annually. Once this was approved, fire boards were funded on a 60:40 split between the council and local bodies, while fire committees got half their estimated expenditure (up to £200) from the Fire Service Council and the rest from local bodies.

The bill was passed with a certain amount of ministerial trumpeting about protection of rural areas. Presented as complementary to the Forest and Rural Fires Act 1947, the Fire Services Act was to provide emergency services that in peacetime would form rural fire protection. In the absence of a 'complete national fire coverage … our aim is to declare fire an enemy of the country, and to make war upon it. It does not matter where there is fire, we shall have the machines so linked we can attack it.' Under section 9, said the minister, all fire brigades in urban areas would be co-ordinated for rural or emergency fire protection, ensuring that units could be assembled, trained to meet rural conditions as well as save buildings and mobilised in an emergency. Under section 12 the Fire Service Council could require a fire authority to make special provisions for forest protection.[64]

The rhetoric was misleading. Urban fire brigades were 'equipped and trained for building fire work and opportunities for effective service in forest fires [were] limited'. That is, the act provided for urban brigades to protect *buildings* in rural areas. (Owners could pay to register these and their fees defrayed any costs.) Moreover, urban fire districts, which were confined to residential and similar closely settled areas with boundaries defined by the local authorities, were not comparable to rural fire districts. However, an amendment to the Forest and Rural Fires Act did sort out the respective jurisdictions of urban personnel and their forest and rural counterparts. A brigade fire officer would be in charge where any rural buildings were burning, but would provide assistance to and work under a rural fire officer in forest and other such fires.[65] The elaborate fire-protection scheme for Taupo township and the surrounding areas, which involved the Taupo Town Board, Forest Products Ltd, the Afforestation Proprietary Co. and the Forest Service as well as the urban brigades at Taupo, Tokaanu, Tokoroa and the Rotorua Fire Board, reflected that balance.[66] Moreover, although the role of the urban fire authorities in combating rural fires did not immediately work as smoothly as the legislation envisaged, by the 1970s there were instances of formal liaison. The Picton Fire Brigade had a minor role in the Marlborough Sounds fire-protection scheme (see Chapter Six) and by the early 1970s Forest Service representatives, local bodies and fire brigades were coming together in some places to improve lines of communication and co-ordination and ensure the brigades were trained in rural fire-fighting techniques and equipment.[67]

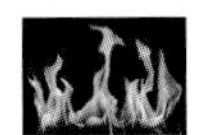

In 1949, when the Fire Services Act was passed, there were 203 brigades, whose membership was overwhelmingly volunteers.[68] That predominance continued. In 1965, 249 autonomous fire authorities were each administering an urban fire district. Each district had a brigade, a chief fire officer with legislated authority, a deputy fire officer and the complement of officers and men needed to provide their district's level of fire protection. Twenty-four brigades were manned by a total of 803 permanent staff. Some 6000 volunteers, who were simply offering their services or at most receiving a token payment, manned the remaining 225 brigades. Some of these brigades were auxiliaries, under a chief probably from another fire district. By 1974 there were 251 volunteer brigades.[69]

Those brigades still faced an uphill battle. The history of the Fire Service Council from 1950 to when it was replaced by the Fire Service Commission in 1975[70] is marked by the problems of meeting the cost of equipment and helping the less financial brigades become efficient. After the government removed sales tax from fire-fighting equipment in 1953, the council imported ex-war light UK trailer pumps and resold them to small communities for £300, a sum within reach of local donors. Two years later ultra-light portable pumps, also ex-war, became available at £80; their sales, which continued until 1959, were seen as helping new services to get organised for minimum outlay. In 1953 new classifications of district fire risk guided small urban areas on what protection levels they should be aiming for and what they should cost.[71] This initiative saw the beginning of fire parties: community-based, voluntary organisations in small rural districts set up primarily to protect property and lives, and who might – but often did not – have local body backing. They were distinguished from the permanent urban brigades whose fire chiefs had legislated authority, as well as from the auxiliary brigades who came under a chief fire officer but were a separate unit probably outside that officer's urban fire district.[72]

In 1963 higher government subsidies went to newly constituted fire brigades in secondary fire districts where the local authorities could not afford the cost of a fire engine or where there was no, or only substandard, reticulated water. Efficiencies were also sought in other areas. The first Fire Service Council training courses, run in 1957, included two fortnight-long training courses for volunteer firemen, and a programme of regular inspections was instituted in 1960.

Organisational initiatives also sought to increase rural protection of buildings. The rural co-ordination scheme, implemented in 1954, extended fire protection to all areas within eight kilometres of the nearest fire station; under agreement that distance could be extended to 16 kilometres. The same year, the reinforcement scheme meant that agreed levels of help were given across districts. Brigades might help on a neighbourly basis and call in reinforcements from further afield; in a national emergency, help could be mobilised on a regional basis. Sixteen regions with regional officers and control bases were established and regulations and detailed codes of practice governed the scheme's operation. In 1959, once the scheme was well established and rural authorities were contributing financially towards fire brigade maintenance, the Fire Service Council claimed that fire cover was 'immediately available in almost all but the remotest part of New Zealand'.[73]

To the men particularly in those volunteer brigades in rural fire districts and under county control, the picture was somewhat different. In the many small or rural settlements fire services continued to be introduced and maintained only through the dedicated effort and commitment of sometimes small groups of committed volunteers and their families. These men were unpaid, they gave up their time for the training vital to developing the team spirit that often acted in place of the inbuilt discipline of the permanent forces, and in later years many studied towards diplomas, at the expense of personal and family time.

Volunteer firemen practising with hoses, Island Bay Training School, Wellington, 1958. Until this facility was opened in December 1957, the practice and training that went into the United Fire Brigades' Association games were critical components in brigadesmen's formal training. By 1961 the school's initial two fortnight-long courses annually had been expanded into regular courses for volunteer firemen, as well as the armed forces.

ATL, EP/1958/0236/F

Even socials (dances and annual balls merged into male-oriented smokos in the 1940s and then phased into social evenings in the 1950s) were taken seriously as the money raised was needed for equipment.[74] Galas and merry-go-rounds, parades and games, growing potatoes to on-sell and doing odd jobs for local citizenry all helped supplement the little money available; hand-me-downs were gratefully received.

The Wainuiomata volunteer brigade, formed in 1943, used a brigade member's Morris 8 car as its first appliance. Two years later the Stokes Valley volunteer brigade donated a motor pump and lengths of old hose, and a ladder came from the Mobil Oil Company at Waiwhetu. The pump was trailed to fires by volunteers' cars or sometimes a 33-seater bus – one of the volunteers was a bus driver. In 1949 the Cust volunteer brigade bought a Pegson Marlow pump, four gorse slashers and four fork handles to make into beaters. A year later a local blacksmith made a trailer to carry the pump, while another offered his stables for storing the watering can, the 18-litre petrol tin, two torches, an 1818-litre square galvanised iron tank originally used to ship crockery out to New Zealand, and a tin of Silvo for cleaning brass and copper items. The Titirangi volunteer brigade, formed in 1949, managed to acquire a 1942 crash tender that had been used in the Pacific war and fitted a pump and an 1818-litre tank. A six-metre wooden ladder (very difficult to get up or down) was bracketed to the side, and their 'mixed bag' of hoses was whatever length remained after other sections had burst under pressure.

The Piha auxiliary brigade, which in 1974 replaced an earlier fire party (the brigade's history does not relate the party's connection to the Forest Service), 'begged and borrowed' basic equipment – a water trailer, Forest Service canvas hosing (which was 'in a shocking state' and burst under pressure), a Wajax 2-stroke, portable heavy pump and hose, and a 'Keystone cops' fire engine that carried about 180 litres of water. The cab had no roof, the crew stood on a plank across the back and the engine needed running repairs (to which the brigade's ingenious members attended). When that truck, too light to steer easily, ended up with its nose in a concrete power pole, members mounted pumps onto a repossessed eel tanker that carried 3364 litres of water.

Uniforms were equally makeshift. The Cust volunteers began with a few second-hand tin helmets from the Addington Railway Workshops. In 1952 they acquired cotton combination overalls, three pairs of ankle-boots, a first aid kit, a few wide belts and fire axes in pouches. Only in the 1960s were cotton overalls and short gumboots issued to all members. The Titirangi men arrived at fires in hand-me-downs from the Central Auckland brigade: heavy woollen serge lancer jackets two sizes too big; trousers whose crutch sagged to the knee, and braces and leather boots that 'were really something – polished up well and great in the garden'.[75] Many other brigade members simply turned up in whatever they were wearing – a white shirt with a tartan tie or shorts and jandals.

The ways in which the Cust volunteer brigade was called out were typical throughout small-town New Zealand. When an emergency call came through, the town's manual phone exchange would ring around the brigade members. The first person to be contacted would then frantically wind the phone handle on the party-line phones and within 30 seconds this would activate a circuit linked to all volunteers' phones. Whoever answered would take down and repeat the fire's location. The first person to get to the station would set off the siren and write the address on the station blackboard, and the engine would leave as soon as there were enough volunteers on board. The latecomers, or the overflow, followed in their own cars.

By the late 1970s sirens were being replaced by bleepers, then pagers, but anyone who lived in a small town or rural settlement in the 1950s and later will remember the siren shattering the night, wheels squealing as the engine left the station, and cars revving as other volunteers followed. Arnold Carr was 'absolutely terrified' as Rawene's policeman, Noel Smith, drove the engine over the Whirinaki hill: 'He whipped around corners … We went like hell.'[76] Titirangi volunteer Trevor Pollard, too young to drive, used to jump out of bed, grab his fire clothes and run in his pyjamas to a crossroads in time to be picked up by a truck loaded with a barrow and tombstones. Because the appliance had generally left the station before they got there, the truck would carry on to the fire while Trevor hung on among the tombstones and exhaust smoke.[77] Wainuiomata fire-fighters and their mates stacked out as cars skidded around corners, collecting the odd gorse prickle on the way.[78]

It was adrenalin-stirring stuff. But as in previous decades, while some brigades successfully instituted well-attended and regular training, others found maintaining a core of trained volunteers problematic. Rawene, in Arnold Carr's time, was never short of volunteers: the fire station had a snooker table, and once a week they took the engine up and down the racecourse, pumped out seawater, 'muck[ed] around in general' and talked. But volunteers came and went, leaving only a small core of committed, experienced men. The Waipara volunteers, great at turning out for fires, were not so good at turning up to the regular training nights, and Mervyn Allaway remembered a number of men in the 1970s found the call of 'Surf's up!' stronger than that of 'Fire!' E. S. Archibald, ex-chief fire officer for Waipara, was not alone in taking the job because nobody else wanted it.

When the Fire Service Council's training courses began in 1957, volunteers' work commitments and the prospect of losing pay often prevented them going. Moreover, for some brigades working on the urban/rural boundaries the usual training – running ladders up buildings, unrolling coils of hose along roads, fighting a static fire rather than one that was moving on a number of sometimes invisible fronts – was not enough. Peter Grant, nominated Piha brigade's first station officer because of his experience, introduced very basic training for his team: 'just plain common-sense stuff because nothing in the rule books covered what we were doing out here'. They worked on starting the portable pumps, finding water sources and ensuring that everyone knew the location of the tracks through the bush. Attending fires in adjoining districts was useful training. But many men in the Piha brigade, as in others, were in their fifties. For them, training 'was about … finding what your limits were and not putting yourself into a position where you're struggling to get to a fire and it's too much for you. You've got to know your capabilities.'[79]

Despite the drawbacks of inadequate and inappropriate equipment and lack of specialised training, rural fire brigades, and whatever gear they had, were increasingly called upon to fight vegetation fires. Sometimes, as we shall see in Chapter 6, this meant working alongside the Forest Service to fight major rural blazes. The Hanmer volunteers, for instance, did 'excellent work' in helping prevent the 1936 Hanmer Forest fire from spreading. More generally the brigades dealt with small vegetation fires within urban fire district boundaries, extinguishing them before they got away. In the late 1930s, as Dunedin suburbs straggled over the hills, Eric Ticey, a very young volunteer in the Green Island Auxiliary, helped to beat out gorse fires with sticks or wet sacks;[80] in the 1950s Rolleston's informal fire force kept sacks up trees or on fences to batter out the fires started by railway sparks.[81]

The list of the Cust volunteers' first implements provides further evidence that their duties went beyond fighting house fires. As the 1950s postwar housing boom and the growing trend towards lifestyle blocks in the 1970s pushed suburbia into bush-clad hills and countryside, and the Forest Service was often actively engaged,[82] such brigade involvement increased. In the 1950s Titirangi residents, concerned about increasing reports of bush fires and the lack of continuous water, formed the Titirangi Volunteer Fire Brigade Auxiliary Committee, agitated for more fire hydrants and put up fire-hazard signs; in the 1970s auxiliary brigades and fire parties were formed to protect west coast beach settlements in the Waitakere Range.[83] Such developments were typical of what was happening elsewhere in New Zealand.

The role of these small volunteer and often rural brigades was confined. Even in the mid-1970s it was generally recognised that 'counties could not be expected to have the resources to fight large fires over long periods', and that the Forest Service had to be ready to assist in any large vegetation fires.[84] While, as we shall see, the relationship between the Forest Service and the Fire Service was not always easy, the volunteer brigades often played a valuable – and taxing – part. In the late 1970s and early 1980s an arsonist set the gorse-covered Wainuiomata hills alight night after night and senior fire-fighter Steve Wilkie 'got about fourteen hours sleep a week' as the brigade battled alongside the Wainuiomata bush fire party.[85] Those were only some of scrub fires that roared over blackened ridges behind part of the town. Some 25 fire appliances from the greater Wellington area were called out to battle one of those fires. In another, a monsoon bucket and helicopter (appliances that transformed fighting rural fires) were brought in to overcome steep slopes and lack of water – a first for Wellington.

Meanwhile, other changes had also increased brigade involvement in rural fires. While the Fire Service had no statutory obligation to fight beyond the urban fire districts, the expansion of the 111 emergency system, first introduced in New Zealand in 1960 as telephone exchanges were being automated, meant that Fire Service brigades were often the first to be alerted, irrespective of whether the fire was burning property or vegetation. As in the past, many brigades had only limited training in rural fire-fighting, but they generally turned out to assist the rural fire authorities, providing at least an initial attack on vegetation fires; if their engines were unsuited to rural fire-fighting, they were invaluable for pumping large volumes of water to fill tankers or monsoon buckets.[86]

But notwithstanding this history of involvement in rural fires, it was not until 1969 that the Fire Service, with Forest Service assistance, ran the first rural fire-fighting course for permanent staff with a rural fire responsibility; a second such course followed a couple of years later.[87] These courses threw the urban brigadesmen into the practicalities of rural fire-fighting. They had to get to grips with the different equipment used in fighting rural fires, construct relay dams for pumping water over considerable distances and up steep slopes, and practise the 'one-lick' method. The course ended with a visit to one of Kaingaroa's forest lookouts, and a lecture on fire research.[88]

In 1975 a further Fire Service Act (effective in April 1976) established a fully centralised fire service. All previous fire authorities were dissolved, and their assets and liabilities were vested in a new Fire Service Commission. The costs of the Fire Service were apportioned between insurance companies (paying to a set formula) and government on a three-to-one ratio. The commission, basically a controlling and largely administrative body, disbursed that funding. It was responsible for promoting fire safety in New Zealand and, in fulfilling that role, ensuring co-ordination between territorial authorities and other bodies. It also dealt with finance, uniforms, equipment, appliances, training, inspection and officer appointments. To enhance efficiencies, the act gave the commission power to gazette any part of New Zealand, including rural areas, a fire region, which could be further divided into fire areas. The Fire Service Commission controlled fire prevention in urban areas; elsewhere fire prevention was to be carried out in conjunction with the rural fire authorities, the Forest Service, county councils and soil conservation councils.

The commission rapidly found that dealing with 277 separate local fire authorities was not conducive to efficiency. If Britain had had the same proportion of brigades to population as New Zealand did, it would have had 4986; if New Zealand had operated with the same proportion as Britain, it would have had two. In addition, many fire committees met only once or twice a year to approve any change or improvements. In 1977 a new, decentralised four-tier structure was established. The Fire Service Commission, in direct and ultimate control, was financed by a parliamentary vote and contributions from insurance companies. It was responsible for finance and national planning and oversight. Below the commission were the regions (Auckland, Hamilton, Palmerston North, Wellington, Christchurch and Dunedin), the main units in which all brigades were administered and co-ordinated. Within the six regions were designated 23 areas (most based on the presence of permanent brigades). Finally, the existing 277 urban fire areas were redesignated as districts with the legal status the fire areas had had. Most (251) of these districts had volunteer brigades.

Many volunteers were concerned about how their brigade would fit into this new structure. But the value of their work, estimated as worth millions of dollars annually, was not overlooked. Helped as needed by larger units, volunteer brigades remained responsible for protecting their own urban area and the gazetted districts beyond. Their chief retained his existing seniority, prepared his own estimates, while also being able to confer with regional and area commanders. The structure also provided for new volunteer forces as towns developed.[89] The better equipment and funding, even if not an immediate panacea, were welcomed by the volunteer brigades – brigades whose members' fraternising, community spirit and adrenalin had provided the 'first organized fire-protection service' from which the permanent Fire Service, almost without exception, had grown.[90]

The Fire Service Act 1975 would also place the volunteer brigades on a better footing to take up a further defined role in rural fire under the new 1977 Forest and Rural Fires Act. Although for the next 10 years the Forest Service would continue to underpin rural fire defences, the Fire Service role, often unacknowledged and inevitably limited, was important.[91] It was also, as we have seen, longstanding.

Chapter Six

The nation's rural fire-fighters 1948–87

During the 1950s and 60s movie-goers stood for 'God Save the King' (or 'Queen'), then settled back into their seats for that week's *Pictorial Parade*.[1] Here they saw fat babies and skinny school children being weighed and measured by Truby King's army of starched uniformed Plunket nurses, similarly attired dental nurses poking in a succession of small mouths, bulldozers and graders cutting into impossible slopes, pushing roads out into the central North Island. Aerial shots swept viewers across the seemingly never-ending Kaingaroa forest, big trucks carted massive loads of logs through roads in densely planted exotic forest, 'new towns for new people who had come to start a new life' were springing up in newly bulldozed land. There was the Kinleith pulp and paper mill, that 'milestone of national importance' that had 'freed' the country from imported wrapping papers; here were logs being processed at that 'new height of endeavour', the Murapara project, the integrated sawmill and pulp and paper plant that was New Zealand's answer to Australia's Broken Hill.[2]

The strange BBC voice heard only in films or radio of the time spoke of achievements that were widely evident to all New Zealanders. The two million trees planted in 1200 blocks, and the 1288 kilometres of road used for access and fire protection, demonstrated progress planned over decades. The Forest Service, using new methods of silviculture, had helped create a new industry. Now people were no longer dependent on slow-growing indigenous species. Now they were reaping the benefit of trees planted only 25 years earlier. 'Man [had] set nature in motion'; trees, a natural asset, a source of foreign exchange, 'a crop whose harvest was homes', were being handled by 'something big ... something only big business [could] handle'. Trees were everybody's business.[3]

Statistics, rather more prosaically, justify the optimism of these films. With the exception of fisheries, the value of the productive sectors rose dramatically between 1947–48 and 1960–61; forestry's contribution more than trebled, rising from £8.7 million ($617.2 million today) to £27.7 million.[4] With the second great planting boom of the 1960s and 1970s (initiated because Entrican, just prior to his retirement, realised that more forest would be needed to meet domestic and export targets) the Forest Service set a new goal of planting 485,600 hectares by the end of the century. It steadily increased its own annual targets. Much of the work concentrated on 'consolidating existing industrial forests in the central North Island and in Nelson' as well as on establishing smaller 'local supply' forests on the Otago coast, for instance, and at Ashley, Berwick, Mohaka and Whangapoua. Marram grass was planted to stabilise vast areas of sand dunes before young pines were established. Forests like the Esk in Hawke's Bay and the Mangatu in Poverty Bay were established to repair erosion and flooding caused by unwise bush clearance. On a modest scale in Westland, in the central North Island and in Southland, cut-over native bush was converted to plantation forest.[5] By 1970 Kaingaroa State Forest was one of the largest plantation forests in the world, at 122,757 hectares, an increase from 100,205 hectares in 1960. By the early 1980s

> each major state forest (and there were dozens round New Zealand) had an officer-in-charge who controlled everything – [land] preparation, planting, pruning, protection. He had 10 to 20 staff and sometimes 100 wage workers. Most forests were being developed flat out – it was the Think Big era – and each forest station could be planting 500 to 2000 acres [200–810 hectares] a year.[6]

Privately owned forests also expanded, facilitated by the agreement that the Forest Service and private interests would share the planting and development work. In 1981 New Zealand Forest Products had 53,000 hectares, and other companies had land-banked a further 118,000 hectares. On a different scale, farmers were planting exotics, spurred on by Forest Service work in farm forestry and the generous forest encouragement grant that returned 50 per cent of costs.

As farmers worried about potential changes in land use, forestry development conferences set increasingly high planting targets. Areas difficult to harvest were planted – as private companies scrambled for suitable land, steep country in the central North Island, for instance, was burned off and planted.[7] By the mid-1980s rising unemployment added new incentives. MPs in badly hit areas often lobbied for additional planting as work creation; on the East Coast forests were again established for long-term erosion control. (These facts were often overlooked in the late 1980s when, within the framework of Rogernomics, the Forest Service was accused of diluting its commercial drive with social concerns.) Mike Hockey, then a senior Forest Service officer, reckoned that 'not a lot of thought went into how the forests would be used commercially 30 years later. Mind you, they were still managed with the best forestry practice of the time to produce a high-quality product.'[8]

Throughout this period of growth and development, the principles developed for fire management and control in the previous decades continued to be applied. The Forest Service recognised that fire was still a pre-eminent land-management tool. Efficient land and forest management could deal with the very rare lightning strike; the real problem was how to manage the fire risk posed by people, particularly to Forest Service land. Towards this end the Forest Service, by 1950 considered the nation's rural fire-prevention organisation,[9] continued with the techniques that Ellis had introduced – good public relations, public education and preventative measures that would enable prompt suppression if there were a fire. National forest-fire education programmes used all available media. Weather information, broadcast on national or local radio, identified specific areas of fire hazard, providing a sound basis for ensuring that logging operations or burning on a wide scale were not unnecessarily suspended or curtailed. Fire patrols, lookout towers in areas prone to long, hot periods and increasingly sophisticated communications methods contributed to early detection. Better access to forests and improved technology increasingly ensured that men trained and equipped to fight rural fires were there to combat the risks.

If none of these control techniques was new, the context after 1947 was. The Forest and Rural Fires Act 1947 expanded the coverage of the Forests Act and put forest and rural fire prevention and control on a national basis. The pattern of steady burning did not immediately diminish after the act's implementation, but the Forest Service developed an increasingly systematic prevention regime as science and technology became ever more sophisticated. A new fire index system for assessing fire risk was introduced. Better pumps became available; bulldozers, helicopters and the monsoon bucket added previously unknown speed and grunt; and chemicals made water a more effective and efficient tool. Fire-fighting training of personnel from the Forest Service and other agencies and companies steadily increased.

From the early 1960s, when the science of fire-fighting was still in its infancy, controlled burns by the Forest Service and private companies for a land clearance contributed enormously to knowledge of fire behaviour. The large private forestry companies took on stronger fire-protection roles within and beyond their own forests. The community, whether as honorary forest rangers or volunteers in a variety of fire forces, also played a part. Increasingly, the Forest Service developed a national co-ordination and facilitation role.

In 1977 the Forest and Rural Fires Act was again amended to address developments over the previous 20 years. But 10 years later, in a new political climate, the New Zealand Forest Service was abolished. The ramifications for fire-fighting were considerable.

The Taupo fires and their origin in wasteland had sent a clear imperative for change in fire regimes. Supply issues and the postwar demand for wood focused attention on developing adequate timber supplies for present and future needs. With the indigenous resource so limited that cutting had to be curtailed, it was all the more important to protect some 202,000 hectares of cut-over land to allow it to regenerate into a timber crop – yet it was under threat from earlier logging, roads and powerlines. The 349 million hectares of highly flammable exotic forests, the Service considered, were therefore crucial to New Zealand's future. Unlike the indigenous forests, they had the productive capacity to support permanent industrial installations and communities. Forests on non-agricultural land were needed to prevent deforestation and consequent impoverishment; their cover was vital in protecting and regulating water. Fire risk in some areas was growing, with the highly flammable gorse, bracken, blackberry, prickly hakea and rank grasses that surrounded holiday resort towns, along with the highland scrub and tussock country, presenting perhaps the greatest dangers.[10]

Following consultation with every possible interested party,[11] the Forest and Rural Fires Act 1947 was speedily enacted. The Crown was not bound, and protection of state forests remained vested in the 1921–22 legislation. But the new act broadened the old one's concept of rural fire districts. Rural fire districts could now include not only forest, peat, flax and sand dunes, but also land planted in gums and other plants. Four different rural fire authorities were set up: the minister (controlling state lands), the rural fire district (controlled by a rural fire committee), county councils and soil conservation councils. All could declare closed fire seasons and requisition manpower. They had to make clear arrangements with local fire brigades to provide apparatus and fire-fighting services; conversely any urban fire board or local authority that had received war-surplus emergency equipment had to make it available as long as this did not jeopardise its response to local emergencies. Levies, or agreements between interested parties, could be used to meet expenses and county councils could charge their fire expenditure to county accounts.

The Forest Service had specific duties beyond those of a fire authority. It was responsible for fire prediction and public warnings over radio and in newspapers. In extreme fire hazards, regardless of any permits, no outside fires could be lit. During such times, if vegetation fires had spread beyond one district, either the Director of Forestry or the conservator had to take charge, with the minister apportioning costs between the fire authorities affected.

The costs borne by individual farmers or companies depended on whether the fire was contained within their property or had spread to others' land. The fire authority was to resolve any arguments about contributions to costs according to the value of the property saved, the assistance each owner had required and other relevant considerations. Finally a tribunal would hear complaints about notices to make firebreaks; requirements to maintain fire-fighting equipment; the levies a fire authority might impose in meeting either its ordinary requirements or the costs of fire-fighting; and the minister's decision on the costs to apportion and to whom.[12]

Inevitably, given the speed of drafting, amendments were needed from time to time to cover gaps or ensure the act worked in a practical fashion. The main changes in the 1948 amendment act gave fire authorities discretion to act to prevent, control and suppress fires in adjoining areas where there was no county council – such as Taupo and the Marlborough Sounds. Any activities on land bordering fire districts and likely to cause fires could be prohibited. Escape routes had to be provided in exotic forests, and, in a provision prompted by forest workers' representations, fire officers could require this in writing. Saving lives was added to the reasons why directors-general could require all commercial millers to have fire-fighting equipment or take precautionary steps.[13]

There were further amendments in 1949 (to the Forests Act) and in 1952. In 1955 a new Forest and Rural Fires Act brought together all the provisions protecting state forests, national parks and unalienated Crown land (as defined by the Lands Act 1948). The Forest Service now controlled rural fire-fighting in all these areas. (General responsibility over all Crown lands did not come until 1977.) The Service's reach did not extend to public reserves under local body or trustee control, land within soil conservation council areas, rural fire districts, or districts controlled by urban fire authorities.

From 1 October to 30 April a closed season now operated in most rural fire or soil conservation areas. Written permits were required from the local fire officer and no unpermitted fires could be lit within 1.6 kilometres of a state forest. County councils acting as fire authorities did not have to restrict fires unless the danger of their escaping was high, but no fires could be left smouldering or burning unless adequate steps had been taken to prevent them from spreading. In extreme fire conditions all rural fire authorities could prohibit any open fire for as long as necessary – and for the first time they could communicate this decision by radio as well as through newspapers. Such bans could apply nationwide or locally.[14] Entrican had at last achieved one of his top priorities.

The 1947 act set out the basic framework and principles. Extensive consultation over the following three years, which must have gone some way to allaying concerns as well as making the provisions workable and enforceable,[15] led to the development of detailed regulations that came into force from the end of March 1951. These provided for possible annual exemptions from fire bans, they dealt with fire signs and who should erect them, and they set out the methods for testing fire-fighting apparatus and how often this should be done. Using wax matches in exotic forests, rural fire districts and state forests was prohibited, with penalties specified for offenders, and (at last) gas-producers could be banned during the fire-hazard months. Other provisions set out requirements for suitable and adequate fire-fighting apparatus for forest employees whose work constituted a fire risk, and specified how the apparatus was to be maintained, stored and identified.

Measurable, observable standards for apparatus and operations established stringent provisions. For instance, steam-haulers were banned from exotic forests; elsewhere they were to carry hose (diameters of hose and nozzle were specified) and a pump capable of operating to direct a nine-metre vertical stream at the end of 61 metres of hose. Over the hazard months, rural fire officers had the power to demand a watchman closely monitor the operating site for 30 minutes after the hauler had stopped. From 1 August to 30 April fire-fighting apparatus had to be carried on all engines. The apparatus required differed according to the type of vehicle: steam-driven locomotives had to carry two knapsack pumps full of water, two shovels and two slashers; a traction engine, tractor, loading crane or logging truck powered by an internal combustion engine needed a nine-litre pump, plus a chemical, foam extinguisher and one shovel. Similarly there were requirements for siting water caches (and keeping them clear of debris that could block pumps), for operating logging and sawmilling concerns, and on conditions and methods for burning refuse or slab over the hazard months (each had different requirements). There were conditions for fires lit under permit. Fire authorities' duties were similarly carefully itemised. They were to provide the director-general with annual returns on fires and offences within their districts; maintain adequate first aid kits, approved by a medical officer; pay for fire brigade assistance and replace or pay for damaged clothing and equipment; and provide for any volunteer fire services. Other valuable but not mandatory actions were establishing and maintaining lookouts, organising patrols and aerial reconnaissance, installing telephones and carrying out publicity activities.

The extensive consultation and industry involvement in drawing up these detailed and rigorous provisions was aimed at ensuring industry agreement, buy-in and self-policing. But within a couple of years, even after shutdowns because of fire risk, milling and logging industries were still apparently failing to appreciate the seriousness of the threat from fire. Many mills still had inadequate fire-fighting equipment and some were not reporting fires or attempting to fight them when discovered. The fires were mostly from steam-haulers or locomotives that should have been replaced, and too often tramways, rather than roads, were being used. Moreover, the protective green belts that were essential in helping to eliminate fire risk from large tracts of cut-over forest were not being established.[16] In 1956 the provisions around the use of spark-arresters and steam engines, dumping ashes and patrolling were revised and tightened.

The rate of loss of forest to fire declined over the next four decades as scrub and fern on forest boundaries were converted to 'fire-safe' farmland.[17] From at least 1955 land-management techniques were beginning to change. Aerial top-dressing and oversowing, converting native grasses to types that burned less readily, and other means of controlling unwanted growth meant fire was used as a last resort, and with other safeguards in place.[18] Thus in 1946, 62 fires in state forests burned 6608 hectares; by 1960 there were only 48 fires (origins mostly unknown) damaging 2031 hectares. Even more impressively, outside state forests, in 1946 the Forest Service helped fight or advised on 311 fires that damaged 232,290 hectares; by 1960 that number had dropped to 82 fires that damaged only 4137 hectares.[19]

This photo of the boundary of the Kaingaroa State Forest, near Rainbow Mountain, demonstrates land-development techniques the Department of Lands used to reduce fire hazards.

ANZ, AAQA, 6506, Box 21, 8/20, 432, M328

From 1960 to 1987, despite the lack of clear figures, it is evident the downward trend continued. An estimated annual average of 602 hectares of grass and scrub and plantation forest was damaged in the 1980s – at a not insignificant estimated cost of $79.8 million (in 2008 values).[20] In indigenous forest between 1965 and 1985 an estimated 1629 fires burned a total of 80,000 hectares. This figure includes areas affected by fires deliberately lit within the 1.6-kilometre safety margin around state forests to encourage deer to browse.[21]

The pattern of burning did not change immediately. From the 1950s newspapers were awash with articles on fires, based on material developed by the Forest Service for its annual anti-fire campaigns. In 1954 the *Dominion*'s article 'Fire danger threatens forest of New Zealand every summer' gave a common message and all too common figures. In the previous seven years an average of 125 fires had burned over 5237 hectares. The 1952–53 fire season saw 182 fires destroying 7657 hectares; 48 had been in state forests, destroying 5237 hectares. The following season the area of 5024 hectares burnt, either by fires from locomotives (9 hectares) or cigarettes (94 hectares), was relatively small. Rather more disturbing were the large number of fires for which neither cause nor culprit could be ascertained (81 fires burning 9803 hectares) and the damage still being done by fires started to burn off land (41 fires over 1367 hectares).[22] In the early 1950s in Southland fires flared a couple of times a month; some were of unknown origin but others resulted from burning off manuka. Dry summers exacerbated the dangers, both within and outside conservancy boundaries.[23]

Although by the late 1950s and early 1960s the Forest Service was attending some 10 per cent fewer fires,[24] some locals found it hard to see an improvement. In 1962 a rural fire officer in Northland vented his frustration over the scrub, gorse and manuka fires that burned once or twice a month:

> Persons unknown have been lighting fires in this area for many years and as they have not been detected in the past I see no reason to believe they will be caught this season ... I would suggest that [the Forest Service] have an organised burn in the area as soon as possible [to ensure private property was not endangered].[25]

The fires, and the mindset that lit them, were more than a nuisance. Over the years water trailers, knapsack pumps, a Fergusson tractor, its driver and discs, bulldozers, cars, fire buckets, beaters and shovels, Hale pumps, gorse spray pumps (spraying even small amounts of water was valuable) and men constantly had to be assembled and got to the fires, often considerable distances away. Fires were left to burn themselves out unless they were threatening property or state forests; tackling those that did need to be extinguished could be challenging and unpleasant. The mining tunnels that 'riddled' the Thames area made night-time fighting dangerous; in the Nelson conservancy fire-fighters battling fire in one of Baigent's forest were reduced to 'tunnelling' underneath high gorse, scrub and fern to get the hose to the fire, while the dense gorse and scrub at times prevented outside fighters from using their knapsack pumps – not the only time such lack of maintenance put men at risk.[26] Costs, sheeted back to those responsible where possible, were assessed on the basis of salaries and wages, including overtime as well as vehicle hire, vehicle running costs and the cost of detergent for 'wet' water (discussed below). The amounts ranged from a few pounds to a couple of thousand – but it all added up.

Money was not the only concern. People died.[27] Valuable conservation land was burned or threatened. On 5 January 1953 'irresponsible hunters' in Fiordland lit vegetation below the bushline on the massive peak of Fiordland National Park's Mt Titiroa, and followed it the next morning with another fire on the open flats in Borland Valley. Rugged and remote country precluded any action. The fire burned out some days later, leaving behind huge erosion-prone scars.[28] Two months later the Eglinton Valley, an increasingly popular tourist area as

Erosion from a fire lit by hunters in the Takitimu Range, Southland, 1964.

ANZ, AAQA, 6506, Box 21, folder 8/30, 435, M10,036

Fire-fighters on the way to the Eglinton valley fire. The men are carrying (from left to right) a portable pump, fuel, an axe, hose and an Indian backpack pump.

Millman collection

the road was pushed through to Milford Sound, was in danger. High wind showered sparks over the defence lines. The fire, threatening to crown, at one stage burned downhill against the wind. Fire crews from Lands and Survey, the Forest Service, Public Works and contract operators battled with bulldozers and chainsaws. Water supplies ran low at times, snags had to be chainsawed down, and mopping-up operations had to cope with smouldering humus a metre thick.

Eleven days after it was first reported, the fire was deemed out. 'Initiative and general resourcefulness' had won the day. Entrican congratulated everyone on averting a potential 'major disaster', and senior fire control officer William Wright pointed to some essential fire-fighting qualities in his comments on individual fire-fighters. Wright saw leading hand Linton as 'intelligent, reliable, and keenly interested in his work. He has an even temperament, with a nice dry sense of humour to carry him through sticky patches.' Forest ranger Tom Swale, who had largely directed the operations, Wright considered '[a] good bushman with a fine physique and a great capacity for work with an easy confident manner in dealing with men'. Swale paid close attention to detail, critical in fire-fighting, and knew that men fought on their stomachs. When the manager of the Tourist Hotel, where the men were billeted, failed to co-operate, Swale 'saw that every man was fed and had a bed, and that he left for work after a more-than-man-size breakfast at 5 a.m. – cooked by Ranger Swale'.[29]

Swale was in the middle of that fire when, on 13 March, 'mentally deficient people' (according to the *Southland News*) or benighted hunters coming down the wrong ridge (the Forest Service) lit a fire above Lake Te Anau's south arm.

With the newly rediscovered *Notornis* (takahe) and its reserve threatened, Swale and three others were rapidly landed on the beach below, with a Paramount pump, a hose, a canvas tank, Indian pumps, slashers and axes. They packed 274 metres of hose and the 32-kilogram pump up some 76 metres of lift, dammed a steep rock gully to create an eight-minute supply, and stopped the fire just in time to keep it out of windthrow, which would have boosted it up precipitous slopes into the *Notornis* area. Fast work, earlier fortuitous rain, and energetic leadership were deemed to have saved the day.[30]

Fires on conservation land continued. On 28 March 1970 hunters in Mt Cook National Park's Murchison Valley let a primus flare into overhanging vegetation. By the time heavy rain had doused the fire two days later it had burned 11 kilometres down the valley and destroyed 1214 hectares of 'precious vegetation', including the slow-growing, mid-altitude Hall's totara. Two years later a fire in the Mt Kyeburn area burned 809 hectares of tussock before the Maniototo fire organisation and the army, with its Iroquois helicopter, communications equipment and always welcome rations,[31] arrived to help extinguish it. The following month the Mt White fire burned from 23 February to 13 March, destroying 546 hectares between 609 and 1036 metres; at one stage it threatened Arthur's Pass village and the national park.[32] In the early 1980s in the vast Rotorua conservancy Mike Hockey, then Forest Service fire control officer in the Gisborne district, had to worry about the 'spectacular' fires that hunters in the Urewera National Park lit to encourage new growth and attract the deer. To add to his worries there were

> the scares from the Rastafarians. Aerial and ground patrols went out at night to try to keep tabs on them, to see what was going on. They were burning marae, churches. We were worried they'd torch the forests – there'd have been vast areas going up in smoke! We were bloody lucky there were no major conflagrations up there.[33]

Nor was Forest Service land itself immune. Many of the significant fires that burned, roughly once a decade between 1955 and 1986, were in Forest Service exotic forests.[34] The 1955 Balmoral Forest fire is one of the better known. Between 1 October 1954 and 1 March 1955 appreciable rain fell on only six days. In this exceptionally dry period an old Balmoral station building burned down. The several resulting fires were quickly controlled and deemed safe, but the following morning one of Canterbury's strong nor'westers quickly turned to gale, and an 'uncontrollable conflagration' travelled at high speed through compartment after compartment. An 'extraordinary' number of wind changes, 'every one of them a half gale', endangered crews and defeated attempts to back-burn or establish defence lines. 'It was chaotic,' forest ranger Jack Barber

Exhausted fire-fighters recover from fighting the Balmoral fire, 1955.

ANZ, AAQA, 6506, Box 21, 8/30, 435, M12555

remembered. 'You had possums on fire tearing out of the bush, running across the road and igniting the next bit of forest.' Fire crews from the Nelson and Westland conservancies and from Selwyn Forest, over 200 volunteers, local fire officers, the Christchurch Fire Brigade, and the valuable army detachments who could work 'with the very minimum of instruction and direction', put in 'long exhausting hours'.[35] Before heavy rain finally doused it on 28 November the fire had burned 2991 hectares.[36] Roy Knight's memories of the hard salvage work he did as a woodsman trainee indicate something of the resulting devastation (and the hard work it occasioned):

> I was on log-scaling duties, selling logs to the mills, and we were walking in powdered ash about six inches deep and we had just a bit of a sunhat on, and singlet and shorts and boots and at night you'd go in to the shower block, you'd have all your gear in your hut in the old brown paper bag and you'd just walk straight into the shower with your dirty clothes on, hang them outside and about half an hour later with the nor'west wind they'd be dry for the next day. With winter, powder turned to liquid mud. And cold – I've never worked in such cold conditions. It was one time I thought what the hell am I doing working in the Forest Service, working in such conditions. But you got used to it.[37]

Beating out flames, probably at the Balmoral Forest fire, 1955.

Millman collection

In 1973, the fourth year of a consecutive drought, fires were banned over half the country.[38] On 7 February a 43°C day rapidly put an end to Roy Knight's intention to give his men fire training. A 'screaming nor-wester' came in. A spark from a grader ended in a conflagration that tore through 194 hectares in the

Ashley exotic plantation. Nine months later the weather had still not broken. Yet on Saturday 3 November, with the wind gusting to Force 7 and blowing directly onto a newly and heavily thinned section of Mohaka Forest, a farmer watched an employee light blackberry bushes about 100 metres from the forest. The fire got into the forest and spread fast. Spot fires crowned immediately and jumped up over 800 metres as the dense regrowth of fern within the thinned area ignited. Sixty men from nearby forests were called in. Seven bulldozers working day and night initially prevented the fire from spreading between compartments. Then the wind changed and, with the fire back-burning fiercely, the tractors could only make breaks between critical areas, not ahead of the flames. Not until about 3am on Sunday morning, when the wind dropped and the intensity of the fire lessened, could tractors form breaks around critical compartments. Relievers from Esk, Mohaka, Patanamu and Kaweka forests took over from night crews and checked for hot-spots By 2.30pm, when the fire was deemed safe, it had burned over 260 hectares of pine estimated to be worth $240,000 ($2.6 million in 2011 values). The men responsible were fined $25 each.[39]

Burnt-out *Pinus* *ponderosa*, Balmoral fire, 1955.

Photo: John Johns. ANZ, AAQA, 6506, Box 21, 8/30, 435, 1518

At the end of 1975 the drought finally broke, but on 22 March 1976 in Hanmer State Forest a hot-spot from a controlled burn thought to have been out for about two days ignited in nor'west gales. Hanmer Springs and Waiau volunteer fire brigades were called out; heavy machinery from the Electricity Department and Canterbury Timber Company set to work damming the river for water and making firebreaks. The wind dropped slightly and the army Iroquois could finally arrive. The troops brought supplies but in the wind and darkness their tents could not be erected; volunteers (whose lack of training meant they were not always wise or welcome) did night patrols. In smoke and wind that hampered efforts to determine the front and boundaries of the fire, helicopters did key reconnaissance, hovered above dams too shallow to scoop while the monsoon buckets were filled, and transported men and supplies. Bulldozer drivers, their vision limited because the lights were behind the extra-high extensions on the blades, drove through the night on steep slopes; farmers' tractors, lacking the vital compressors that cleared the filters of soot and ash, sat unused (another bone of contention). Two days later, when the wind had dropped and 200 men had controlled the blaze using machinery worth hundreds of thousands of dollars, the fire was declared out, though it was patrolled until there was a downpour on 9 April. Some 526 hectares of forest carrying 141,500 cubic metres of sawn logs, enough to supply the Canterbury timber industry for six months, was gone.[40]

The Forest Service was not alone in its fire losses. On private land, for instance, in January 1980 a helicopter mistakenly ignited a controlled burn on a Carter Holt plantation at Willow Flat at the bottom of a steep incline rather than the top, which would have created a slower burn down the slope. In the uncontrolled blaze that roared up the slope, the safe area provided by the previous year's burn was insufficient for the three men at the top. Two lost their lives and the third was seriously burned.[41]

A year later conservator Peter Maplesden wrote his account of the two-day Hira Forest fire to 'bring home to those of you who have not experienced a major disaster the trauma associated with fire'.[42] Existing fire danger had meant that logging crews were already working mornings only and had been brought in closer to headquarters. But locals who did not want exotics planted in an area of regenerating bush had opposed Nelson City Council's planting of exotics in the Waahi Taakaro area. The council's fire protection was not up to scratch and Maplesden himself had been reluctant to press for gorse and scrub to be removed in case protesters misinterpreted his actions.

On 5 February 1981 a small fire in its primary stages was reported in the area. It grew rapidly: 'Large quantities of gas from intensely heated vegetation rolled ahead of the burn, exploding into flame when sufficiently diluted by air, setting fire to more scrub'[43] and then moved into the Hira forest. In an understorey of dry gorse with thinning residues of older trees, a strong southerly that prevented helicopters flying, and steep terrain that inhibited fire-fighting action, '[t]he fire was not only out of control, it was uncontrollable'.[44] Soon three uncontrollable fires were burning in different areas, and the possibility of fire sweeping north to the top of Whangamoa Saddle and beyond into the Rai Valley and Havelock was only too real.

Ross Hamilton, a Nelson City Council rural fire officer, can still see the fire jumping from Tantragee Saddle in the Brook Valley into the Hira Forest in the Maitai Valley, at least 2.5 kilometres away. 'Many times the heat was so great that men could not get close enough to play water on the flames. Roads and firebreaks ... could have become death-traps. The steepness of the country and the large quantities of dry fuel added to the intensity of the conflagration.'[45]

Maplesden was in charge, working with the Fire Service brigades, local rural fire-fighting forces and 100 soldiers. Volunteers were not even considered, so vital were experience, leadership and discipline. Over two days, and as the wind gradually dropped, helicopters could get into the area and the fire was brought under control. Although 10 men had been in serious danger, the absence of casualties and serious injuries was 'one of the bright spots in the action'. In total, 1972 hectares had been burned, but the long business of mopping up could begin.

A common theme in the response to these fires was that they were directed by Forest Service men and were fought (largely) by Forest Service staff using Forest Service equipment. Certainly contractors and their increasingly heavy equipment might be called in, trained and disciplined soldiers might assist, local volunteer fire brigades, with a primary responsibility for protecting any threatened homes or buildings, might battle alongside the Forest Service men or use their heavy pumping equipment. But by 1951 the Service was responsible for 3.84 million hectares of indigenous bush as well as some 184,500 hectares of exotic forest. At a head office level that meant responsibility for protecting rural lands on a national scale: 'In all matters of rural fire protection, the Forest Service is looked upon as the authoritative source and pattern for equipment, methods, and organisation.'[46] It was the nation's rural fire-prevention organisation.[47] It gained this status partly simply because it was there. At Ashley Forest, for instance, where Roy Knight was based, 'there was no volunteer force within one hour's travelling time. So who would you call on? The Forest Service!'

As the 1950s moved into the 1960s and as they gained experience in fire behaviour and fire-fighting the huge controlled burns of the time, Forest Service staff officially and unofficially saw themselves as 'the experts, and that was that ... We could look after things.'[48] Over the years, odd comments filtered through about particularly the incompatibility of Forest Service radio with that of other agencies (an ongoing problem for today's fire-fighting agencies[49]), but otherwise there is little to suggest that the Service carried out its role other than thoroughly professionally.

The work lay not only in fire-fighting. All the earlier ingredients of the Forest Service public-education programmes continued in full force. 'Warnings and prosecution' were one approach. Another lay in the decades-old principles of opening up the forests for 'healthful recreation and relaxation' and creating educational opportunities to develop 'a feeling of shared ownership' and 'a diminution in vandalism', and thus continued to be strongly espoused and acted on.[50] What was needed were initiatives that could give city-dwellers, who had

Developing public tracks and bridges and running open days for the public to view forestry operations and machinery provided opportunities to show the Forest Service in a positive light.

Wally Seccombe, private collection, image no. 192

widely varying recreational interests and were largely out of touch with forests and forestry, a 'Scandinavian perspective' and respect for forests and forestry.[51] Forestry huts designed for Forest Service hunters were opened to the public, tracks became ever more carefully signposted, and by the 1970s the wire suspension bridges that led trampers across rivers and above ravines had been expertly designed and constructed.[52] In 1965, 15 years after Entrican had first mooted the idea, the exotic forests were opened to the public.[53]

In most of the nation's newspapers, reports on fires and fire statistics, on fire bans and the rationale behind them were all part of the public relations campaign and formed a constant summer theme. By the 1980s lookout services were generally discontinued or used only in times of high fire hazard – hard bottom-line accounting meant the Forest Service could no longer afford to have men immobilised for long periods.[54] But newspapers, particularly in areas where lookouts were situated,[55] saw lookouts as great copy – though they did not report on the alcoholics and recluses who used their time in the tower to dry out for six weeks, then went on a blinder; or Buckshot, with dysentery and nappies to solve the problem of being caught short in a tower.[56] Instead they focused on the more respectable types like Joe Fruish at a Golden Downs

A lookout (this one photographed about 1947) provided a vantage point to keep watch over the central North Island forest – a forest so large that it took two hours for a plane to fly over it, and two days for a car to drive around it. While towers were not used everywhere (Southland and Otago did not have the long, hot periods that necessitated them), these protective measures attracted interest 'even in Australia', noted the accompanying text.

ANZ, F1, 13/11/6, Pt. 1, Newspaper clippings

lookout in the 1950s. As he told the *Nelson Evening Mail*, he was awake and on the job from 4am. He didn't like time off: 'It can make you feel unsettled and put you right off the beam.'[57] 'AOL' found 'contentment' in the seven months he spent isolated in a lookout. If, he advised, you can 'live without the company of fellow pals and gals, "Go bush, my man! Go bush and be a lookout man!"'[58]

Joe Fruish, lookout man.

Seventy Years of Forestry, p. 121

Snappy phrases for catching the public imagination were posted in all over the country: 'We want no butts about it' (considered an outstanding success), 'Don't walk out on old flame', 'Chaperone that cigarette – Don't let it go out alone'. The 'Keep New Zealand Green' campaign, developed in the 1950s, warned generations of the dangers of forest fires.[59] Mervyn Taylor's highly crafted woodcuts were run in daily and weekly newspapers; displays were designed for different locations and mounted in hotel bars, lounges and lobbies, campgrounds and on highways. Half-moon fire danger signs began to be erected and are still in use today. Film-strips were prepared in conjunction with the Education Department and material was published in school journals. As a wider ecological consciousness dawned, the backs of schoolbooks were enhanced by a drawing of a kiwi and the accompanying text: 'Without flourishing forests, there would be no kiwis. We sometimes forget that each forest is a community … Each animal has its own role to play.'[60] The public relations firm Dormer-Beck advised on effective media placement[61] and later on the best time-slots for TV and radio.[62] By the 1970s conservancies were producing displays tailored to their own areas.[63]

'Keep New Zealand Green', a poster designed for broad appeal.

Dominion, 6 January 1947 (in ANZ, F1, 12/1/12/2, Pt. 1, Fire prevention/Newspaper clippings)

A sign at the entrance to the popular Mt Holdsworth track in the Tararua State Forest. No opportunity was missed to drive the message home.

ANZ, AAQA, 6506, Box 32, 12/16, 945.2, M1315

From the 1950s National Film Unit assistance was enlisted. *Forestry Fire Season* used Golden Downs to enact the drama of a fire 'that destroys our nation's wealth'. Men, alerted by sirens, emerged from the forest at full tilt, scrambled onto still moving trucks and were off to battle the fire.[64] (An anecdote suggests the fire, specially lit, took three days to put out.) *A Fire-conscious Public* gave clear options: look after your fires in well-constructed fireplaces and be thanked by the Director of the Forest Service or leave that small billy fire to turn into a raging inferno, creating 'waste where there was wealth'. As to land-management practices, top-dressing planes and planting pines were modern ways of overwhelming gorse and scrub. Why use fire, that outmoded tool?[65] From the mid-1950s the themes of fire and forestry were integrated.[66] *Nobody Ordered Fire* presented trees as material for everybody's home; a natural asset that was everybody's business, that could be planted, tended, harvested and processed – or easily destroyed by that casually flicked butt or untended fire.[67]

The 'good neighbour' policy was still rigorously pursued to enhance locals' co-operation. Not wanting community relations to sour, the Southland ranger decided not to fine one struggling family; Tapawera School's education board, 'extremely obliging' in lending its hall and generating plant for film shows, was not charged the cost of extinguishing a fire.[68] There were odd disputes, which it was important to resolve quickly – in many cases the Forest Service might be interested in buying that land for new plantings and getting on with the community, and the possibility of arson had always to be reckoned with.[69] The occasional farmer had a 'real hard-nose, red-neck attitude', but generally the low-key approach of offering to

Mervyn Taylor, a wood engraver, painter, illustrator, sculptor and designer, was one of New Zealand's finest artists. The lengthy explanatory text in this Mervyn Taylor poster reflects US practice at the time, though the public relations firm suggested this sort of wordy publicity should be given relatively limited placement.

ANZ, F1, 12/1/12/7, Pt. 3, Fire prevention/Screen advertisements 1952–54 help with a burn, or just standing by as an insurance, 'went down really well,' and provided good training for Forest Service men.[70]

Fire-prevention advertisement 1952–54.

F1, 12/1/12/7, Pt. 3

Scientific and technological developments continued to contribute to forest protection. In 1982 the first computerisation of fire control data considerably speeded up retrieval and analysis.[71] Fire prediction, too, became more precise. However Entrican might have enthused about his early weather instruments, staff were unenthusiastic about doing the essential regular readings with the old array of instruments, the information was insufficiently detailed to give a reliable picture and, though raw data were collected, there was no research to systemise their use. In 1963 a new system based only on wind speed and moisture content of wood gave 'rise for optimism'. The real breakthrough came in the 1980s. In 1981, after national trials, a Canadian fire weather index, which allowed very specific estimates of the relative dryness of a range of fuels, was adopted.[72]

This system increased the accuracy of fire-risk estimates: for instance, peat will dry out in a long, dry period and requires considerable wetting or moisture to make it less flammable. A period of rain may not reduce the risk as much as might be assumed. Fine fuels, which present a particular ignition risk, can also be closely monitored.[73] In 1980 locally available hardware successfully transmitted the collected data from remote weather stations via a VHF-AM land mobile radio channel; from 1981–82 these stations were progressively installed across the country and eliminated the laborious collection of data by hand.[74] However, the system, adopted because it was developed for fires at similar latitude, had first to be calibrated to New Zealand fuels – *Pinus radiata* and gorse, which burn at a terrific heat. Moreover, initially no one fully understood how the system worked, so only the end fire-risk figure was used. The system's potential was fully exploited only when Martin Alexander, a Canadian researcher working in the field, came to New Zealand in the early 1990s, and 'opened [people's] eyes' to the finer details.

Today the indexes and the acronyms – the FFMC (fine fuel moisture content), the DMC (duff moisture content) and DC (drought code), ISI (initial spread index) and the BUI (build-up index, indicating the total amount of combustible material available on any given day) – roll off knowledgeable people's tongues. The fire weather index system (along with the training and research that have contributed to its appropriate use) has led to significant advances in assessing highly localised fire risk, understanding fire behaviour and developing fire-prevention readiness and response measures.[75] Today, says Kerry Hilliard, the Department of Conservation's national fire co-ordinator, these stations are 'the key that drives fire management in New Zealand because from that [data] you can set up trigger points and all sorts of things that you can tie into management practices. They are really what have changed things.'[76]

But fires had still to be put out. Developing technology again created efficiencies. By the mid-1950s telephone switchboards had been installed at forest headquarters, ending the use of overloaded party lines. Conversion to metallic circuits continued.[77] The more reliable and mobile radio eventually supplanted the phone for field communications. In 1948 a former army mobile dental surgery vehicle, newly outfitted with a two-way shortwave radio, other equipment and bunks for the crew, set off for its new job as a mobile radio headquarters. In 1954 there were 54 fixed and 94 mobile transportable and portable stations, and new radio installations were in place at Glenbervie, Tuatapere, Mamaku, Taurewa and the Murupara logging scheme. Retired forest manager John Ward remembers the huge sets of 12-volt batteries sitting in the trays of trucks. These powered the 'twisted pair' of centre-fed half-wave antennae which, supported on a short portable mast, gave a good performance for up to 48 kilometres. A lower-powered version whose extreme range was 32 kilometres was also available, though its performance was reduced substantially if used in a truck or if hills and trees stood in the way.

But it was all war-surplus equipment that the Forest Service was by then desperate to replace.[78] In 1957 New Zealand-made mobile radiophones improved and extended radio communication to the point where a communications engineer was appointed.[79] From then on the number of both fixed and mobile stations grew steadily; the sets became more reliable and user-friendly and their range increased; and another New Zealand manufacturer developed a pack-type portable set for forest use.[80] By 1960, with staff ever more widely dispersed, the Service was developing and operating a high-quality radio network that virtually covered the country. This extensive reach enabled co-operation between Search and Rescue and the regional emergency organisations, and was considered vital in fire protection work and forest workers' safety.

New lightweight portable sets purchased in 1963 gave increased coverage and enabled fire-fighting positions to be co-ordinated.[81] Nine years later VHF radio was introduced, representing 'the greatest advance in many years'[82] and, in

A radio truck reporting back to HQ.

ANZ, AAQA, 6506, Box 22, 387, H1815, Forest engineering/ Telephone communications

pre-cellphone days, an important way of communicating rural emergencies. By the mid-1980s 'there were VHF repeaters on all major hills throughout most districts',[83] though by then, under an international agreement, all VHF radio was being converted to SSB. The frequency, too, was moved – from AM to FM.

Radio technology was exploited for other purposes. In 1971 an alerting device on the mobile radio-telephone system to summon gangs for fire-fighting was trialled at Kaingaroa, then progressively installed in other forests. Now men could be posted for both fire protection and effective production work. Two years later personal paging systems began to be introduced.[84]

A base station radio being used on Fire Course no. 12, September 1955.

ANZ, AAQA, 6506, Box 22, 387, M1565

Progress with vehicles and fire appliances was slower. Senior Forest Service fire control officers L. H. Bailey and William Wright, for instance, toured the various conservancies doing field inspections, assessing fire risk and the fire preparedness of conservancies and county councils as well as reporting on recent fires and their management. Their tortuous and time-consuming journeys on public transport and local conservancy vehicles led Bailey, in 1948,

> to place on record that the lack of a motor vehicle of any kind renders field inspections of £90,000 worth of equipment at various supplying stations, instructional tours etc, a matter of extreme difficulty ... The amount of trouble experienced in carrying out field supervision of fire protection resources ... is such as to militate against the success of the job.

In part because of postwar exigencies the matter was not quickly resolved. In 1953 the Forest Service's responsibilities for fire protection had been considerably enlarged by its direct responsibility for national parks, scenic reserves, and all fire districts administered by other government departments.[85] A year later Entrican wrote to the Director-General of Post and Telegraph asking if public service vehicles could be made available so his staff could get to fire emergencies around Wellington as 'we still have not sufficient transport of our own'.[86] The replacement programme for the standardised station vehicles that met the particular Forest Service need likewise progressed only slowly. By 1974 the exotic forests were open to the public and fires had be contained quickly; heavy fuel-efficient crawler tractors were sought.[87] In the 'very strong bottom-line era of the 1980s',[88] however, the fleet of ageing vehicles, graders and crawler tractors could only be expensively maintained, not replaced.

A portable radio-telephone set being tuned to a fire frequency. Within a decade the size of sets had been considerably trimmed down.

ANZ, AAQA, 6506, Box 22, 38, M7774

By the mid-1950s the Quads had had 12 years of rugged use as both personnel carriers and tankers. By then the old vehicles were no longer being made – and they were 'blimmin' noisy. Those Quads were really difficult to drive, really heavy in the steering and once you got off the road, they were very hard to get back on. But they could go places, those old fire engines, they could go places the Internationals [the replacement engines] couldn't.'[89] By 1951 a limited number of trailer pumps that carried water, pumps and hose, and slip-on tanks that went on other station vehicles were being developed to replace the old Quads.[90] In 1958 a five-year replacement programme, starting the following financial year, was planned to end such piecemeal replacement.[91] The newly available, moderately priced Internationals were water-carriers with pumps, lockers and places to sit; built of light alloy, they were two-thirds the weight of steel, longer-lasting and tougher. Investing in such machines was not done lightly. Loaded down with a 2273-litre tank, lockers for almost 610 metres of hose, equipment and hand tools, and two Wajax pumps, an International was extensively tested on all sorts of terrain. At one point it was raced over eight kilometres against a Quad: the International arrived 10 minutes before the local machine – quite a pick-up over that distance. Internationals had a comfortable road speed of 80–97 kilometres per hour, about 29 kilometres per hour over rough country, and were easy on the driver. Fitted with a safety-belt, good for rough country, and having a cab designed to protect both instruments and driver from dust, they were 'a real pleasure to drive after those other things', Ward remembers. 'We thought we were made when we got those.'[92] When Quads were thoroughly clapped out (or in the department's words, when

A Quad and its tanker, Waipoua Forest, 1952. These extraordinary vehicles clearly caught the public imagination and were constantly being photographed.

ANZ, AAQA, 6506, Box 21, 8/26, 432.3, M61

An International fire engine. These were initially built by G. N. Hale and Son in 1958 to the specifications of the Forest Service Engineering Department. They came complete with wire meshing protecting the headlights, flashing amber lights and backward-sloping shelves for the tool locker (at rear).

Collection of Roy Knight

maintaining them was no longer 'economically practical'[93]) they were used for spraying weeds in the forests.

The 1970s saw further innovation. The versatile and economical smoke-chaser was developed and manufactured at the Forest Service experimental workshop at Golden Downs run by Jack Calder. This vehicle, designed to overcome the limitations that heavy appliances faced in rural conditions, was 'a small, say one-ton Toyota multi-purpose utility truck with probably 100 gallons [455 litres] of water on the back and a small pump [and] nipped a few fires in the bud'.[94] When not fire-fighting it was available for other work.

The 1970s also saw the Internationals replaced by Bedfords. Then, in 1984, 14 new Isuzu fire trucks replaced the Bedfords.[95] Equipped with the new power-take-off Darley pump, which was driven by a device integral to the truck's engine and a Wajax pump as a spare, and carrying 2728 litres of water in a mild steel tank, 'these [were] to be the fire engine for the 1980s'.[96]

By the end of the 1950s the Canadian Paramount pumps had had their day. This six-stage low-volume centrifugal pump could develop pressures of over 29 kilopascals with a maximum output of approximately 455 litres per minute. It came with a two-wheel steel body trailer, which also carried hand-tools, extra fuel, a 273-litre canvas dam for open relay work, and 305 metres of hose that was 39 millimetres in diameter. Alternatively the pump alone could be carried by backpack. Selected because they were the only centrifugal pump whose impellors could withstand the abrasive action of the suspended pumice in many of the central North Island streams, the pump and backpack weighed 33 kilograms – a killer to carry, especially in rugged country.[97] They were also highly temperamental: 'If you started them right with the first pull you were okay, but half an hour later you could still be trying to get them going and the fire was getting bigger.'[98] A handle on the side had to be perfectly positioned for starting so 'you had to know where you had your tongue poked and your feet placed! If you didn't know the animal you were working with, it was really frustrating.'[99] In the mid-1950s some Hale pumps were available but, while deemed reliable, they needed two men to carry them so were only used where access was reasonable and the country not too rugged.[100]

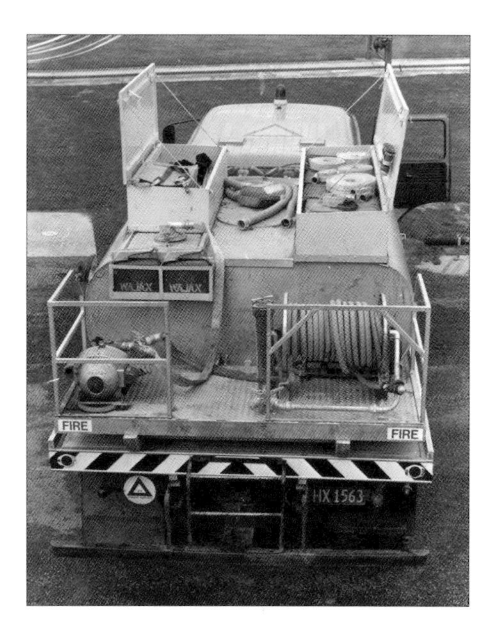

The 10.2-tonne Bedford tanker truck, with its slip-on tanker appliance, Like the smoke-chaser, it could be used for other work. However, although versatile, the truck was deemed slow and underpowered.

Collection of Roy Knight

An Isuzu fire truck, the make still used today.

Collection of Roy Knight

Brian Helms adjusts the controls of a Paramount pump during a controlled burn on a firebreak in the Hutt Valley, 1954. Although these pumps were not entirely satisfactory, something of their value (and of the extent of the Forest Service role in rural fire control) can be gauged by the Forest Service's decision to demonstrate their use to a variety of outside groups.

ANZ, AAQA, 6506, Box 21, folder 8/25, M1069

By 1958 the Wajax one-man pump was on the market. Originally a two-man chainsaw, then adapted as a fire pump, 'it was head and shoulders better than its predecessors', Barber recalls, but its 26 kilograms on your back 'made heading up to a dam hard work'. Able to deliver 2100 kilopascals and pump water by hose 152 metres, the Wajax was 'very noisy – you certainly got earache after working with those'. But it was more reliable and simpler to operate than the Paramount and, because it could be adopted everywhere, training and maintenance were simplified.[101] The army, air force (and prisons) also used it, bringing a welcome uniformity to the scene. That many of these Wajax pumps are still used today, fitted with a Robin 17DFF motor, is a tribute to their design and construction.[102]

The new equipment needed good water supplies, highlighting the issue of water storage, particularly at Kaingaroa. Better and bigger dams were steadily built throughout the country's exotic forests.[103] At the same time other developments assisted with reducing water use. In 1959 rubber-lined hose finally replaced the canvas hose and new, controlled nozzles were introduced. The new nozzles did more than spread the water better. When Roy Knight started work with the Forest Service in 1952,

Hose being carried in a backpack during a controlled burn of a firebreak in the Hutt Valley, November 1954. A meticulously detailed manual gave instructions on flaking the hose in a three-ply 'former' before lowering it into the backpack and removing the wooden guide. The object was to ensure the hose fed out smoothly as the walker moved forward. Bearers were instructed to carry the backpack as high on the shoulders as possible to prevent any sideways movement.

ANZ, AAQA, 6506, Box 21, 8/25, 432.3, M1071

The Wajax pump. This photo was taken by John MacDonald in the mid- to late-1950s. The Wajax put out about 318 litres per minute. Like the Paramount, this portable high-pressure and low-volume pump was useful because it could work from limited water supplies in isolated areas but also pump long distances in forested areas.

ANZ, AAQA, 6506, Box 21 8/25, 432.3, M9966

> we only had brass nozzles and canvas hose and your nozzle wasn't controllable and if you had a hose line going down the hill you'd lose all the water out of your hose – you could stop it down at the pump but you could lose 50 or 60 gallons down the hose line, just running out into nowhere because you couldn't turn the nozzle off.

Over time the aluminium nozzles and branches, prone to corrode then seize up if used with salt water, were replaced by the current plexon ones. These nozzles and shafts, which have hardened plastic exteriors with a brass insert, are largely trouble free.[104]

As Knight struggled with his nozzles and hose, chemicals began to be used to suppress rural fires and inhibit combustion. 'Wet water' – water with a detergent added – was utilised from the late 1950s. (Saturnal, a solid-form detergent, was initially used. In later years Ian Millman went around garages collecting Teepol, a commercial detergent, in wholesale lots.[105]) A hydroblender, comprising a 1.2-metre tube of fibreglass fitted with alloy caps at both ends and a removable cap at the top, was used to insert 'a big soap capsule in the hose line that reduced the surface tension of water. The soap made the water spread faster, and water treated like this tended to stick.' It penetrated both wood and charcoal significantly better than plain water and, because it evaporated significantly less, 'it was used a lot in monsoon buckets where the cost of water delivery was extremely high'.[106]

A Wajax Mark IV pump, 1981.

ANZ, AAQA, 6506, Box 21, 8/25, 432.3, 12803

Chemical fire retardants had long been used in fire extinguishers on static fires. In 1958 calcium sodium borate mixed with water was trialled as a vegetation fire retardant. Although nothing seems to have come of this, in the mid-1970s Fire Trol and the more effective – and possibly cheaper – Phos Chek were being used. The retardants were made of fertiliser-type salts, red dye, corrosion inhibitors and flow conditioners. The salts, heated by the approaching fire, reacted with any organic matter to make it unburnable and retard the fire spread.[107] They were not necessarily simple to use. Different ratios of chemical to water had to be mixed for a crown fire (1:85), a fire in fern or ti tree (1:10) or a grass and tussock fire (1:15).[108] But used properly retardants were highly efficacious, reducing water use and acting where water alone was ineffective.

There were some teething problems. The retardant tended to settle in the

Dams were not always permanent. During a sawdust fire at Conical Hill in 1952 bulldozers made this emergency water tank.

ANZ, AAQZ, 6506, Box 21, 8/27, L6403

heavy drums in which it was shipped from Indiana. John Barnes, in rural fire for more than 40 years and in 2012 manager of rural fire in Christchurch, remembers himself and two others at the Golden Downs fire depot 'experimenting a bit on a venturi system that would flow water through it and suck the retardant up out of the drum – an easier way of getting it out. Geoff Hildreth [later Golden Downs fire station officer] and I were asked to got to Taupo to demonstrate this at a fire controllers' workshop'. But, lacking the right pressure, the pump acted in reverse, covering him and Hildreth in red retardant.[109]

Other technology helped make water on fires more effective. By 1955 aerial top-dressing had already done much to promote soil conservation on the hill country. Could fixed-wing aircraft also help? Extensive trials in 1955, 1957 and 1959 carried water in discharge trunks that were fitted inside the hoppers usually used for top-dressing. Sufficiently encouraged by the results, the national Soil Conservation and Rivers Control Council organised two service operational trials in the Waitaki and Selwyn fire districts over the 1960–61 summer. Although the Forest Service called the results 'impressive', it had concerns about cost and the technique's limited applicability (in any really hot fire water was vaporised before it reached the ground) so it did not explore the technique further at that stage.[110]

However, the Forest Service did pay aero clubs to survey sometimes vast areas.[111] The Geyserland Airways and Taupo Air Services Cessna 172 patrolled up to 50 hours a week, flying up forest roads, circling logging areas for smoke, monitoring farmer burn-offs and checking that stationary cars were not carrying illegal entrants or loiterers. When the *Rotorua Post* reporter went up, not once was the 'wing-mounted loud speaker' used to order people out of the forest.[112] The National Airways Corporation (NAC) continued to report any fires (and Air New Zealand, with which NAC merged, still sometimes does). Moreover, although the capital cost of replacing ground crews with aerial ones was too large to contemplate, within two years of the aerial fire-fighting trials the Forest Service realised it was economical to charter commercial aircraft to carry fully trained and well-equipped ground forces and four or five 1136-litre collapsible water tanks to a fire. 'Wet water' was also used.[113] Some fixed-wing water-bombing was also done in Kaingaroa.[114] In the mid-1980s forest manager Mike Hockey used fixed-wing aircraft initially to keep a lookout for breakaways, then later as

> bird dogs, a control platform for the fire boss. If you're on the ground in a fire, close to the action, it's not very easy to see the whole picture. From the air you can see the whole thing, make dispassionate decisions. We had excellent comms, so could control other aircraft, advise crews on the ground etc. To

A forest-fire patrol plane passing the Kakapiko lookout in the Whakarewarewa Forest, 1954.

ANZ, AAQA, 6506, Box 21, 8/22, M459

> some extent we pioneered the method in the Gisborne district.

A bird's eye view of the immense Kaingaroa Forest, looking south. The huge firebreak was just over 100 metres wide, but tree encroachment had reduced it to 90.5 metres when the photo was taken in 1971. In the centre of the photo, on the road verge, is a water tank for fire-fighting purposes. Such tanks were placed at 3.2-kilometre intervals along the highway.

ANZ, AAQA, 6506, Box 21, 8/20 432, M11,342

In 1968 helicopters were first actively involved in fighting open fires.[115] The Forest Service scorned some urban authorities' continued use of the fixed-wing aircraft to drop water – 'a most expensive bucket-brigade system'[116] – but rapidly recognised helicopters as highly effective. By 1972 New Zealand-made monsoon buckets were becoming standard fire equipment, both within and outside the Forest Service.[117] In the late 1970s their essential contribution in controlling major fires was recognised and 100 helicopters were relatively well distributed through both main islands. New Zealand Forest Products, Tasman Pulp and Paper Co. Ltd and Fletchers owned some; the RNZAF and some counties owned others.[118]

It is possibly easy to overstate what helicopters did, and can, contribute to fighting a fire. Some early models needed to be shut down for up to an hour in summer heat to cool off, and not all could be immediately restarted. Even today adverse weather limits their effectiveness, they are expensive, they need good water supplies and they need to be used in tandem with ground crews.

A helicopter laying hose, 1970s. While Forest Service staff gained experience in working with helicopters, fire brigade personnel and local fire authority staff had an opportunity to evaluate the technique. Forest ranger R. A. Cohen, who organised the event, stressed the value of the helicopter's speed and economy, especially when used with red fire retardant, which made it possible to target with water accurately.

Nelson Evening Mail, 26 September 197? (in ANZ, AANI, W3219, 12/11/7/3, Pt. 1, Newspaper clippings/Fire-control methods

On the other hand they are highly valuable for carrying water, fire-fighters and equipment, and they can carry out hose-laying and reconnaissance work. They can tackle overnight fires or fires in full flight, and pilots can work closely with ground parties, dropping water and retardants on hot-spots or ahead of a fire to limit its hold.

Helicopters became much more effective when, in 1972, New Zealand-made monsoon buckets became standard fire equipment. At $800 each in 1971 ($10,125 in 2011 values), monsoon buckets were not cheap.[119] And considerable flying skills were needed to concentrate the water and maximise its use: 'You had the fire going in one direction, the chopper going in another and the water coming out of the bucket probably going in a third. Fifty per cent of water evaporated before it got to the ground.' Steep, hilly country compounded the difficulty of concentrating the water.[120] Monsoon buckets were initially considered valuable for particularly light fuels and for mopping up,[121] because their low volume (at that stage they carried only 282 litres) made dropping water through a bush canopy useless. Nor did they scatter water, as a hose does, so that for burning ground, snags, spot fires or single trees, 'the monsoon bucket is out on its own'.[122] They also provided 'the only way you can get near the fire. We beaters can only beat out the bits that flare up after the fire has passed through. Even then the heat is fantastic.'[123]

At first an assistant pulled on a lanyard or rope to activate the bucket's release mechanism. After a lot of experimenting, an electric solenoid withdrew a pin to release the water; this technique allowed an extra 73 litres to be carried and the pilot could put the water where he wanted 'rather than have to shout' at his assistant.[124] Work continued. In 1977 fire control officer Bill Girling-Butcher, with typical gentle mockery, reported that the Forest Service had modified the monsoon bucket, activating it with an outboard-motor mechanised tilt and ram – a 'fail-save mechanism that may even appeal to the Australians'. (The Service had also developed a lip that further extended the bucket's capacity from 545 to 909 litres for when more powerful helicopters were available.)[125] Two years later a hydraulic-operated system provided a totally reliable release mechanism.[126] Today the release system is generally pneumatic.

The same year as monsoon buckets were introduced, infra-red hand-held devices to identify hot-spots were issued to all conservancies. Described then as 'promising' for ascertaining when an area could safely be left, the initiative seems to have stalled, but Canadian-approved equipment was bought and tested for New Zealand conditions in 1978. This rapidly proved its worth over the time-honoured way of testing – putting your hand on the ground.[127]

The helicopters and monsoon buckets were contracted in and the other tools were owned by the Forest Service, but there was no rational basis to the way in which they were distributed among the conservancies, leading to discrepancies. In the early 1960s equipment was 'allocated to individual forests on the basis of opinion rather than opinion supported by a systematic analysis of the risk', and perceptions of risk could diverge widely: the Forest Service spent 9s 3d per acre on one area of high risk and 16s 6d on another of moderate risk. In an attempt to address this issue a points system of assessing fire risk was proposed, based on human activity in and around the forests. Given the human tendency to ask for more than strictly necessary, the problem was not easily solved.[128] To further complicate the matter, particularly heavy machinery was accumulated over the 1960s and 1970s, when thousands and thousands of hectares were being broken in annually.

Technicians at the various stations were trained in tool maintenance but standards were inconsistent. By the late 1950s, and as maintenance became more skilled, much of the South Island equipment was sent to the Golden Downs workshop. There was a similar depot at Palmerston North, headed by the unforgettable Johnny MacDonald who had been equipment officer at Karioi in the 1950s. He is credited with ensuring gear was of a uniform standard and suitable for the job and with 'dragging the … Forest Service out of the knapsack and shovel era and into the modern high pressure pump and technical equipment' era.[129] MacDonald, 'one of the most conscientious public servants' Ward has ever met, had a significant influence on men and equipment. Knight remembers that

> MacDonald tested everything before we got it, tested the construction. People used to say in Palmerston North, 'Who's the silly joker in the Forest Service depot going round the streets, dragging a bit of hose behind him?' That was an abrasion test. Then he'd put a blowtorch to a hose to see how long it'd take to burn with water in it. First time I went on a fire equipment course, about 1958, he had an abrasion thing like a big file rubbing across the hose. [He left nothing to chance. When the plexon nozzles were introduced] he started throwing them on the concrete floor. He said, 'If you're up on a bank, and the hose line's going up a gully and you want a plexon nozzle, how are you going to get it? You'd throw it to the joker down amongst the rocks.' He said, 'Well, what happens

A fire hose scrubbing device. Technicians throughout the Forest Service were highly ingenious. The device being demonstrated here was made by staff at Otautau.

ANZ, AAQA, 6506, Box 21, 8/25, 432.3, M2783

> if it hits the rocks? It'll shatter.' I learned a lot from Johnny, and so did a lot of the other jokers.[130]

MacDonald's role was formalised in 1959 when he was appointed chief fire technician. He and an assistant provided central control over purchase, checked components and investigated equipment breakdowns.[131]

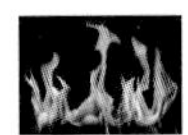

Bill Girling-Butcher, 1969. Girling-Butcher's correspondence and memos are characterised by a humour and irreverence that makes them a joy to read.

ANZ, AAQA, 6535, Box 27, M10,527

Training in fire behaviours (then still an inexact science) and fire responses became an essential component of fire protection. Just after World War II Entrican's Category A scheme recruited ex-servicemen such as Bill Girling-Butcher, son of Ron Girling-Butcher. Remembered as 'one of these great characters; a rugby man, for a start', he was astute, wise and infuriating, and let people get on with the job. He brought a certain informality to his position. Doug Ashford, a forestry trainee at the end of the 1960s and now president of the Forest and Rural Fires Association of New Zealand, remembers doing uphill water trials with Girling-Butcher 'in his jandals and beach shorts and we were all done up in overalls and boots'.[132] Category A men were sent to the Tapanui or Rotorua training centres for intensive training. The scheme was a 'godsend', bringing in men who all went to the top of their field. From the late 1940s to the 1970s these men were introduced to a 'field-training regime that was the envy of overseas counterparts'.[133]

From 1953 the Forest Service held regular annual training courses for representatives from the larger forestry companies and from local bodies and government departments. Here administrative officers gained practical experience in developing fire plans and in the latest fire-fighting methods, along with a working appreciation of the value of aerial fire patrols and meteorological instruments.[134] This work was extended in 1969 when the Forest Service and the Fire Service Council jointly ran a course for urban fire authorities with fire responsibilities. Forest Service staff, too, continued to update their own training. A range of international journals, correspondence with their overseas counterparts, and attendance at Australian courses brought the latest developments in fire control and management to their attention.[135]

By adopting this posture and wearing full-length clothes, this firefighter is managing to shield himself effectively from radiated heat.

If caught in the flames throw yourself flat on the ground, face down. Proper clothing will protect your body, and you can get added protection from the blade of your shovel.

The Forest Service also drew on Australian CSIRO publications in producing small booklets to inform a wider public about basic fire-fighting practices. The accompanying text deals with the fluid loss (up to 3.4 litres an hour) in fire-fighting and measures to prevent it.

New Zealand Forest Service, *Safety and Survival in Forest Fires*, 1974

For those on the ground, something more was needed. From the 1950s US-imported booklets such as *Forest Fire Fighting Fundamentals*, illustrated with cartoons, set out basic knowledge on the principles of fire-fighting, the tools used and the most common means of suppression – why fire spreads, how different sorts of fuels impact on it, what you can do to suppress it, and what you use where and when. *Water vs. Fire* provided the same sort of good advice more colloquially: use water sparingly, aim low ('Whether water knocks the flame/ Depends a lot on how you aim'), use spray, don't swear at the coupling. In the 1960s the fire section of the New Zealand Forest Ranger Certificate had extensive material on fire – fire plans, basic fire-fighting methods, fuel types and behaviours, fire-fighting organisation. Over the years the Forest Service put this sort of information into various booklets, combining it in different ways with different emphases for different publics.[136] Booklets such as the Forest Service's *Forest and Rural Fire Precautions 1973* were published to help contractors observe rural fire prevention regulations and were updated over the years. From the mid-1980s Ministry of Agriculture and Fisheries leaflets such as *Burning Scrub (or Bush Cover): Requirement and methods* gave practical advice on burning and its legalities to farmers.

That was desk stuff. Rangers, woodsmen, and the wageworkers and work gangs who in Barnes' view were the real fire-fighters, received practical training, practising the one-lick method, cutting and digging fire-lines, using the fire equipment, spreading dirt, beating out small fires, and map reading. Stations in Canterbury were required to drill every fortnight, regardless of the day's conditions;[137] elsewhere, half a day a week or fortnight was generally put aside, or training was carried out when fire conditions shut down the forests or when operations were suspended in the early afternoon. Over-frequent practices could mean 'you cried wolf too much'.[138] From 1949 competitions similar to the Fire

Fire training in Tarawera Forest on a steep ridge at the base of Maungawhakamana, 1982. Competence with the Wajax pumps was essential, and drills were held in forestry camps or, as here, in the forest itself. Not everyone is identifiable, but Aussie Kohl is the training officer beside the pump. From left to right behind him are Jason Wetini, Phil Peters, Sam Ngataki and, on the far right, Marcel Van Westbrook. The frame on the ground was used for carrying the pump. Spare shovels, fire hoses and nozzles and radio equipment also had to be carted up.

Wally Seccombe, private collection, image 268

Brigade games, and with ever-more complicated rules, involved growing numbers of keen Forest Service employees, other forestry workers and, for a time at least, county fire officers. They became known as the Wajax competitions; head office sent the winners telegrams; shield and trophies proliferated.[139] They generated a great spirit and enthusiastic practices, and helped develop a quick response.

Fire-fighting exercise, Kaingaroa Forest, September 1955. The exercise was held in logging slash. Here the nozzle-man is progressively damping down the edge of the fire-line. The second man is ready to couple on a new length of hose to allow the nozzle-man to advance. Note the backpack and the hose.

ANZ, AAQA, 6506, Box 21, 8/25, 432.3, M1556, Fire suppression

Top fire-fighters from the Rotorua conservancy, about 1970. Rear, L to R: J. Langedoen, Wayne Morgan, W. Hamiora (coach). Front row: Warren Morgan and D. Galvin. The team had to make a 70-metre dash carrying fire-fighting equipment and pump water to a 'fire' in the shortest possible time – a feat they accomplished in a winning time of 1 minute, 1.8 seconds.

ANZ, AANI, W3219, 12/11/7/3, Pt. 1, Newspaper clippings/ Fire-control methods

Outside its forests the Forest Service was often involved with vegetation fires in semi-rural county council areas where the local volunteers needed backup and assistance. And it was ready to bear the costs. Its actions it construed as being in the public interest, it considered it gained good will and that there was educational value in being able to demonstrate how effective a small, well trained and properly equipped crew could be – and the work was valuable fire-fighting training.[140]

From the 1970s Forest Service training of fire brigade members, council workers and others increased participants' skills. The training also gave further opportunities for the Forest Service managers to develop and hone fire-fighting exercises and skills that they could subsequently use in training their own men. But, as the 1976 Hanmer Forest fire demonstrated, none of this training prepared men for dealing with large, ferocious fires. Simulated exercises, the more realistic the better, Mike Hockey remembers, with all sorts of problems built in to them, were systematically developed and used from that date.[141]

A fire-fighting exercise, 30 November 1971. Here, in a co-operative training venture, Forest Service employees K. Kotau and D. Gay extinguish a controlled burn while members of the Nelson Fire Brigade look on. The absence of protective clothing is interesting. The Forest Service began to issue cotton overalls in the early 1960s, and dress would change markedly over the next decade with the development of flame-stop materials.

Nelson Evening Mail, 30 November 1971 (in ANZ, AANI, W3219, 12/11/7/3, Pt. 1, Newspaper clippings/Fire-control methods

However, even more than those exercises, controlled burns, ever-increasing in number and size, provided a major training ground.

Forestry managers had long used fire to burn scrub and fern on firebreaks or, like farmers, to clear land prior to planting.[142] Particularly after the 1955 Balmoral fire, they also started to burn slash (in spite of some conservators' initial opposition) as a protective measure.[143] However, before the 1960s, a station's '200-acre [81-hectare] planting programme per annum was regarded as formidable'.[144] That changed with the planting boom. Whether in local figures or national aggregates, the amount of land burned off was enormous. In the Kaweka Range alone, in January 1969 more than 1416 hectares of wind-eroded land and kanuka, scrub and fern regenerating from decades of earlier fires were burned to create employment for Hastings' unemployed. Burning continued with similar intensity into February, and the activity was even greater in the following years.[145]

Nobby Reekie, officer in charge at Wairapukao, a 50,000-hectare subdivision or management unit within Kaingaroa Forest, remembers that from 1974 to 1982 about 40 fires a year burned off some 4000 hectares on the forest's flats as poor species were converted to *Pinus radiata*. The men were pushing the trees' altitudinal limits.

> A good cleanup' could increase ground temperature by 5–6 degrees Celcius – all the difference between having a live crop or a dead one – and ensured that

> vegetation could not impede the earth absorbing the sun's heat during the day, nor trap the air as the ground cooled, thus encouraging frost. The summer season brought additional fires. '[T]here'd be about 70 or 80 fires within the Kaingaroa Forest boundaries. We were very seldom able to burn before Christmas – it wasn't dry enough – but from about mid-January to the end of March (we once fluked it as late as ANZAC Day) there's always be someone burning.

In 1974 the Forest Service was involved in 260 controlled burns over almost 25,000 hectares, and the Meteorological Office helped with developing an index to check that smoke dispersed properly and did not cause what Girling-Butcher euphemistically referred to as 'temporary inconvenience'.[146] The burning was by no means confined to the central North Island, nor to the Forest Service. Private forestry companies, large and small, were all involved in such controlled burns as well. Don Geddes, working for Fletchers in the mid-1960s, remembers burning 400–600 hectares annually.[147]

The size of the burns varied greatly. The big ones burned with huge force and intensity. Barnes remembers the fierce fires at Kaingaroa burning 'tons and tons of residue that was left on the forest floor'. If they got into unburnt areas they became uncontrollable. Barnes remembers 'you'd pray for them just to die out. I remember another one – like a waterspout, but it was a smoke spout' that seemed to go 'thousands of metres into the sky'.[148] The high fire load that each fire carried threw 'out tremendous heat and great palls of smoke, and creat[ed] clouds like stratocumulus, which caused rain'.[149] Some firms approached the

The fire in this burn-off behind Lake Rotoiti in early 1980s generated the huge columns of smoke typical of such fires.

Wally Seccombe, private collection, image no. 87

task responsibly; for example, Carter Holt established trees on old farmland covered with second growth,[150] with other farmland as buffers. However, much riskier burning was often carried out next to (or in) established forest.

The force generated by the forest burn-offs carried logs up into the fire's vortex, then hurled them to the ground. This small fire-thrown log was embedded to a depth of about 60 centimetres in a metalled road. Norm Dunn is standing by.

Nelson Forests Ltd, in Ward and Cooper, *Seventy Years of Forestry*, 1997, p. 127

But there was no doubt that burning was highly efficacious. In 1968 what would have cost $32 a half-hectare to fell was cleared by fire at 55 cents per half hectare.[151] In 1981 procedures were developed to speed up approvals.[152] But opposition was growing: housewives in Gisborne did not want to have to worry about soot on their washing;[153] Rotorua and Auckland residents' ire, raised by the blankets of soot and smut as fires burned over thousands of hectares, cannot have been mollified by the promise of another 10 years of the same; and from the 1970s some groups were speaking out against the destruction of the bush and its flora and fauna.[154] The Forest Service's failure to deal with changing attitudes was perhaps part of a growing complacency that would contribute to its demise.

The burns were of course carried out in the fire season, when the fuel was dry. Head office recognised everyone's 'limited experience of safe and effective techniques', but officers in charge planned carefully. Head office then approved the plans (some months in advance), and they were then carefully followed. By the late 1960s increasingly complex and detailed plans gave information on all aspects of the terrain, fuel types, the sort of equipment to be used (bulldozers, fire engines, helicopters), where and how the men were to be deployed, how light-up should proceed, and plans for containing any fire breakout.[155] Initially four classes of burns, each requiring a different level of approval, were identified, but from 1967 Girling-Butcher vetted the plans of all but the most minimal burns. 'If he ever altered anything, it was "Ah, now why didn't I think of that?"'[156] By 1971 weather charts issued on the basis of 15 years of previous readings were available. This information, used as a guide, became more sophisticated a decade later with the Canadian fire weather index.

Prior to burning, areas to be burned were carefully prepared. Don Geddes remembers that 'the existing vegetation was crushed and killed by scrub cutting,

A gravity roller crushing the second-growth bush and scrub behind Lake Rotoiti, preparatory to burn-off, early 1980s.

Wally Seccombe, private collection, image no. 87

rolling with heavy cleated rollers, and killed with herbicides. Boundaries were prepared with bulldozed firebreaks and access tracks.' On the day, local knowledge, particularly of wind[157] and the availability – or otherwise – of experienced men and equipment from other forests to act as back-ups, played a major role in any decision to go ahead. The men were carefully briefed to promote maximum safety, and light-up went ahead anytime between 2pm and 8pm, when (it was now recognised) wind strength and direction would have settled for the day. Burning techniques became more sophisticated, with different areas and vegetations requiring different methods. Initially, the boundaries were back-burned, then the outer perimeter lit; this technique encouraged the fire to travel across the area, creating convection currents that drew the wind and fire in from all directions so that all the fuel was consumed.[158] Speed was essential – and not always achieved in the early days because some perimeters took so long to get around.

Early methods of lighting up were rudimentary: 'cotton waste soaked in diesel on the end of a pitchfork'; later diesel blowtorches that always left their operators surrounded by smoke and covered in diesel; even a plumber's blowtorch.'[159] Difficult terrain or gullies could complicate matters:

> You picked probably three or four of your most experienced guys and fittest guys that could run the fastest, and sent them down into these gullies to light up. They would disappear and all you could see of them

> was the smoke that they were creating as they lit up … Sometimes you wouldn't see those guys for a couple of hours and [with line-of-sight radios only] once you got to the gullies you lost contact.[160]

In the early 1970s incendiary devices began to transform the scene. Where it was too dangerous to hand-light, delayed action incendiary devices (DAIDs) were available. Like 'big wax matches' they were fired from a rifle that looked more like a dart gun. 'As they went into the air they burned along the wick and exploded as they hit the ground.' They could also be dropped from helicopters. Hockey remembers that as great fun, and the stories generated moved into Forest Service legend. With 500 DAIDs to a box, it became a highly repetitive task to put one against a heater element to light it, drop it down the tube into the scrub and reach for the next. In the process someone inadvertently put a lit one back into the box. Fortunately he had the wit to kick the box out the helicopter door before the whole lot exploded.

But DAIDs did not always light well (even if helicopters hovered to fan the unwilling flames) and it was difficult to achieve a continuous line of fire.[161] Aerial fire incendiary devices (AFIDs, also spelt APHIDs) were first trialled in September 1977. Using AFID fire-lighters – (a drum that dripped alumagel (napalm) and was slung below the aircraft, for which Malcolm Mead at Ashley

A young John Barnes demonstrates an incendiary rifle, early 1970s. Peter Maplesden, who was in charge of the Hira fire, is on the extreme right.

ANZ, AA

A helicopter and AFID lighting up a controlled burn in Ashley Forest, 1979. With helicopters, says Don Geddes, light-up patterns could be changed. Heat from fire rises, creating vertical convection winds. Fast light-ups in the centre of the burn area would generate convection winds strong enough from all boundaries into the centre. Next all the boundaries would be lit, and the fire drawn away from all areas outside the burn. As a result, the practice of lighting slowly from the lee boundaries could be abandoned so men no longer needed to spend extended periods in smoke and heat. Burns were able to be accomplished very quickly, making them cleaner and safer, and often saving the trouble of working into the early morning. Containing burns was much easier.

ANZ, AAQA, 6506, Box 21, folder 8/31, 436, M12505

Forest must take prime credit),[162] helicopters flew only about four metres above the ground at 10–15 knots to light the back-burn. Once this was safe, the main burn was lit by laying continuous lines of lit alumagel. 'There would be helicopters moving around, dropping the napalm,' remembers Bill Studholme, former CEO of the Selwyn Plantation Board, 'and then – *whoosh*! The whole hillside would explode into flames. It was very exciting.'[163] The pilot (or the operator) controlling the drops could also drop blobs of alumagel from over 50 metres above the ground, eliminating the need to fly down into gullies; unless complicated by irregular ground or gullies, burns of up over 400 hectares could

Checking light-up gear, early 1980s. Larry Bennet (left) and Don Geddes are inspecting the burner. The equipment had to be tested every so often to assuage Civil Aviation fears.

Wally Seccombe, private collection, image no. 122

be lit in about 40 minutes.[164] Because of their speed and efficiency in lighting up these 'minor fire storms', AFIDs were rapidly adopted. Civil Aviation imposed increasingly stringent regulations on flying and on the lighting equipment.

Early burns had their quiet moments. On night patrol, Knight would 'go out and rake some of the ashes out and put some spuds in for a meal'.[165] But there was always risk. Ward was nearly caught several times when hand-lighting, and had to take cover, sheltering in creek beds or any other less flammable area. The best-laid plans could always be disrupted by an unforecast wind shift; human error was not unknown.[166] Nor was pre-burn stress:

> Some of the big burns, when they were burning cutover, got a bit dodgy. Sometimes you'd really sweat, you'd worry for days about weather conditions (long-range forecasting became critical), about backup, whether you'd covered all the bases. The timing had to be good to ensure that the main fire drew correctly. Especially with helicopter light-up, once you'd started, there was no going back.[167]

Finishing the job was always a relief.

Inevitably there were breakaways. Don Geddes, senior ranger in Fletchers' Kawerau forest, recalled:

> We expected minor escapes to occur, but we had the experience, trained fire-fighters and the equipment to extinguish them. There was an understanding that if we had contained all escapes with the resources that had been allocated to a burn-off, then we had control. Managing minor fire outbreaks outside the designated burn boundaries was part of the process.

By the 1980s the big burns helped focus attention on 'previously neglected' safety issues. Instructions were circulated: to ensure the light-up was properly controlled, highly trained helicopter operators with ground/air communications were to be used. In hazardous conditions, the instructions noted, it might be better to hand-light. Training, supervision, experience, good firebreaks, access and good, well-tested and well-maintained equipment were essential. Resuscitators had to be available at all fires, controlled or otherwise. Safety clothing – bulk-purchased fire-resistant balaclavas, overalls and gloves – had replaced the pre-1970s shorts and singlets and fire-fighters were instructed to wear cotton next to their skin, with a wool overtop, not trendy nylon. Safety was paramount.[168]

The value of the controlled burning went well beyond achieving cheap land clearance. The burns were consciously used to provide excellent training. Fire-fighters could 'get smoke in their lungs' – gain highly practical, real experience of back-burning, of using the equipment and using it well. There was nothing

Fire-watching in the smoke, about 1981. The phrase 'getting smoke in the lungs' represented a literal as well as a figurative reality. The job was physically demanding, Wally Seccombe recalls, with the heat and often thick, acrid smoke affecting the eyes, nose, throat and lungs.

Wally Seccombe, private collection, image no. 114

like a real fire 'to sharpen the mind', keep fire plans up to date, test their prescriptions and keep their mobilisation schemes alert.[169] The burns helped develop understanding of fire behaviours and of local weather conditions that affected them. Honing skills in a way that training exercises could not, the burns helped create a body of trained and skilled fire-fighters who were available, more or less nationwide, for attacking rural fires.

The Forest Service was the most significant player in rural fire. But from the 1950s 'the largest company forests' were by then 'reasonably well organised for fire protection, and … included in the Forest Service radio and weather networks and aerial patrols'.[170] New Zealand Forest Products, for example, was involved in a wide range of fire responses.[171] Along with other major interests in the area, it was part of the Tokoroa rural fire district, which covered forest and farmlands around the Kinleith mill.[172]

The company established a clear fire regime that had many parallels with that of the Forest Service – partly because some Forest Service staff moved to private companies and took the thinking with them. It graded its fire plans in relation to the level of protection needed for the existing risk and steadily upgraded its protection strategies and practices. Specifically nominated high hills that gave vantage points across the forests had initially served as lookouts; during the late 1950s and into the early 1960s Forest Products set up six high towers, equipped with radio, direction finders and so on. The lookouts were manned, at first full time over the fire season, but only by day from the 1970s, when transport and access had improved. After the Balmoral fire, and again after the 1977 Ashley Forest fire, the equipment in the towers was boosted.

Until 1946 Forest Products' fire-fighting equipment was mostly hand-tools and backpack pumps. From the early 1960s the company had four-wheel-drive Bedford fire appliances and Wajax pumps, as well as high-powered pumps whose greater output was valuable for filling the monsoon buckets (the Wajax put out a smaller volume of water but under high pressure). Bulldozers constructed firebreaks; when helicopters came on the scene the firm hired them but had its own monsoon buckets. All vehicles had radio-telephones – large, unreliable and high on maintenance. The company also had equipment to measure fire risk, old instruments acquired from the Forest Service and not sufficiently sophisticated to provide the necessary information from specific localities to reflect the wide variations that are possible over small areas.[173]

Setting up a fire district sign, Tokoroa, August 1954. The signs were made of galvanised iron and were 1.2 metres square. They were taken down at the end of one fire season and re-erected just prior to the next for maximum impact – and so that they would not be used for target practice.

ANZ, AAQA, 6506, Box 32, 12/16, 945, M703

From the mid-1960s the company control-burned to create firebreaks and clear flammable fuels from its boundaries. Training was well organised. In the late 1960s and into the

A forest fire danger sign, 1954. Today such signs are commonplace in both urban and rural areas. Their novelty then is suggested by the small notice: 'Fire Danger Indicator Sign. This sign indicates the existing fire hazard for the Taupo–Oranui–Reporoa area. The pointer is set twice daily in accordance with fire weather readings of temperature, relative humidity, wind & fuel moisture content as measured at the fire weather station of N.Z. Forest Products Ltd at Tahorakuri Forest.'

ANZ, AAQA, 6506, Box 21, 945.2, M703, Signs/Fire prevention

1970s silviculture and forestry crews would do up to a day's training, along Forest Service lines, once a week during the fire season. The firm's highly competitive Wajax competitions at Kinleith involved all its forestry crews – and those crews were waged so turnover was low and experience retained.

The company also worked to develop fire awareness among its logging crews. Patrolmen went out to all logging crews to do the required inspections of the fire equipment and check that it was operational and up to standard. Machinery and tractors were washed down to remove any vegetation and minimise the potential for overheating. Spark-arresters were used on all heavy machinery and mesh shrouds were put round engine compartments to keep out any leaf matter that might subsequently ignite.

Forest Products worked as a good corporate citizen to foster good relations with its neighbours, conducting burn-offs on farmland adjacent to the forests. It had the skills, equipment and manpower to operate protectively, and the burns gave crews experience with fire behaviour and opportunities to apply their skills. Taupo Totara Timber Company was one small company that called on Forest Products' assistance in a large burn, using the company's equipment, engines and manpower.

Significant spending went on public education.[174] By 1964 the Tokoroa Rural Fire Committee (of which Forest Products was part) ran a local essay-writing competition, was developing a national fire protection symbol and was posting fire signs. The firm also began developing a 'fire character' (although, along with the symbol, it did not eventuate).

Locally, Forest Products, like other companies, worked to build and maintain productive relationships with neighbouring forestry firms. It was also involved with regional initiatives that helped improve the administration involved in protecting forests from fire. Before the early 1980s the Forest Service had met annually with fire authorities' personnel – rangers and other staff from the Forest Service and staff from the district councils and from the forest companies – to update them on changes that affected them. But in the early 1980s, within the Bay of Plenty and South Waikato area, a regional fire group comprising the Forest Service, Tasman, Forest Products and Fletchers was established under Forest Service auspices. The group created an early zoned-response system, shared equipment and offered back-up regardless of whose land the fire was on. Once the Forest Service resolved early communications problems by providing means of accessing its radio system as an initial response, this system worked well.

When it came to fire, the companies had to be allies, however competitive they were in the marketplace. Any big costs were met by the company on whose land the fire started but otherwise costs lay where they fell. What operated initially as a gentlemen's agreement eventually became the Regional Co-ordinating Committee under the 1975 Fire Service Act. Run by the Ministry of Forestry after the Forest Service's demise in 1987, the initiative was carried over into the post-1990 regime. In the Bay of Plenty and South Waikato area the Rotorua Co-ordinating Committee, which also evolved during the 1980s and continued to operate after the Forest Service's demise, met regularly to achieve consistency in the fire seasons set by forest companies and the district councils and to help co-ordinate fire protection and response measures.[175]

Company administrators also provided a means of protecting pockets of Tasman and Fletchers forests that was novel at that time. Rural fire districts had previously consisted of contiguous forests, usually owned by one company or owner; company administrators proposed that these scattered forests be included into a single rural fire district. The Forest Service did not greet this departure from normal practice 'with unequivocal enthusiasm' but, as Don Geddes says, there was nothing in the legislation to forbid it. The resulting Matahina rural fire district covered three main forests around the Kawerau area, and scattered blocks stretching from Tauranga to East Cape. Interestingly it excluded local authority rural fire districts, on the basis that the forest owners 'had most of the wildfire protection and fire-fighting skills and equipment, and did not wish to accept responsibility for areas outside their forests'. The principles of that district later informed the establishment of a combined rural fire district at Golden Downs in the early 1990s, administered with significantly lower costs.[176]

Volunteer field day, Ashley Forest, late 1970s. As time went on the conservancies became ever more sophisticated in their volunteer training.

Collection of Roy Knight

Don Geddes at Fletchers 'plagiarised' North American practice to turn fire plans into working practical documents. Critically, his use of existing management structures in fire management organisation meant that outsiders could join a fire management structure with each man well cognisant of his job within it. The Tasman Inter-District Fire Plan was later adopted and published by the Forest and Rural Fire Association of New Zealand for use throughout the country.[177]

The forestry companies and the Forest Service were not the only ones involved in rural fire matters. From the early 1950s volunteers, in a range of structures, assisted. Honorary forest rangers continued to patrol and inspect places inaccessible to forest officers, and conferences were held to motivate them and reiterate their duties.[178] Even if constrained by inadequate tools, by the late 1940s the traditional community response to fires was facilitated by better transport. In

Lower Moutere, John Barrington remembers going as a small boy with the volunteers when the phone rang to announce a fire and looking on as events unfolded. Volunteers crowded onto the tray of his family's farm truck along with the thoroughly wetted sacks, scythes, gloves and other minimal safety clothing; as they drove, more trucks swelled the numbers making for the fire. A short briefing warned about the dangers of wind changes. (Barrington particularly remembers being told not to climb a tree to escape any fire.) Then the work began, attempting to clear firebreaks and beat back the fire at the edges.[179]

In 1956, facing enlarged responsibilities under the 1955 act, the Forest Service began facilitating more formal volunteer brigades to be first-action crews. A group was set up that year at Te Anau to protect the eastern fringe of Fiordland National Park and Eglinton Valley where, with the new road, the number of tourists and cribs was increasing.[180] Shortly after, in a then unique organisation, eight permanent residents in the Marlborough Sounds who had telephone or radio communication were appointed as fire officers. Under the Commissioner of Lands in Blenheim they organised local fire crews (trained and equipped by the Forest Service) for first-action work on structural or vegetation fires. The Picton Fire Brigade provided a crew that could be dispatched by launch to operate Forest Service pumping equipment that the brigade housed.

The arrangement extended through Queen Charlotte, Kenepuru and Pelorus Sounds and out to D'Urville Island. By 1965 some 30 fishermen, farmers, freezing workers, tradesmen and launch-owners fought fires each holiday period; wardens in remote areas issued fire permits and held gear.[181] In the mid-1970s a floatplane provided speedier access to remote fires.[182] Under the 1977 Forest and Rural Fires Act the two fire authorities for the Marlborough district, the Forest Service and the Marlborough District Council, became the first to merge as they created the Marlborough North rural fire district, to ensure uniformity in fire protection across the wide and geographically challenging area of the Marlborough Sounds.[183] The annual Sounds fire conference engendered a great atmosphere where senior Forest Service staff met and trained with the volunteers and wardens, caught 'a few fish and had a few drinks'.[184]

We have already noted the active

A monsoon bucket belonging to the Canterbury High Country Fire Team. Acquiring one's own equipment and supplies was (and is) an essential part of successful volunteer effort.

ANZ, AAQA, 6506, Box 21, folder 8/25, 432.3, M12642

involvement that the Forest Service had with small rural volunteer fire brigades and, from the 1970s, with what became known as fire parties. (As discussed in Chapter Five, those groups dealt with vegetation and structural fires; the bush fire forces were dedicated vegetation fire-fighters.) The Service provided training and equipment so the volunteers could become first-strike forces for vegetation fires, with the Service there as a back-up. The volunteer role grew steadily. In North Auckland from the mid-1950s the Forest Service had

A hillside of gorse, just above the road over the hill to Wainuiomata, offered cover to regenerating natives. This photo also points to the blatant disregard for Forest Service publicity.

ANZ, AANI, 6506, Box 21, folder 8/30, 435, M11,357

agreements, formal and informal, for the Fire Service local brigades to attend scrub fires; it thought flexibly on ensuring potential additional manpower, and was relaxed about which force provided equipment as long as the fires were put out.[185] By the 1970s in Canterbury, depots with Forest Service trailers and Wajax pumps written off to the local volunteers were dotted all around the Rakaia River.

In the late 1970s John Ward was in charge of Tairua Forest, one of many Fire Service managers who collaborated closely with the local volunteer brigade. That brigade, generally because of proximity, would arrive at the fire first and leave only when the Forest Service men had taken over and no longer needed them. (Ward, conscious that volunteers had paid jobs to go back to, developed a policy of earliest possible release.) The Forest Service station at Eyrewell Forest did not have many men on hand in the weekend so the Oxford brigade had a couple of packs of forestry hose on appliances; the Rangiora and Woodend volunteer brigades had hose on two main appliances (and retained it after the Forest Service was disestablished because they were a first-call force in protecting Department of Conservation reserves in the Ashley River bed). Volunteer brigades at Kekerengu, north of Kaikoura and at Springs Junction, which first came together decades ago, are still active – and their ex-Forest Service gear, now 40 or 50 years old, is still in use, as it is around Springfield and Sheffield (though now owned by DOC).

The Forest Service's relationship with volunteers was not always entirely comfortable. Agreements and understandings did not always work smoothly. Some volunteer brigade groups, more committed to a professional approach, were simply better than others; at other times the Service had issues with volunteer management of personnel and recognition of lines of command.[186] Moreover, the Fire Service Commission's reluctance to fund fire parties after 1975 slowed their formation until 1990, when financial incentives helped persuade the local authorities of their worth and fire party numbers rose steadily.[187] The 'vollies', as they are affectionately known, were (and are still) 'an integral part of rural fire control' in a number of areas.[188] 'They're diamonds, those guys,' commented Barber, 'and they don't get paid.'[189]

In 1971 forest ranger Bob Collier and conservancy officer Dave Saunders turned some 90 Christchurch-based fit volunteers into the Canterbury High Country Fire Team. Extensively trained and equipped by the Forest Service to assist rural fire authorities and to deal with vegetation fires only, the team rapidly became a specialist elite force, available to be helicoptered in and out of 'the back of beyond' with gear and supplies. They arrived at fires in white overalls, with special insignia to identify 'in charge' personnel. The 1980 Mt Oxford fire illustrates their role and their capacity. There was snow and the fire was jumping from ridge to ridge. The team was there for about six days putting out hot-spots, cutting three helicopter landing pads close to the fire scene, patrolling, and felling the spars that were giving off 'sailors' (showering sparks and flaming vegetation). Today DOC runs the team. Some of the original members are still among the (now depleted) muster, and many have earned the venerated Fire Service long-service medal.[190] Concerns over rank vegetation resulted in a similar team being set up in Marlborough in the early 1980s.[191]

In other areas locals took the initiative to address growing environmental concerns to conserve bush and keep gorse as a protective cover for regenerating natives. (Today other flammable vegetation is recommended in place of highly flammable gorse.) By the late 1960s some Wainuiomata residents recognised that the local brigade was insufficiently manned and equipped to deal with major bush or scrub fires.[192] In February 1970 scientist Bill McCabe called a public meeting, and the local beautifying society was transformed into the Wainuiomata Bushfire Service. Its 24 members, who devised their own equipment and methods,[193] were the first dedicated volunteer rural bushfire force (as distinct from a volunteer rural fire party, which attended structural and vegetation fires). Girling-Butcher ('and if you weren't co-operating with [the Forest Service] you weren't anybody'[194]) stretched the provisions of the 1955 act to provide the group with a legal basis and, later, a legal definition under the 1979 regulations. Now known as the Wainuiomata Rural Volunteer Bushfire Force, it is still very active.

At about the same time Arnold Heine set up the rather more ad hoc and short-lived Hutt Valley Conservation Society with an aim of protecting regenerating bush on Wellington's gorse-covered Eastern Hutt hills,[195] a 'vigilance committee' in the Haywoods Hill area kept tracks open and worked to develop local school children's regard for bush. The Forest Service approved, provided training and equipment, and in 1971 chaired the ad hoc Wellington Hills Fire Prevention Committee, formed to co-ordinate its own activities with those of local bodies, fire brigades and volunteers in preventing and controlling fires around Wellington.[196] By 1977 similar volunteer groups had been set up in Eastbourne and the Taupo area.[197]

Fighting and preventing forest fires and administering the fire districts did not come cheaply. Forest Service annual costs grew steadily from £8149 in 1948 to $17,831 in 1966, then rose sharply to $79,867 in 1973 (representing a growth in costs in 2011 values from $561,698 to $870,046).[198] In the 1985–86 financial year the Forest Service spent an estimated $5.9 million ($10 per hectare) and private forest owners an estimated $3.8 million ($7 per hectare) – a total of $9.6 million.[199] The implications of such costs in relation to efficiencies and commercial loss are discussed in the final chapter.

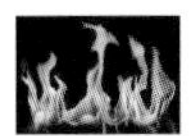

The Forest Service used volunteers in part because aspects of the legislation had not always worked as envisaged. Implementing the 1947 act initially seemed to meet with success. The creation of fire districts was never going to solve the whole problem, but they did put the responsibility where it should be: on forest owners. Government departments, county fire authorities and catchment boards generally appeared to be co-operating.[200] Forest Service staff were gaining representation on the Soil Conservation Council, the Rivers Control Council and 12 of the county catchment boards. Seven years later the Forest Service reported that county authorities were carrying out their responsibilities under the act more capably as their experience increased, taking greater care in issuing permits to burn and mostly carrying out inspections before issuing the permits. By 1955, 75 rural fire districts covering 1.82 million hectares had been constituted under the new act, and local bodies, county councils and rural fire committees had appointed rural fire officers. The Esk Rural Fire Committee was formed in 1963 from local Forest Service staff and representatives of forestry interests in the area. It met monthly to work together – on a shared-cost basis – to eliminate or reduce fire hazards, issue fire permits and develop fire-suppression plans. The Esk fire in 1968 gave the committee further motivation. The Maniototo county's rural fire control system was a model of what could be done to attack and control a fire in its early stages.[201]

After 1955 the Forest Service stopped regularly reporting on fire district formation. Although large exotic forests continued to be covered,[202] by 1977 only 17 rural fire districts remained. Many soil conservation and territorial authorities had only a limited commitment, if any, to their fire-prevention role and had done little or nothing to organise it, often over some considerable time. County councils were failing to take substantive moves to properly equip their local brigades for vegetation fires and were not producing fire plans; narrow agreements precluded local brigades from participating in anything other than property fires. In short, they were over-reliant on the Forest Service and did not appreciate the significant role they themselves had. In many instances the problem was that councils were reluctant to rate citizens for expenditure that produced benefits in terms of what had *not* happened (devastating fires) rather than what had, but forest owners, also unwilling to incur fire-protection costs, 'added their voices to the general cacophony' of council meetings to prevent effective fire protection.[203]

Soil conservation councils, whose members were generally farmers, often ignored conflicts of interest and made decisions on the basis of mateship. They tended to issue permits to burn that were valid over a number of years, they did not inspect the properties, and did not require intentions to burn to be notified. Nor, says Ward, summing up a general perception, were small forest owners always 'as fire conscious as they should have been in terms of looking after external firebreaks or keeping roads open'.[204] Hard-pressed private forest managers on land adjacent to state forests in particular were likely to rely on the Forest Service and its self-interest in providing assistance and guidance.[205]

Whatever the reasons for the Forest Service continuing to take the front line at all fires, the expenses mounted. The Service fought many smaller fires without sending out bills. Even when the costs were sheeted home (the Service became increasingly hard-headed at fighting fires first and extracting money later), there was additional expenditure involved in doing the costings and chasing payment, and in removing men from more productive work.

By the late 1960s the Service was clear that the 1955 legislation needed amending. With the increase in planting by state and private companies, forests were growing as an economic investment just as fire district numbers were reducing. Land use was changing. Cities and towns were expanding into the countryside and demanded more intensive control; retired land was seen to increase the high-country fire risk. Nor was the risk only from fire – vegetation needed protection from erosion and animal pests. The rural fire district operation needed to change to enable soil conservation areas to be gazetted as rural fire districts, thus facilitating better administrative and financial arrangements.

A new bill was introduced in 1967. But, as Girling-Butcher reported a year later, the bill '[i]n the homely figure of speech ... aimed to patch up a few leaks in the old tyre in order to keep us going until a new tyre could be secured, but quite a number of people told the Committee what kind of new tyre there should be, and that they did not like the old tyre'.[206] Farmers in particular construed the Service's attempts to control fire in the public interest as inimical to individuals' freedom to work the land, and Federated Farmers launched a tirade of essentially anti-forestry statements.[207]

There was further consultation. A redrafted bill was introduced in 1971 but 'a completely unexpected collision of interests left the Select Committee, the Minister and [the Forest Service] in a most embarrassing position'. The Forest Service maintained its optimism that 'common rather than sectional interests [would] prevail'.[208] Moving cautiously so as to avoid further confrontations between farming and forestry interests,[209] it introduced another lengthy process of consultations. The Mt White fire inquiry, with representatives of the Soil Conservation and Rivers Control Council, Lands and Survey, the Forest Service, the Counties Association and Ministry of Works, helped to secure a broad agreement on the direction of the legislation.

A further redrafted bill, introduced on 17 June 1977, received only 12 submissions. The Defence Department believed it needed to be exempted from a ban on activities such as firing live ammunition and lighting fires during fire bans, and it had technical concerns about how restrictions on spark-arresters would affect its Sioux helicopters.[210] Local bodies pointed to a number of mainly technical issues, and Federated Farmers raised further objections. But, in what has to be seen as a tribute to Girling-Butcher, the farming lobby noted its pleasure in participating in 'a model exercise in legislative consultation'.[211] Further amendments accommodated the Defence Department and some of the farmers, and the Forest and Rural Fires Act 1977 passed into law in November 1977, although it did not come into force until April 1979, to allow regulations to be developed.

The act was to provide practical measures to detect, control, suppress and extinguish any vegetation fire. The basic features of the earlier fire regime remained. Fire districts continued to be administered by fire authorities: the Minister of Forests in relation to Crown lands; the Ministry of Defence for the gazetted Defence rural fire districts (used for training exercises); local bodies (borough and district, as well as county councils); and private owners in specially formed rural fire districts. Other elements that remained were the permit system for burning within fire districts and Forest Service responsibilities for weather prediction. However, vegetation was now widely defined to include all plants, whether alive or dead, and whether waste, rubbish or peat and fossil fuels within 20 metres of the surface. Careful exemptions to bans on open-air burning (to exclude abattoirs, crematoriums and barbecues – newly on the market) were developed.

The act applied to all land except that under the care of the Fire Service Commission. County, rural and state fire districts would provide locally based fire control. Specific provisions applied to particular areas. Owners of large private forests had to do their own fire-fighting under fire authority supervision; unusual historical or geographical features were given special protection; a safety margin, generally 1.5 kilometres wide, applied around all forest areas and state areas, where permits to burn had to be sought, and in which the forest owners were responsible for fire-fighting; and in periods of extreme hazard, bans on all burning could be notified for affected areas. By arrangement, the act could also be applied to urban vegetation.

The rural fire authorities appointed fire officers within their districts. Co-operation between fire authorities was given statutory backing, and arrangements could be contracted with Fire Service or industrial brigades to supply equipment or assistance. Rural fire mediators (effectively arbitrators) were a new development. Envisaged as a sort of local ombudsman nominated for the occasion, they would be available to decide disputes promptly, informally and more cheaply than going through the courts.

General expenses were usually met by a levy on each fire district. Fire-fighting costs were to be paid either by the person responsible for the outbreak (often defined as the person on whose land the fire started) or by a levy on the property protected. Limits on levies (periodically reviewed) were imposed to reflect the value of the property. There was also state assistance. Other provisions required the public to help, removed liabilities when fire-fighting and dealt with regulating.

The act reinforced four principles that are still important today in differentiating the tasks of rural and urban fire-fighters. First, the new act was concerned with the safety of vegetation, not buildings. As vegetation fires spread fast, mobility and special protective clothing were essential. Second, rural fire safety was not simply about suppressing a fire: it involved 'fire control', including weather-watching, training and equipment, public education on fire danger, supervising firebreaks and ensuring public safety. The approach that Fire Service Commission brigades brought to their job was inappropriate for rural fire. It was important, therefore, that the two groups, with their own specialities, remained separate (though the act enabled each to come to the aid of the other, whether by longstanding arrangement or in an unforeseen emergency).

Finally, the act sought to balance the centralisation essential for achieving the wide measures needed to control rural fire with independent, decentralised local control. Having mediators on hand who were required to act in the public interest was envisaged as a means of balancing state and local control.[212]

The 35 pages of highly detailed regulations, around which there had been 'cordial, extensive and practical' consultation,[213] came into force on 1 April 1979. Many were updated from the 1955 regulations – the safety of apparatus, the regularity of its testing, the use of spark-arresters for equipment ranging from diesel-powered motor vehicles to hard-fuel stoves. Fire-protection measures in sawmills were mandated. The capacity requirement of chemical fire extinguishers was increased and the situations in which they were to be used were specified. People without reticulated water had to keep their supplies clear of anything that could block pumps, or suffer considerable fines. Provisions for gas-producers were now in force for the whole year. Wages for fire-fighting service reflected attendance at improved fire-fighting courses. Fines for offences were adjusted for inflation.

Among the new provisions were definitions of three different types of fire (which previously had not been defined at all). Sundry fire-control measures now deemed that adequate firebreaks included roads, green crops, top-dressed hill and high-country land, water, sheer walls or rocky faces clear of vegetation, river flats, tidal estuaries and mud, unless a rural fire officer or specified fire plan directed otherwise. New provisions for the hill and high country set out the Forest Service's co-ordination role and responsibilities, as well as prescribing how fire authorities were to compile fire plans for these areas and co-operate with catchment authorities, and what their responsibilities for small pockets of rural populations were. Rural fire authorities were to assist, train and finance volunteer rural fire-fighters (carefully distinguished from the volunteer and industrial forces registered under the Fire Service Act 1975). Further provisions addressed volunteers' training and protective clothing.

The act's implementation proceeded fairly smoothly, in part because of the extensive prior consultation, but also because forest managers were getting better at balancing the letter of the law with its practical implementation. They now had more precise weather data, along with a greater appreciation of the fact that rural fire could be anticipated, unlike urban fire, which could start anywhere at any time. And land preparation, helicopter weed-spraying and track-making with heavy machinery had contributed to a broad knowledge base.

In the mid-1970s Fletchers' Taupo forest protection manager, Ian Millman, was able to balance his fire responsibilities with production deadlines over the high-risk months. He, like others, worked to reduce the hazard and minimise the risk. No chainsaws operated in bush after lunch because logging crews had started at 4am so by then they had worked a full day. Tractors could not work after 3pm, but they had already had time to pull the logs out. Loading and cutting on the skid could go on until 6pm, and someone stayed back an hour to ensure the site was safe.[214] At Kawerau, Don Geddes enlisted New Zealand Railways' co-operation, expecting them to comply with legal obligations to spray their line only if Fletchers had notified a high-hazard period.[215]

Other parties responded as expected to both the legislation and the regulations. B. J. Jurlina of Northland had 'never heard of anything so silly' as what the legislation had set up,[216] and when the regulations came into force a Waikato county chairman 'who loves to see his name in print' said he 'should not be expected to check gear or ensure fire fighters were properly dressed'. Girling-Butcher took him in hand.[217]

More substantive were the county authorities' concerns over costs and the farmers' anger over the levies. They could be countered, the director-general hoped, by arguments that they no longer needed to contribute to the New Zealand Fire Service, that if they complied with the regulations they should experience a reduction in costs through improved efficiencies and co-ordination, and that ultimately they stood to gain.[218]

Farmers worried about insurance and liability. In urban areas insurance levies financed urban fire authorities, whereas in country districts many farmers paid both insurance and a local authority levy but still did not get the same protection. In addition, where there was fire on their land, farmers could be covered by their own fire insurance. But they were also liable for fires that escaped from a contiguous forestry block – and any insurance they might get to cover such circumstances would have impossibly high premiums. Counties' willingness to make proper provision for rural fires (which, given the distances involved would include small trained helicopter crews) was also treated with a certain scepticism.[219] Why not have a central fire-fighting fund, the farmers demanded.

The same issues arose when the act was reviewed in 1981. Should responsibility for rural fire protection be a national issue with national funding rather than a local concern? Should the general taxpayer help fund the substantial costs of modern machinery? The system allowed small companies and other landowners to 'free ride' because larger companies funded the rural fire committees. Some local authorities were not developing fire plans, at a time when the fire threat was growing because of greater public use of rural lands, silviculture practices that thinned to waste, leaving trunks to rot, and the increasing use of industrial equipment. Forest owners were already paying $5000 for essential publicity. Should others contribute?[220]

A rural fire-fighting fund was established to take effect for the 1985–86 summer to at least partially resolve those issues. Funded from insurance levies on farm buildings, it was a way of channelling back that levy to those households without any Fire Service assistance.[221] But by then other forces were at play. The Labour government's election in 1984 heralded in a period of profound economic and social change, and the Forest Service came under heavy criticism. Conservationists disagreed with how it managed native bush; financial analysts and businesspeople saw disappointing returns from exotic wood sales and criticised the Service's planting and employment programmes as not being 'core business'. Economic thinking dismissed a role for government in business. If it were to be there, it should be so only at a remove.[222] On 31 March 1987, without adequate thought being given to an effective rural fire-fighting structure for the future, the Forest Service was disestablished.

Chapter Seven

Fragmentation, re-formations and the future: 1987 to the present day

New Zealand's weather can originate far from its shores. In the eastern Pacific a warming of ocean temperatures heralds in El Niño conditions. Thousands of nautical miles away, New Zealand's prevailing westerlies blow with greater intensity than usual, and drought hits the eastern parts of the country. Canterbury, in the rain shadow of the Southern Alps, can be badly affected. In 1987, the year the Forest Service was disestablished, El Niño conditions brought drought to the region, and the dry, hot, windy conditions continued into 1988. By that summer the fire danger was extreme.

The fires over those two years – at Oxford, Woodend, Dunsandel and Bottle Lake – are in the annals of New Zealand's rural fire-fighting history. Dunsandel was a big fire; the threat the fires collectively posed demonstrated the extent to which, with the loss of the Forest Service, fire-fighting co-ordination and resources were gone. Initially the newly formed Department of Conservation (DOC) lacked the resources to in any way replicate that role; nor did it – or Timberlands (charged with looking after the commercial forests) or the Ministry of Forestry (basically a policy agency) – have a mandate to do so. Suddenly local authorities, with no Forest Service buffer, were faced with the full burden of their responsibilities as rural fire authorities – a role in which, generally speaking, they had not excelled in the past.

The fires resulted in *A Review of Rural Fire Services in New Zealand*, Gerald Hensley's politically astute, widely admired and practical report on the country's rural fire services. Taking advantage of a territorial authority restructure, building on existing structures and tweaking existing initiatives, he addressed the key issues of standards, co-ordination and funding. The organisation of rural fire-fighting today is largely a product of his work.

Today, partly because of changes in forestry management practice, rural

Wainuiomata women's team, 2007, Wajax pump competitions.
National Rural Fire Authority

fires damage mostly grasslands, not forests.[1] But 5 per cent of the fires still cause 95 per cent of the damage, and people are still responsible for most of them. Fire-fighting techniques have changed little over the past 30 years, though protective gear gets better, equipment more ingenious, the vehicles more varied and versatile and the big machinery bigger.

However, the environment in which New Zealand's 3000 volunteer and part-time rural fire-fighters work has changed. The New Zealand Fire Service, rather than the rural fire bodies, now provides that first-strike capacity, putting out 90 per cent of rural fires. The possibility of integrating urban and rural fire services in the interests of efficient emergency management across the country has been debated, but the government has not pursued this option.[2] Consequently 'the urban and forest and rural skill set and the built versus land environment remain, from a policy perspective, like chalk and cheese'.[3] Meanwhile, to improve the efficiency and safety with which people are deployed, and keep equipment well maintained, the National Rural Fire Authority has introduced new training regimes and monitoring systems and is pushing for the amalgamation of rural fire districts, a proposal seen by some as striking at the heart of the volunteer forces. There is another even more controversial issue

– the role of fire as land-management tool in an increasingly carbon-conscious world. One thing has remained constant – the work of the volunteers, largely unpaid and unsung.

In 1987 Canterbury, long known to have one of the country's highest continuous fire risks over the summer, struggled with hot, dry and windy conditions. By the beginning of February 1987 the Canterbury forests in particular were tinder dry. Where they were open to the public, rigid fire bans were in place. Forest operators were stopping work early, staff training was intensified and fire engines were brought in.[4] On 2 February a burn at Burnt Hill got out of control. It was suppressed, but perhaps a hot-spot reignited because at midday on 4 February another fire, fanned by gale-force nor'westerlies, was sweeping through farm properties at up to 100 kilometres an hour, spreading great clouds of smoke over Christchurch's northern suburbs and threatening Eyrewell Forest.

The Selwyn Plantation Board, which 'had adopted the role as the de facto local fire authority', did not send its men because they were fully occupied with another relatively small fire at Burnt Hill for about a week.[5] Fire brigades from Oxford, Cust, Sheffield, Coalgate, Kirwee, Darfield, Rangiora, Dunsandel and Burnham were there in force, but some had already been battling grass fires near Darfield and Greendale. The size of the fire and the already stretched resources left little option but to call in the Fire Service Christchurch area commander, Michael Burke. Regional command posts (to deal with the risk to Eyrewell Forest) and local command posts were quickly established. Eyrewell Forest resources were focused close to home; elsewhere the Fire Service, unable to actually stop the fire, had appliances moving around the periphery, updating the situation so that Civil Defence could declare a state of emergency if necessary. The wind made helicopter operations out of the question, and dust reduced radio communications to intermittent.

Burke set up his control headquarters at the Oxford Fire Station. There the Red Cross, Salvation Army and the army provided food and refreshments. But as convoys of water tankers and 300 army, air force, St John and Red Cross personnel assembled, enthusiastic volunteers, whose cars clogged the roads, endangered themselves and refused to leave except under instruction from uniformed police. As the situation 'bordered on the chaotic', nature intervened. The wind drove the fire up against the Waimakariri River, swollen from rain in the Southern Alps, and that flank was quelled. By night the wind had dropped. The army and air force came in with shovels and knapsack tanks, and exhausted volunteer fire-fighters could be stood down. By 1am the next morning it was over, bar the dampening down. One farmer was in hospital with burns, homes and fences were lost, some 40 hectares of farmland was destroyed and sheep, trapped in their paddocks, were burned in heaps. A relief fund was set up.[6] It was like going back three-quarters of a century.

Burke considered that the tactics in this fire – having the Christchurch Fire Service up front and in command until the emergency was over – constituted a satisfactory operating procedure. He was going to need one. The following year would be 'the record windiest, sunniest, hottest and driest year'.[7]

On 17 October 1988 a Canterbury nor'wester cut off hundreds of hectares of crops at the roots or sand-blasted them with soil and debris. At Woodend, not far from Christchurch, a flare-up from an earlier fire suddenly raged and crowned. The Woodend, Rangiora and Kaiapoi volunteer fire brigades were soon outclassed. As some 200 people evacuated their homes, the Christchurch Fire Service was again called in. Burke worked to instil confidence in Woodend residents, then arrived at the fire to find 15 appliances from the Fire Service, DOC, the local authority and a four-wheel-drive club, and six water tankers. Once again the Red Cross and the Sallies were on hand with food and drink for some 150 people. A large swimming pool and a water-race provided a water source, from which hoses formed a fairly weak ribbon along a road that served as a fire-break. As darkness fell and the fire roared towards the fire-fighters, the situation seemed ominous.

Then the wind swung to the east, the fire front dissipated and an effective attack could be mounted. Fire-fighting went on all night but by morning the Fire Service was able to withdraw, leaving the principal rural fire officer of the Rangiora district in charge of mopping up among the 60–70 hectares of standing pine through which the fire had driven.[8]

Worse was to come. By December, Dunsandel Forest was dying back from drought. Bill Studholme, then Selwyn Plantation Board's chief executive, remembers that they 'had worked really hard to minimise any risk and had spent six to eight weeks sitting on the edge of our chairs' lest the slightest spark be fanned into fire.[9] On 8 December an electrical storm hit the area and a fire quietly kindled. Extraordinarily, although the fire was seen in the morning, it was not reported until 2.15 that afternoon. By 3pm local volunteer brigades, the Defence Department, counties' and contractors' earthmoving machinery, plantation board staff and four helicopters were on the attack. Local farmers, too, turned out: 'We had about 150 at Dunsandel, and even if they only had a couple of back-pumps, a few sacks and a slasher and they turned up in gumboots and teeshirts, you could send them off knowing they knew what to do.'[10] By 11pm the fire was confined to some 25 hectares. Over the next few days, with the fire apparently well contained, bulldozers created windrows out of the burnt wood and drove in a firebreak.

The high temperatures and low humidity persisted, and by 12 December the situation was explosive. By mid-afternoon 50-knot nor'westerly gusts were blowing fine dust from the previous firebreak work into a dust storm in which men could not move. An hour later a sudden sou'west change, heralded by 50-knot gusts, rekindled the blaze. Five hours later fire was burning over 160 hectares, having 'hopped over five-chain [100-metre] firebreaks, got into the forest and started to move towards Dunsandel. One line of fire could have jumped the Main South Road, got into a ripening crop and threatened Leeston.'[11]

On the morning of 13 December it was clear the situation was beyond local capability. The Selwyn Plantation Board called on the Ministry of Forestry and sought other assistance. Headquarters were set up at the board's office in Darfield, and a field base upwind of the fire, near a major water-race that would become the main water supply. A fire strategy meeting determined to fight the fire under the Forest and Rural Fires Act rather than the Civil Defence Act, allowing those with local knowledge and a better understanding of rural fire-fighting to be in charge (in the event the county engineer, who was the principal rural fire officer, delegated the position to Studholme). The decision also took the weight of responsibility from the local elected representatives, who were 'a little daunted at the prospect of declaring the civil emergency needed to invoke the Civil Defence Act'.[12]

A set of fire orders was drafted, the fire area was broken down into sectors through which daily written instructions could be relayed, and a command structure was established and quickly disseminated through the sectors. Murray Dudfield at the Ministry of Forestry had the Meteorological Office fax four-hourly forecasts specific to the Dunsandel area. County overseers and Christchurch City Council staff took turns at being the 'machinery boss', providing that vital co-ordination of resources. Helicopters were contracted in, though the high wind meant they could sometimes be used only for surveillance. DOC offered the vast quantities of fire suppressants and retardants needed to get the fire under control; like the Ministry of Defence supplies, these resources had to be freighted overnight from Auckland. Four times the number of heavy bulldozers, diggers and water carts as had been used in the first flare-up had to be kept operative in spite of the dust. One Christchurch City Council mechanic had this responsibility; another did oil checks and minor repairs. Fuel and oil had to be called in; a borrowed DOC mobile repeater provided good communications between headquarters and those on the ground. The army was there to fight the fire and the Salvation Army was there with food – 'indeed, without [their] efforts, the Dunsandel fire would have been a much more miserable place'. The Red Cross also provided food, drink and companionship.

This picture has been captioned 'Blokes standing around'. But despite the urgency that attends fires, not everything can be done at full speed; updates need to be given and briefings confirmed.

Bill Studholme

But as that new fire flared, the situation got more complicated. 'The volunteers and farmers had to get back to work and to their farms, and although the army had been commandeered in, they had a big military manoeuvre over January and a lot of their men were going on leave.'[13] The situation was so serious, the board decided, that it had to be handled at a political level. Bill Studholme

> stirred up Jenny Shipley, our local MP. She organised Peter Tapsell [Minister

Feeding fire-fighters is an essential task but a huge logistical problem at any fire. At the Dunsandel fire, says Bill Studholme, the Salvation Army and Red Cross undertook this vital service spectacularly well.

Bill Studholme

of Forestry] to come down, and I visited Mike Moore (the duty minister) at his home. Moore obviously impressed [Prime Minister David] Lange (away at the time) with the seriousness of the situation and Tapsell realised he had to move. He got us a number of new military fire appliances that were sitting in Wellington and which we badly needed for fighting the fire.[14]

Treating the hot-spots is always a laborious task. The difficulties of fully extinguishing fire were exacerbated by the particular situation in the Dunsandel Forest. An earlier forest had been harvested, the slash had been windrowed and the area stumped before second-rotation forest had been planted in that area. Although the still-standing trees were badly burned and scorched, the fire there could be dealt with by monsoon buckets. In the windrowed area, the still-burning hot spots could be exposed and extinguished only by heavy bulldozers pushing the slash back. It was, Bill Studholme says, 'an excessively expensive operation'.

Bill Studholme

On 22 December the fire was finally contained. Then on 23 December it rained. 'You can hear the relief in my voice,' Studholme told the press.[15] But 164 hectares of timber, valued at $440,000 had been destroyed; the Fire Service alone had contributed 4265 staff-hours over the two weeks – and it was not involved subsequently in detecting hot-spots.[16] For over three weeks a helicopter, with the Timberlands infra-red camera mounted on its skids, took off at first light each day, dropping ribbons to mark the spots for ground crew to extinguish. 'Mopping up', says Studholme, 'is as demanding as actual fire-fighting. Nine-tenths of a fire can be underground and in some weather conditions hot-spots can last a very long, long time.'

There were other fires that summer. Bottle Lake Plantation flared on 21 December 1988. Seven days later, at Worsleys Spur in the Port Hills, 'one of the most threatening fires of the whole summer season' was fought by Fire Service units, a hose-layer, command and control units supported by helicopters, DOC high-country fire teams, 100 Civil Defence workers, the army and air force, volunteers and the seemingly indefatigable Salvation Army and Red Cross. It took 26 hours to get it under control. On 13 January 1989 a further fire on the Dyers Pass Road again took considerable resources. The helicopter alone cost $25,000; fortunately for the Christchurch City Council, the air force waived charges.[17]

In April 1987 the Forest Service was split among three government agencies: the Forestry Corporation (ForestryCorp), the Department of Conservation, and the Ministry of Forestry. ForestryCorp, a state-owned enterprise, managed the state's exotic plantations; DOC managed the indigenous forests. The Ministry of Forestry, a policy agency dealing with compliance, protection and enforcement, administered the Forest and Rural Fires Act 1977, its 1979 regulations and the 1980 rural fire regulations (to be replaced by the 2005 regulations), which outlined the roles and responsibilities of rural fire district entities. The ministry promoted fire-control measures, declared regional fire emergencies and co-ordinated annual fire-control meetings. It could advise and assist fire authorities, including providing fire training courses for them, preparing and gazetting notices and ensuring that authorities could buy 'suitable' literature on preventing fire. There was also the usual range of governmental functions, in this case around monitoring rural fire authorities, advising on fire mediators' appointments, and assisting the New Zealand Fire Service vet claims on the recently established rural fire-fighting fund.[18]

The ministry had no hands-on rural fire-fighting function. That was left to the rural fire authorities, rural fire committees and rural fire parties. Urban brigades, under the New Zealand Fire Service, had no statutory obligation to help fight rural fires but often did so under formal or informal arrangements.

Some 165 territorial authorities were the rural fire authorities for their respective areas; the Ministry of Defence was the authority for eight Defence rural fire districts; DOC was the authority for its 13 conservancies; and ForestryCorp was the rural fire authority for its own forests. Rural fire committees could establish rural fire districts for additional protection. In the Waitakere Range, for instance, the local council wanted to protect land for water catchment, recreation and conservation; in the dry Moutere Hills orchardists were conscious of high fire danger. Some private forests were declared fire districts.

Rural fire parties could deal with small local structural and/or vegetation fires; they had various sources of funding and some were registered under the Forest and Rural Fires Act. The New Zealand Fire Service, because it was alerted to all types of fire by 111 calls, and because it had agreements with various rural fire authorities, often provided at least an initial attack on some vegetation fires and could call out helicopters and bulldozers.[19] The major private forests were considered sufficiently equipped to manage fires within their own boundaries and neighbourhood.

However coherent that structure looked on paper, it was highly flawed because it failed to recognise the extent to which the Forest Service 'had, in effect underwritten the whole system for decades'.[20] The Service had provided standards, training and manpower; its policies of writing off serviceable equipment enabled local authorities and fire parties to meet those standards; and its workshops maintained fire-fighting equipment from all around the country. Whenever possible it responded to all fires, irrespective of whether they were in a state area.

Studholme's experience was not atypical. The Selwyn Plantation Board had not created a fire district for its forests but had worked instead on helping develop a local resource capability whose manpower and equipment were otherwise constrained. Yet if there was 'any reasonable-sized fire ... Joe Levy, the regional Forest Service conservator, would be on the phone – "What do you need, how can we help?" If you said, "We're all right on our own," he'd come back, "Well actually I'm standing outside your gate and we're coming in now!"'[21]

When the Forest Service went, so too did the structure that had provided back-up, co-ordination, standards, equipment, training to many local bodies and forestry-linked organisations, and a means of meeting costs. This was not a role that the new government agencies, the rural fire authorities or the Fire Service could fill. The Ministry of Forestry employed some experienced fire managers and fire-fighters, but with reduced staffing and funding and a circumscribed role its options were limited. It advised local authorities especially on cost recovery, established a small number of rural fire co-ordinating committees to help fill the gap, and could, as at Dunsandel, send a senior officer to a bad fire – though the ministry had no equipment or manpower.[22]

ForestryCorp, because day-to-day activity made its forests more fire prone, had received about two-thirds of the Forest Service's fire-fighting equipment. But the 5500 Forest Service men available to fight fires in 1985 had dropped to 2600 under ForestCorp. By 1988 contractors were tending to employ fewer, less experienced men, and fire-fighting experience and 'organisational culture', Hensley considered, were likely to decline further when ForestryCorp was sold. (If the newly created SOEs were to act as businesses, the government needed – among other things – to establish the value of their assets to gauge an appropriate commercial return. Disagreements between government agencies about that value, and budgetary pressures, resulted in ForestyCorp, among others, being offered for sale to allow market mechanisms to determine value. The sale process began about 1989.)

The decision during the Oxford fire to have Eyrewell Forest's men guarding that forest was perhaps not surprising. ForestryCorp, under its new commercial imperatives, was required 'to act as a business' and preserve its plantations. It was 'not equipped and not prepared to underwrite' local fire-fighting capacity.[23]

The Department of Conservation's lack of equipment and suitably trained personnel hamstrung it over 1987–88. In 1986 Kerry Hilliard, the Forest

Service national fire equipment inspector (and in 2012 DOC's national fire co-ordinator), had to 'go through all the Forest Service fire equipment assets and identify what should go to [DOC] and what to the Forestry Corporation'. Acting without any instructions but using his extensive knowledge of what was where, he tried to allocate equipment according to fire risk. But whereas DOC was still to be established, 'some in Forestry Corp had been identified in their roles and when it came to their forest giving up some pumps and ancillary fire equipment to be allocated to DOC, no way were they going to bloody do it'. Pleading at a ministerial level was to no avail. DOC got sufficient equipment only because Hilliard quietly forgot about the substantial fire equipment at the Forest Service bulk store in Palmerston North, turned a blind eye to the senior ranger who hid a specialist fire engine in the bush to ensure it went to DOC, and acted fast on a tip-off from Kaingaroa Forest headquarters to commandeer equipment from the fire station where an understanding senior fire officer had left the door open.[24]

That manoeuvring went on until about 1988. It took a letter from Girling-Butcher to ministers seeking additional technical support for DOC before two senior fire officers (John Barnes and Terry Stuart) were appointed in 1987 to help train DOC staff, sort equipment and set up policies and fire plans. (The positions were lost in the 1991 review.) Initially most field staff had little or no background in fire, though as fit, experienced outdoor operators they quickly gained the necessary skills.[25] 'By necessity, [DOC also] established volunteer fire parties and formalised agreements for mutual assistance with some other rural fire authorities.'[26]

The territorial rural fire authorities had always differed widely in their capacity and the extent to which they met their obligations. Too often the job was left to the county engineer, who didn't know about fire and didn't want the responsibility. Not everyone had the wit, as one man did, to sign blank sheets so an experienced outsider could issue the orders. Forest fires 'scared the local authorities to death'. Elected councillors did not relish voting for an 'unpredictable, unbudgetable activity' that messed up their budgets. The all-too-frequent result? Inadequate rating provision, the county water tanker masquerading as the fire engine, lack of equipment standards or ways of checking adherence to them, and substandard gear that was sometimes incompatible with that of the neighbouring authorities.[27]

Furthermore, without the Forest Service co-ordinating the multiplicity of rural fire authorities there were no rules to cover what happened when fires crossed boundaries and this became problematic. In the past, 'if either the Forest Service or the Selwyn Plantation Board could drive to someone's fire in half a day, then we would. It wasn't all altruism – we were all well aware that any of us might need a helping hand one day.'[28] Now, as the authorities argued over who should bear costs (sometimes delaying critical action) and their fire-fighters fought fires 'with a hose in one hand and a calculator in the other', that previous system of 'local self-help' was further eroded.[29]

Nor could the Fire Service provide more than limited assistance. Under the previous regime the relatively flexible boundaries between urban and rural fire-fighting had enabled the Fire Service to provide pumping services or back-up; by 1987 better road and communications had extended its reach and it had responded to some 4500 calls that year.[30] But that assistance depended on mostly informal local arrangements rather than formally assigned roles. Moreover, the Fire Service's heavy protective gear and breathing apparatus, its vehicles designed for on-road, not off-road, use, its own urban responsibilities and the limited rural fire-fighting training its men received inevitably restricted its role.[31]

Over 1987 and 1988 Bill Studholme recorded that he and the Selwyn Plantation Board were concerned at the government's 'dismantling of the national forest and rural fire capability' and its 'inability to replace it with anything else'. He made submissions and raised his unease in various fora.[32] Yet his was a lone voice. Why did he get such a poor hearing? Why had those in charge of policy not thought more about how the change would affect fire protection?

Peter Berg, then leader of the Forestry Corporation establishment unit, says that 'in the early part of 1986 there were a whole lot of other things to focus on – people, commercial issues such as keeping the factories supplied with wood. These were paramount. There was not a lot of time, but there was also no mandate.' From the establishment unit's perspective little had changed:

> The big forest owners ... already ensured good levels of protection. The small forest owners and the local authorities were affected, but their forests constituted only 10 to 15 per cent of the total forest resource. DOC and the Forestry Corporation had a primary responsibility to ensure things were OK in-house.

From the ministers down, players did not appreciate what the changes and corporatisation implied: 'The expectation was that the agencies had been set up with the same capabilities. There was little understanding that business had to act as business.' Understanding was not helped by central government's failure to 'communicate with local authorities as they probably should have. I can't recall a single consultation.'[33] Nor were the local authorities the only ones who had no opportunity to have input into the new structure. 'Senior Forest Service personnel tended not to be asked to attend meetings, so the organisation could not engage with the discussions,' says Hilliard. 'It was hell bent on "[fixing] this bloody Forest Service who weren't really making much money".' Some saw those who were shaping the new structure as 'going to go away and show the government that they were business-orientated'. Hensley and Hilliard both remember that in that environment, so dominated by Treasury and its economic model, it was difficult to raise, let alone debate, what might replace the social functions the Forest Service had fulfilled. The fires were 'an unintended consequence'.[34]

The Canterbury fires, however, highlighted the deficiencies of the situation, in relation to both the immediate and the broader problem. The weather had been critical in quelling the fires. Without that luck, fires the size of those in Australia or the US could have erupted, creating situations that needed huge commitment of resources and expertise for which the country had neither financed nor planned.[35]

Two other initiatives had picked up on issues lost with the Forest Service's demise. In 1986 at an annual conference senior fire control officers expressed their concern about the impending loss of 'a carefully nurtured rural fire capability' and of senior officers' experience once the Forest Service was gone. On 2 February 1987 the Forest and Rural Fire Association of New Zealand was set up to maintain and improve efficiency in the rural fire-control field. The names of those involved – Neill Cooper, Murray Dudfield, Brian Ely, Kerry Hilliard, Alan Fifield, Don Geddes, Morrie Geenty – are a *Who's Who* of rural fire-fighting. At the time the association was an 'important mechanism that kept people in rural fire in touch with each other and kept them in the business';[36] today 'it is the one voice for the rural fire-fighters that can take their concerns to the top and be assured of an excellent hearing'. Although it is totally independent from government, its good working relationship with the minister has on occasion translated into financial assistance for specific projects. The association has also set the standard for the 'Rural Fire New Zealand' uniform that volunteers and staff of rural fire authorities across the country are adopting, and that the National Rural Fire Authority supports for use in international deployments.[37] In addition, the Forest Owners' Association, established in 1920 to lobby on members' behalf on matters such as tax and fire protection, took on responsibility for nationwide fire publicity.

But the problems were wider than either group could address. The Bottle Lake fire in the summer of 1988–89 proved to be the tipping point after which Prime Minister David Lange told Gerald Hensley, co-ordinator of Domestic and External Security, that something had to be done.[38] On 10 February 1989

Cabinet established a committee, chaired by Hensley and representing relevant government agencies, to identify the best structure for rural fire-fighting, the role of volunteers and the relationship of rural fire-fighting authorities with the Fire Service; and to address funding, standardisation of equipment, training, public education and the resources needed to realise the recommendations. Just over six months later the committee produced a report that essentially set up the rural fire-fighting framework we have today. The process of widely publicised public meetings and submissions had elicited views from groups as diverse as high-country farmers and isolated rural fire-fighting groups, then ascertained responses to a draft report, and Hensley's own diplomatic skills, which guided 'brisk' discussion and which the report's phraseology reflects, ensured wide acceptability.[39]

The committee became well acquainted with the fragmentation and the resulting difficulties that beset particularly small rural local fire authorities, and with the difficulties of the urban–rural interface. It heard, too, of funding problems that the local authorities faced when contending with periodic and costly conflagrations.[40]

The rural fire-fighting fund had come into being in the 1985–86 summer. Nominally it met fire-suppression costs above $2500 for fires where liability could not be established, or where a permitted fire that had met all conditions of the permit had nonetheless got away. Applications could only come from a county or district rural authority, or a local or regional authority rural fire committee responsible for districts with conservation, water catchment, recreational or similar non-commercial activities. However, by 1989 it was clear the fund was not resolving the funding problems of rural fire authorities. It was, the committee heard, too tightly constrained: of the $500,000 set aside for that financial year only $102,000 had been paid out. Nor was the system responsive. Economic pressures made 'all involved … more cost-conscious and more argumentative about getting their money back'. If fires crossed boundaries, authorities strapped for cash tried to apportion costs or dithered about getting in the big machinery while the blaze developed around them, becoming even more expensive to put out. Contractors, after being unpaid for months, proved reluctant participants in subsequent fires.[41]

The committee developed some basic principles for dealing with fragmentation and funding. It declared that fire was one of many land use/management issues, and should continue to be dealt with under the hitherto effective decentralised structure. Co-ordination at local and regional levels enabled the authorities to adapt their resources to all sorts of emergencies and fully utilise that flexibility and local knowledge. An effective, functioning and fully fledged system of volunteers reinforced local responsibility and, given the intermittent nature of rural fires, was cheap and effective. Responsibility at a national level was needed to develop standards and ensure inter-operability of equipment across the country; to ensure periodic audits on statutory obligations; to develop clear arrangements for apportioning costs and their recovery; and to undertake fire awareness publicity, maintain the fire weather index and organise training. These changes were to be fiscally neutral, with each tier meeting its own expenses. However, a small office of recognised experts would ensure that standards were kept up and the system maintained or adapted as needed.[42] 'We held our breath over the 1990–91 summer,' says Hensley.[43] But the system that the committee proposed worked.

At the heart of the system are local involvement and responsibility. Links to resource management legislation (since 1991) and local government, Civil Defence and emergency management (since 2002) mean rural fire is tackled on a wide front.[44] Some 3000 volunteer and part-time (including seasonal) rural fire-fighters employed by forest and rural stakeholders tackle the country's rural fires. They belong to 86 rural fire authorities (responsible for non-state land that is outside rural or urban fire districts), 11 DOC conservancies (responsible for DOC land and the one-kilometre fire-safety margin around some boundaries), eight Defence Force rural fire districts, and the 31 specially gazetted rural fire districts administered by committees whose members are mostly forest owners. Regional rural fire committees contain representatives from each rural fire authority. Linking local and national effort, these committees promote regional co-operation and assistance, and advise and assist the National Rural Fire Authority, the national body that co-ordinates rural fire management work in New Zealand.[45]

The National Rural Fire Authority is itself a division of the Fire Service Commission. Murray Dudfield was appointed national rural fire officer in 1990 and continues in 2012 to lead the 10-person, non-uniformed team, all with practical fire experience, who constitute the authority. This structure avoided the expense of having separate rural and urban authorities, recognised the particular nature of rural fire, the inclusion of portions of rural land in urban districts and the limited Fire Service experience in dealing with rural vegetation fires. The Fire Service Commission provided the strong and financially viable institutional framework that allowed the National Rural Fire Authority to develop the necessary expertise. It also meant that rural and urban services could progressively develop their existing co-operation and perhaps become more closely integrated.[46]

The National Rural Fire Authority sets minimum standards for training and equipping fire-fighters and for response times. Since 2005 it has used a business excellence approach to audit fire authorities' compliance with these standards and evaluates their performance against set criteria to ensure equipment works to certain levels. It publishes technical advice and assistance to fire-fighters, guidelines on rural fire management, and information for the public on assisting with rural fire prevention. It has a role in fire investigations. Responsible for the fire weather index, the authority monitors fire danger conditions throughout the country. It administers the rural fire-fighting fund and it provides grants to rural fire authorities.

The National Rural Fire Advisory Committee, which incorporates a fire research advisory committee and working groups on equipment, training and the management code of practice, provides strategic advice and support to the National Rural Fire Authority. Its membership is drawn from the authority itself, DOC, Federated Farmers, the New Zealand Forest Owners' Association, the Fire Service and Local Government New Zealand. It is therefore another forum from which the authority can gain ideas and feedback from major interest groups and practitioners.

Since 1995 the Fire and Rescue Services Industry Training Organisation has been responsible for setting industry standards for all aspects of fire and rescue. 'Vegetation fire-fighting' training comes in a series of New Zealand Qualifications Authority unit standards that are modularised into various levels. After one to four days of training, the trainees' competencies are judged, and the

A training exercise. Rural fire-fighters use a fire-rake and a mattock to construct a firebreak, from which they can burn outwards. Note the safety clothing: helmets, protective eyeguards and overalls. Once made of Nomex, overalls are now less expensive cotton impregnated with Proban, a chemical that makes the material fire resistant. The troops wear yellow overalls, the officers orange.

Department of Conservation

Wellington DOC teams training for a Wajax competition.
National Rural Fire Authority

assessments are moderated. Students can also sit national certificates and diplomas; subsidised training programmes funded from Vote: Education are delivered through registered training providers, helping rural fire authorities with the training costs.[47] In another attempt to address the lack of operational experience that they see as the result of less frequent burning, the Canterbury rural fire authorities hold large-scale exercises in which hundreds of people participate in practical training situations, in competitions and in games aimed at putting laughter and sport into serious learning.

Paul Devlin from the Christchurch City Council and Trevor Bullock from DOC helping out at the Victorian bushfires, 2006. The fire crowning in the background underlines the need for expert training.
National Rural Fire Authority

International experience adds a further dimension to training. In 1998 New Zealand, following overseas practice, adopted the incident management system that grew out of the 'horrendous' US fires in the 1970s. The system designates roles and responsibilities that can be used in a range of situations. 'Because we're

New Zealand fire-fighting volunteers at work in the horrifying 2006 fires in Victoria, Australia.

Department of Conservation

talking the same language and using the same principles', emergency services both here and abroad find it easier to co-operate and co-ordinate. In 2000 the National Rural Fire Advisory Committee was therefore able to turn longstanding study tours to the US and Australia into opportunities for international deployment.[48] The tasks vary from country to country. In Victoria, Australia, where links have long been in place, 'you take your own crews who are tasked out into the fires. Victoria *actively* uses our men. In the States they bring in prisoners and the military but need experienced managers. New Zealanders and Australians fill those roles.'[49] There is hard work, there is fun, and the intense focus of working together as tight teams, says Jock Darragh, Wellington City Council's principal rural fire officer, means that 'those people in six weeks will probably get more fire experience than they would in New Zealand in 20 or 30 years'.[50] They also raise questions. Equipment overseas is used on a scale New Zealand cannot match. Yet 95 per cent of New Zealand's fires are over within the first six to 12 hours, says Ross Hamilton, Marlborough's principal fire officer and emergency services officer. Are we inappropriately modelling our response to our fires on Australian and North American responses to theirs?[51]

The crews for the initial international deployments were initially picked on reputation and availability but a more systematic method was clearly needed.

Controlled burns still provide fire experience and training. Here, in smoky haze, a medium-sized fire appliance patrols for spot fires outside the area of a planned burn-off.
Southern Rural Fire Authority

Three national multi-agency incident management teams were formed in 2002, and are now available for use either across New Zealand or overseas. This initiative also provided a way of maintaining services in the face of reduced numbers of skilled people in the country, as has the system of appointing seasonal fire-fighters who can be paged or phoned and got quickly into the field (a system that apparently developed from overseas experience rather than the similar practice of our own high-country teams.) 'You don't need thousands and thousands of people,' says Dudfield. But the training is essential. 'Ideally, if I have a maximum of 10 or 20 trained, dedicated people, there is not much we couldn't do,' notes Darragh.

The National Rural Fire Authority also provides training for fire investigators. This role is growing in importance given the high costs of using today's heavy machinery; and, in the inevitably more litigious climate, people are too often inclined, says Barnes, to 'wander around with a GPS saying, "Hey, this is on your land." '[52] As the national authority has to recover costs when a rural fire authority claims from the rural fire-fighting fund, determining responsibilities is important. Key resources available to investigators are the 121-page 2006 publication *Wildfire Origin and Cause Determination Handbook*, a New Zealand edition of an international publication, and the courses based on industry unit standards that Barnes runs for the Rural Fire Authority. In addition the investigators themselves bring considerable expertise to their job – a law enforcement background seems to be particularly useful. The results are often impressive. After the 6000-hectare Withers Hill fire, investigators 'got down to a metre from where it started'.[53] Cases still go to court, although settlements are generally reached. They may, however, lack the humour Girling-Butcher typically found, as in the 1991 case in which a Christchurch QC, acting for plaintiff, had over-imbibed: 'The Marlborough Law Society decided to honour the presiding judge with a bibulous evening immediately prior to "sum up" proceedings. We had to awaken and dress [the QC] and then pray [Chief Justice] Tom Eichelbaum could interpret his slurred summation.'[54]

Publications with simple, direct and straightforward information are aimed at a range of readerships. Some are operational texts for regular fire-fighters and managers. The pocket-sized *Rural Fire Management Handbook* provides information on everything from safety to fire behaviour and fire investigations. The double-sided, laminated 'Ten standard fire-fighting orders' fits in a breast pocket. *Air Operations Information and Checklist*, also pocket-sized, has colour-coded checklists and procedural guidelines. Farmers and other rural dwellers can turn to the colour-illustrated *Farmers' Practical Guide to Rural Fire* for information on preventing and suppressing rural fires and on the agencies concerned; *A Landowner's Guide to Land Clearing by Prescribed Burning* aims for safe and effective burns. Fire authorities putting health and safety programmes in place can use the *Guidelines for Managing a Rural Fire Health and Fitness Programme* to reinforce key points.

For the broader public, *FireSmart Partners in Protection* helps individuals and communities recognise and reduce fire threats, *FireSmart Home Owners' Manual* advises on reducing the risk to homes from 'interface' fire; bookmarks feature TV cartoon figure Bernie, a national rural fire-fighter; and pictures of rural fire operations stress the partnership model in rural fire management. A raft of information and advice is also available on the web.

Increasingly the material is informed not only by New Zealand experience but also by research. In 1992, after a 13-year hiatus, Martin Alexander and Grant Pearce launched a new research drive with their work at the Forest Research Institute (re-branded in 2005 as the Crown research institute SCION) on adopting the fire weather index to New Zealand

Bernie helping to keep New Zealand green. The slogan, like the concept of a public image associated with fire safety, goes back a long way.

National Rural Fire Authority

conditions. Experimental burns on New Zealand vegetation – gorse and slash pine in 1993 and 1994 – continue, while a retrospective study of a grass fire in 1991 was the first wildfire study. In the following years SCION and other researchers have published on the flammability of different species, on how fire affects different vegetation types (valuable for land-management purposes) and on aspects of the fire weather index. Since 2003 the fire researchers have worked with the Australian Bushfires Co-operative Centre. Such cross-Tasman work is relevant given how much the two countries have in common: similar weather patterns (though the weather events here are shorter), the potential for 'catastrophic' fires, and the need for information on smoke toxicity and its management. Although there is some criticism of the nature of the post-fire investigation reports required by the Rural Fire Authority,[55] they do contribute valuable data on fire behaviours and suppression. A relatively new research focus, including for SCION, is on human factors in fire control: how physiological and psychological factors affect fire-fighters' critical decisions, how poor decisions may be offset, and aspects of teamwork and team management. It is in these areas, not in technology, Dudfield believes, that the advances in fire control will now be made.[56]

The Rural Fire Authority does not work unaided. Plantation forests cover about 7 per cent of New Zealand's land area. The Forest Owners' Association, which represents about 75 per cent of the total plantation forest ownership (about 1.3 million hectares out of an estimated national estate of a little over 1.8 million), responds to its members rating fire as a top priority. It contributed to developing the Bernie figure; alongside the authority it runs an annual $500,000 fire-prevention campaign. It covers the fire signs over the winter to give them renewed impact when unveiled in the summer; its forest fire training policy demands that everyone working in a forest owned by an association member must be trained to industry standards; and it gives high priority to the ongoing research programme it contracts with SCION on fire indicators, fire behaviours and new technology.[57]

Many of the bigger forest owners, too, especially when part of a rural fire committee, self-insure by having trained personnel on staff and their own fire-fighting equipment: over 1600 trained staff (60 in full-time positions) and some 130 vehicles contribute to local efforts and international deployments. Excluding the costs of major fires (those over 200 hectares can attract suppression costs of over $1 million per fire), these forestry owners in 2005 spent $8–10 million on protecting the forestry estate. (This compares with local government spending of $6–8m, DOC's $10m, the Ministry of Defence's

$3m and the National Rural Fire Authority's $3m.)[58] Based on the total area, these levels of protection, which reflect the assets' value, are above average, and indicate the key role that forest owners play in rural fire management.[59]

Local efforts contribute to the overall effect. In the Nelson district, for instance, the Waimea Rural Fire Committee, which represents all the big forestry interests, and the Rural Fire Network have 'done a wonderful job of training, equipping, helping, encouraging, keeping the enthusiasm up over years'.[60] Canterbury went through a period of many fires, with prevention hampered by ongoing issues with Railways, such as overloading and poor maintenance due to a narrow focus on profit. In 2008, however, the rail companies, heritage companies, DOC, the Fire Service, the National Rural Fire Authority and local government put together best-practice guidelines for the whole country. Composite non-metallic brake systems were installed on some locomotives to reduce spark risk, and KiwiRail, which took over the country's rail network in 2008, sprays vegetation along the tracks to reduce the risk. Its staff attend fire-training courses. More maintenance is being done, and locomotives known for causing fires are moved to less hazardous parts of the country. Some are having spark-arresters installed. (Where there is a fire the companies must pay recovery costs, a factor that doubtless contributes to their willingness to co-operate.)[61]

That 90 per cent of fires are extinguished by the Fire Service illustrates its critical continuing role, particularly through its volunteers, in rural fire control. Although in a rural area the Fire Service's initial response is to save life and protect property, considerable co-operation now exists between the managers of the urban and rural sectors. The Fire Service's capacity for immediate response makes it ideally suited to putting out small vegetation fires; even in large fires, once its primary responsibility is fulfilled, it can identify local water sources and carry out dry fire-fighting tasks if safe. In addition, although its equipment is unsuited to front-line work, its tankers, pumps and hoses are used to carry volumes of water to where the smaller rural fire hoses are linked in. The many memoranda of understanding signed between rural fire authorities and the Fire Service underline its integral role.

The New Zealand Fire Brigade and rural crews tackle a fire in Taraka Gorge, Johnsonville, in about 2001 – one of countless examples of the brigade working out in the country alongside a rural fire team.

Wellington Rural Fire Force

However, the Forest and Rural Fires Act places the responsibility for rural fire prevention and control firmly on the rural fire authorities;

its 2005 regulations spell out even more precisely than Hensley's committee's recommendations how the authorities are to promote and carry out fire-control measures. Voluntary rural fire forces have to be registered with the National Rural Fire Authority; members' duties have to be spelt out, as do details of ongoing training and the housing of equipment. Grants are made to this end. Compulsory fire plans are no longer merely operational documents. They are now structured around the components of the co-ordinated incident management system: reduction, readiness, response and recovery.[62] *Reducing fire risk* demands details on establishing fire seasons, carrying out public awareness activities and issuing fire permits. *Readiness* requires details of training, equipment, and personnel, of any agreements with other fire forces, and of the rural authority's fire seasons. *Response* covers operational planning, ranging from dealing with calls for assistance to ensuring that all parties can communicate effectively. *Recovery* policies and procedures deal with the post-fire situation – health and safety considerations, debriefs, post-fire investigations and, after major fires, repair of any damage caused by putting the fire out.

The Hensley report's recommendations were implemented to take advantage of local government amalgamation. The bigger and better-financed councils, the committee reasoned, would better understand and resource rural fire. Now two grants provide additional sweeteners. By 1990 DOC, faced with an annual bill of $300,000– 400,000 for suppressing fires on state lands, did not have the finance to call in heavy equipment it needed. In 1991 an augmented rural fire-fighting fund enabled rural fire authorities to claim costs beyond the first $5000 (increased from $2500) and 95 per cent of the remainder of any significant fire.[63] This gave the rural fire authorities a way of striking a rate for those rare events. They could confidently hire the necessary helicopters, monsoon buckets and bulldozers without worrying about a budget blow-out.

Helicopter support at an unknown bushfire on the West Coast, in about 1980. The sheer volume of water and the speed at which they deliver it make helicopters a vital tool in today's rural fire-fighting.

National Rural Fire Authority

Although a small percentage of households are beyond the reach of the Fire Service and so pay the fire levy without receiving any service, the centrally controlled, enlarged fund (today worth $1–1.5m annually) achieves at least a 'rough equity'. Charges lie where they fall, whether on local authorities, insured property owners or forest companies. The fund also provides an important carrot. To

The new Ford appliance purchased by the Wainuiomata Bushfire Force in 1971. It was manufactured in Canada in 1940 for the air force, bought by the Stokes Valley Fire Brigade in 1946, and converted to a double-cabbed appliance with a front-mounted Colmoco pump and 455-litre tank. Eastbourne bought it in 1967, and onsold it to Wainuiomata, who purchased it with community funding. The truck was stored in the Wainuiomata Fire Brigade's new station, and further overhauled to provide room for forestry hose packs and side lockers (precluded by the tank). A roadside control magnetic board was installed, and further minor changes were made over the years.
Gavin Wallace

lodge a successful claim against the fund, rural fire authorities must meet minimum National Rural Fire Authority standards. That incentive works towards achieving the inter-operability needed for efficiency and for today's increasing focus on 'fire-fighter safety and situational awareness'. Accountability, too, reinforces self-policing: volunteers with the wrong footwear may find themselves directing traffic, not battling the fire.[64]

The second form of funding, introduced in 1992, was a grant assistance scheme that gave bona fide rural fire authorities a 50 per cent subsidy towards the costs of their fire equipment and personal protective clothing. Today subsidies can be as high as 75 per cent. Protective clothing is still covered; equipment covers the bright yellow rural fire appliances familiar to many small town and rural dwellers, mobile radios and the fire-danger-monitoring hardware. Remote weather stations feed data via computer links to the National Rural Fire Authority towards a national overview, while regional fire authorities use the data in their own localised weather monitoring.[65] The subsidy ($2 million in 2010) is paid out on evidence of purchase: the local authorities' own contribution, which 'demonstrates a need, not a want' is 'the litmus test'.[66]

We don't have 'a Rolls Royce service', says Dudfield, 'but it is working' – and working efficiently. Since 1987 the area covered by plantation forest has increased almost 40 per cent (from 1.3m to 1.8m hectares). Significant high-country areas have also been returned to the Crown, which, for conservation reasons, manages fire there less intensively.[67] Yet while these factors have created

This new, purpose-built, medium-sized yellow rural fire appliance, with its specialised lockers for equipment, was bought in 2010 with assistance from the rural fire-fighting fund.

National Rural Fire Authority

'increased fire risk and potential fire intensity', there has been, fascinatingly, almost no change in the economic cost of wildfires between the per annum cost in the 1980s (estimated at $97.8 million in 2008 dollar equivalents) and that of the 2002–07 period ($97.7 m).[68]

Other figures have changed, though, demonstrating greater efficiencies and different spending priorities. The number of vegetation fires has risen significantly (probably because of better reporting) from close to 1000 in 1987 to almost 4000 in 2007–08. But while humans and their land-clearance burns continue to be responsible for most of these, the area affected today (an annual average of 5000–6000 hectares) is less than half of what wildfires burned in 1987. Grass (54 per cent of all fires) and scrublands (40 per cent) are now predominantly affected; forest fires account for only 6 per cent. Against the overall increase in fire-prevention costs of some 30 per cent since 1987, the suppression costs have remained somewhat lower. However, pre-suppression costs (for public education, forest management activities, patrols, data recording and monitoring) were considerably higher – if they fell, suppression costs would probably rise.[69]

A tussock burn in the Avondale/Mossburn area, late 1990s.

Southern Rural Fire Authority

What does all this mean for the way rural fire authorities interpret their role? We will look at two quite different fire authorities. First we consider how DOC manages a co-ordinated rural fire response across the 11 conservancies that make up its 'nationally organised fighting resource'.[70] Then a principal rural fire officer details his (slightly atypical) management of a large urban fire district (one aspect of which differs from the management practices of other districts).

DOC, which some see as an industry leader because of its ability to combine depth of knowledge and the use of modern technology, is responsible for about one third of New Zealand's landmass, for scenic areas that abut urban areas and major highways, for high-country tussock and for deep bush in remote mountain valleys or on inaccessible coasts. But, says Kerry Hilliard, DOC's national fire co-ordinator 1987–2010, managing fire is just like managing any other threat, such as possums. You just need nationally consistent rules to maintain a well-equipped, well-trained fire force, to which higher-level assistance can be easily provided if needed.[71]

In their everyday work, field staff read maps, operate radios, drive all-terrain vehicles and use chainsaws – or aircraft. Some 'know the back country like the back of their hands'. Those are all valuable fire-fighting skills, which are carefully and formally honed. DOC ensures its conservancies have the mandatory fire-fighters by requiring all field staff, men and women, to signal in their terms of employment that they are their prepared to fight fires either on the front line or at a managerial or support level. Staff choose the level at which they want to be involved and then take a relevant fire-fighter fitness programme. All new trainees also do the six or so unit standards of basic NZQA training that DOC requires, and then may carry on to crew-leader level.

In the summer all field staff are involved in dummy exercises. There is an art, says Hilliard, in working the pumps, running hose lines, using radios and getting in and out of helicopters; it sounds easy but regular training and familiarisation are essential. In high-risk areas one afternoon a month is devoted to practice. In addition, in high-fire-risk regions, DOC has volunteer rural fire forces registered with the National Rural Fire Authority. These volunteers receive the same training and gain the same competencies as DOC staff.

Over and above that, each region has its incident management team, composed of people across the conservancy. It is a very considerable resource that on a couple of occasions has assisted the national team and has even gone on international deployments. But while, if the need arose, Hilliard could 'drag [staff] from one end of the country to the other', DOC prefers to retain a primarily regional response: 'We didn't want to seem like the Forest Service again.' In dry areas DOC staff, no matter where they go, have to take their fire kit with them – gloves, overalls, helmet, water, high-energy bars

– so they can turn out to a fire and be self-contained for the first 12 hours.

DOC's structure is used in deploying its fire-fighters. Each conservancy has three or four area officers who are responsible for all DOC functions within that area, and each area has at least one crew leader and some fire-fighters. When a fire is called in on 111, the Fire Service Communication Centre takes the call and informs both the closest brigade and the relevant fire authority. Often the neighbouring fire authority reaches the fire first. 'The important thing is to get it out, and you can sheet the costs home later.'

Because delays are inevitable when calling staff in from the field, DOC has memoranda of understanding with other rural fire authorities. However, it can generally get to a fire within an hour or so of the first notification. Most fires are cleaned up within 10 hours, 'though that's a bloody serious fire in New Zealand'. If a fire in its area of responsibility is still 'a runner' after four or five hours, DOC implements response plans and despatches additional fire crews and equipment. If the fire becomes even larger, Hilliard co-ordinates the response from other conservancies and, beyond that, can call in people from other conservancies if all neighbouring resources are tied up. If the mop-up goes on for a long time, most of the fighters will be DOC staff, because others will have been turned back.

Those big fires have their own dynamics:

> Believe it or not, they seem to happen late Friday afternoon, so it's utter chaos for first 12, 24 hours. The first poor buggers in that night get no sleep even though the rules say you're meant to stand them down. But [unless it is dangerous and you have to pull them back] you just can't, there isn't anyone else, you've got to keep on going.

Then the incident management team arrives and its managers begin finding accommodation, setting up meals and finding the community hall, school or whatever local facility can provide telephone lines and computer links. They also build the incident control plan setting out the strategy, tactics and resources for managing the fire. A key for resource deployment is to identify an assembly point where all resources and personnel are signed in. The control is tight:

> You don't leave that point to go to the fire until you're tasked … Fire can be like war, sitting around and waiting. It can be frustrating. Fire-fighters have had a lot of training, they can see the smoke, all they want to do is to get up and put it out but you've got to say, 'No, no, you're staying here.' They can find it very hard to understand why they're not out there.

Finance influences what equipment DOC calls in. Unlike other parties involved, whose costs are paid from the Fire Service levy and who pay excesses on the rural fire-fighting fund, DOC, by contributing to the fund, pays the full cost of the fire. A high-cost option such as bringing in helicopters, DOC knows, is 'out of [our] pocket. We're a bit more frugal and realistic on what we're going to do.'

Beyond fire training and suppression, DOC staff involved with compliance and law enforcement related to flora and fauna (some of them former detectives) make very good fire investigators. The 'incredible amount' of equipment maintenance services that DOC's six regional fire depots lease (at a charge) to other fire authorities gives DOC a role in maintaining national equipment standards. The department also hires out a high-tech infra-red camera to assist in mapping hot-spots when mopping up; rural fire officers in each conservancy office issue permits or, in really dry periods, proclaim no-go areas; a vast publicity resource is directed at users of public conservation land; and DOC staff work to engender good relations with their neighbours. And, as with other rural fire authorities, the National Rural Fire Authority regularly assesses its performance.

In Wellington City the scale is different but the principles underlying the overall practice are similar. Jock Darragh, principal rural fire officer and operations manager of Civil Defence, describes Wellington as one of the highest fire-risk areas in New Zealand. That risk, he says, is compounded by 'hills and wind and houses abutting rural land that's often covered with volatile gorse and

This scrub fire behind Porirua in 2000 illustrates the rural–urban interface in which the Wellington Rural Fire Force often works.

Wellington Rural Fire Force

Members of the Wellington Rural Fire Force battle a scrub fire on Cape Terawhiti, April 2011. Crew boss Paul van Wamel wets down a hot-spot that Colin Robson, controller (above), has dug up. The 'hoody' to the left is the person who allegedly lit the fire.

Wellington Rural Fire Force

other second-growth scrub', and coastlines that lack road access. Like many other councils, Wellington City Council deals with fire as part of its Civil Defence emergency management. Because Darragh received his fire-fighting training in the British army and in DOC as a technical fire manager, his approach differs from that of his counterparts in other districts. He is, he stresses, an operational manager. Where other authorities often have an administrative principal officer whose deputy controls the fire forces, Darragh himself does the administration, deals with his public, attends fires and arranges controlled burns. Moreover, working in an environment where he sees the science of fire evolving daily, he argues that evolving technology is a critical ingredient in gaining efficiencies in fire-fighting into the future. Modern technologies, consequently, play an atypically large role in his district management and in the training his rural fire force receives.

Pre-suppression activities, in which technology is used alongside more traditional approaches, form the other major part of Darragh's job. He monitors the remote weather stations daily via the National Rural Fire Authority's web page: the nature of the fire rating determines whether he has to request forest owners to shut down operations early. As other principal rural fire officers will attest, being highly visible and available is a key ingredient. As he works alongside rural dwellers to manage fire as a tool, he helps build the image of

Nick Westwood, a member of the Wellington Rural Fire Force, at an open day. The hose-handling/water-squirting is very popular, Darragh says. He also takes all opportunities to engage with the public. He tells the children about how hard it is to be a fire-fighter, about fire danger and about what happens if they play with fire. He also targets their parents, giving them pamphlets about fire safety or defensible spaces around their homes.

Wellington Rural Fire Force

'Jock Darragh, our rural fire-fighter'. By burning off for local farmers, his team can study fire behaviour and develop fire plans. It also gives him an opportunity to battle the complacency that comes with farmers who say they 'have been burning for 50 years', and to talk quietly to them about the $50,000–100,000 he might need to recover if a fire where his team is not assisting gets away.

Through the rapid number property identification system he has developed and for which Civil Defence, not the property owner, pays, property locations are identified precisely. Effective fire management demands knowing your area – Darragh knows every track, stream and firebreak. He produces the requisite annual fire plans. To create a positive context for his fire-safety material (developed from US examples), he posts it out with the fire permits rather than the rates. He talks to schools and 'squirts some water', and in return receives children's letters with drawings of rural fire tankers. His safety bulletins on local radio over the summer, as well as some 30-second TV slots, are indicative of his good relations with the press. His public addresses to rural women energise them into starting work on fire plans, and recruitment posters feature Lesley Porter, a 53-year-old crew boss, to reinforce that fire is women's business too.

The 25-strong rural fire force of which Porter is a member was established in 1993 as a bush fire force dealing only with vegetation fires. It now tackles external fire-fighting on structures (though not the interior). Sparks, Darragh points out, can easily ignite the surrounding area, and vegetation fires can equally easily threaten homes. The volunteers (five are women) range from the self-employed to senior public servants and company managers. Porter got into the work almost by accident and loves the adrenalin rush. While she acknowledges an element of community service in her motivation, 'For me it was a way of finding my thing … I'd never have dreamed that a 45-year-old woman could have chucked packs on her back and gone up into the hills …' The people she works with are a critical ingredient:

> The guys are fantastic, so accepting, and even if our team is mostly men, they're really supportive. They call me the nana of the team. There's real camaraderie, which is partly due to Jock [Darragh]. We'd follow him anywhere, we trust him with our lives and he looks after us. I wish I'd known about this years ago.[72]

The force is structured into the operational fire-fighters (17 men, three women) and five support personnel, who may not want to actually fight fires or who may not have the necessary fitness. Their work in the command vehicle, preparing maps and photos and communicating by radio during operations, enables the fire-fighters to operate knowledgeably and safely.

Darragh insists that members have a medical and fitness check. 'It's a matter of safety. It is a dangerous environment, and if you're not fit enough to ensure your own safety you shouldn't be there.' Those who last six months become probationers and thereafter operational fire-fighters, or members of the 'remote area fire team' once they are sufficiently experienced and trained and able to be winched into a remote area and left to get on with the job. The final step is to become a member of the heli-tack team, which requires at least five years' experience, along with training to operate with helicopters, set up heli-bases, load cargo and passengers, and be left overnight. Unique in New Zealand, 10 'rappellers' are the most specialised group, rappelling or abseiling in from 150–300 feet to cut heli-pads, set up camp and be self-sufficient for three days.

'There is no democracy on a fire-line,' says Murray Ellis of the Wainuiomata Bush Fire Force.[73] Certainly Darragh runs his team with military precision. He and his volunteer deputy, Colin Robinson, liaise as controllers. Three crew bosses are each in charge of seven or eight people, and each crew boss has the senior fire-fighter as squadron leader. The crews develop tentative plans as they go to the fire and discuss what they are likely to meet. What are the terrain and vegetation like? Where are the water supplies? How do we attack? The crew is numbered off – No. 1 on the branch, or hose, with support from No. 2. No. 3 is the dogsbody: pulls hose, takes messages and helps No. 5 (pump person) set up the pump. No. 4 is the crew boss. When they arrive the crew bosses are briefed by Darragh and Robinson. Any changes are signalled and the safety briefing is

Gavin Wallace, then deputy fire controller (now fire controller) of the first-formed and still operational Wainuiomata Rural Volunteer Bushfire Force, sits exhausted during an Eastbourne fire in about 1985 that saw two fire-fighters hospitalised. Such experiences do not diminish his enthusiasm for fire-fighting. The task, he says, 'is pretty personal – you against nature' – and it poses 'clear challenges'. It is 'focused – you're not aware of time, you're up in the bush running water on flames and getting out. There's none of the crap from modern life. It's pretty hard to pull someone off the line – most want to stay until the fire is out. But it's not as if it is day after day.'

Gavin Wallace

given: keep in touch, follow the hose line down if there's a problem, get into a burnt area in an emergency, use the branch to spray you and the ground around you as a last resort. Then the fire-fighters are off. Porter, as crew boss, does not actively fire-fight but keeps in contact with her crew, tasks people and liaises with Darragh and Robinson about things like water pressure. She reminds people to take water to drink. It is very easy, she says, to find that 'two or three hours have gone by and you've not had a drink'.

It is hard, unglamorous work. Hoses, adapted from what the Forest Service brought in from Canada in the 1960s, now come in two types. The percolating ones, made from natural and synthetic fibre, weep water and do not burn so can be moved across burnt-over areas. The non-percolating, non-friction hose, which gets water to the fire at strong pressure, is used from waterside to fire edge and is then attached to the percolating hose. But however ingenious the technology, the hoses, with a diameter of 25 or 41 millimetres for ease of handling, constantly have to be untangled from the small stubs of gorse that remain after the fire has gone through, then dragged on up the hill; the lightest pumps weigh about 30 kilograms. There is smoke and hard work – and potential danger, especially from heat, as Ellis observes:

> The thing that really strikes recruits immediately is the sheer blast of heat, which invariably far exceeds what they expected. Even from quite small fires the radiant heat gives people a bit of a shock. It is one of the things one needs to understand: that working next to serious fire you can get very heated and you sweat much more than in other situations, so there is a danger of dehydration that can happen much more rapidly than normally.[74]

Porter's recollections of her first fire, at Porirua, bear out Ellis's words. They rounded the hill at Johnsonville 'and there was this huge plume of smoke, dirty brown smoke that means it's very hot. "Fuck'n hell," said the driver.' By the time

Members of the Wellington Rural Fire Force and Wellington City Council Parks and Gardens Fire Crew mopping up after a fire at Makara Cemetery. The absence of any vegetation (fuel) indicates the extreme heat the fire generated. Crew in the foreground are using a Pulaski hand-tool, which combines a grubber and an axe. In the background fire force members are soaking the ground with water. Each group is systematically covering every centimetre of the fire to douse all possible hot spots.

Wellington Rural Fire Force

they got there 'it looked as if a mountain were on fire'. But there is also euphoria when the fire is out. 'Everyone's cold, tired, filthy, and everyone's got a grin on their faces.'[75]

The six-hours-a-week training that Darragh puts his rural fire force through mixes high tech with the inevitable hard slog. He has designed it to challenge the team and develop skills useful in any Civil Defence emergency. Team members receive radio and medical training; the swift-water rescue course in the middle of a freezing July equips them to draft water from rivers safely; and rapelling practice, planning and controlled burns are constant. But, skilled as his team may be, Darragh is a pragmatist:

> In an ideal world you'd have trained and equipped people all the time in place waiting to battle fire. But it doesn't work like that. I

Kathryn van Wamel, Lesley Porter and Marie Brosnahan from a Wellington Rural Fire Force crew eating and resting during the 2005 Makara Road fire.

Wellington Rural Fire Force

> know of work gangs in forests, in silviculture crews, and they've got big bellies and holes in their jeans and these are our fire-fighters and we have to use them. If they're there to respond to a forest fire they can save thousands … you make do with what you have.

The fire force does not work unsupported. When, in a pattern that is repeated across the country, the New Zealand Fire Service takes a 111 call for a vegetation fire, the 'urban boys' fight for an hour or so. They then call in the rural fire force and move to a back-up role, using their 70mm hoses to relay water at pressure to the fire force pumps and hoses.[76] Darragh, who chairs the regional rural fire co-ordinating committee, has developed memoranda of understanding with the other fire forces in the region and they all help one another out.

The equipment and gear on which Darragh unashamedly 'spends money to save money' gives the team an edge. Helipro, New Zealand's most experienced fire-fighting helicopter company, get Darragh's team to any coasts around Wellington in 15–20 minutes, where it would take two hours to drive: 'It is the essential fast response – hit it hard, keep it small.' Helipro's Wes Cam thermal imaging camera, a gyro-stabilised TV that comes in at a mere $500,000 and rather upstages what DOC can offer, identifies hot-spots days after the fire, and allows the helicopter's position in relation to the ground to be monitored. There are also the well-equipped four-wheel-drive tanker, three smoke chasers and small four-wheel-drive utes that carry all the small stuff to the fire scene.

'You have to love and breathe fire. My job is my passion. I just love coming to work each morning.' Darragh's passion (a term many interviewees used) and his approach seem to pay off. 'I have 100 per cent success in fire-fighting; there are no fires burning out there, are there?' Not that he allows himself to become complacent.

Wildfires may not impinge as dramatically on our lives as they once did, but they cost the country some $98 million annually.[77] Yet, while the Forest Owners' Association still affords fire the highest priority, regularly reviews its fire-protection activities and seeks greater efficiencies, fire-practice routines in about a quarter of the country's plantation forests still need improvement. This is partly because of patterns of ownership. Where once a single company owned 50 per cent of New Zealand's exotic forests, there are now dozens and newcomers regularly enter the market. Typically they own small to medium-sized forests. Such fragmentation does 'not lead to good overall co-ordination of resources or effort', and the suitable resources are often simply not to hand. Nor is there the erstwhile interest in helping in fire-protection organisations.[78]

Companies with fragmented ownership or small companies established as shell companies are seen as exacerbating the problem. These companies are often totally dependent on contractors with sometimes questionable levels of training and an inexperienced workforce that turns over regularly because of stringent drug testing rules. Not all are represented in the Forest Owners' Association or bound by its requirements on fire management. They may also be based outside a fire district, opt out of fire control, and freeload on the fire protection available from a neighbouring larger company or rural fire authority. If there is a fire, those coming to assist may lack the resources to cope with poorly designed and poorly maintained roads and firebreaks, tank traps dug to deter trespassers, and other dangers such forests present.

Questions of how rural fire-fighting can be made more efficient and effective remain unresolved. In 1993 a ministerial review first proposed an amalgamation of the Fire Service and the rural fire-fighting service, provided that there was no additional cost. Principally for a range of industrial relations reasons to do with the Fire Service, the proposal did not progress. However, 2003 saw the publication of the first of several papers on achieving efficiencies in fire management, the role of fire services in emergency and first-response work, ways this role might be 'enhanced' and the services' gradual transition to a fully integrated service.[79] Discussions, a large number of submissions and a change of government protracted the proceedings. In 2009, with opinion on the virtues of amalgamation divided between the urban services (in favour) and the rural services (against), the minister abandoned the proposal but did direct the National Rural Fire Authority to reduce the number of rural fire districts.[80] The resulting strategy, at the moment being implemented on a voluntary basis, aims at the long-term reduction of the current 80 districts to fewer than 20.[81]

Enlarged rural fire districts are not new. In 1992 the Waimea rural fire district was formed from three different rural fire district committees, DOC and the Nelson and Tasman councils in the first major amalgamation after the 1987 reforms. Today the district reaches across to include Golden Bay. Such districts have been formed elsewhere too. The current drive to enlarge rural fire districts is about facilitating a proactive focus and providing broader perspectives and better fire management and protection. The intended result is more equitable funding, better management of resources and fire management and protection, and an increased capacity that can 'reduce the consequences of wildfire and facilitate it as an effective rural land management tool',[82] as well as ensure consistency of permit practice across contiguous areas and better public understanding.[83]

Those opposed to the initiative argue that boundaries will move well beyond communities of interest that critically underpin the volunteer system – a system that is at the heart of New Zealand's rural fire management structure. Volunteers will walk away and the relationships and years of work will be lost. Ross Hamilton, citing his own experience and recent research, questions whether councils in the bigger structures will be able to fulfil effectively their 'moral obligation to ensure the community is back to normal as soon as possible'.[84] Even supporters of larger districts can be equivocal. They criticise the National Rural Fire Authority's selling of the idea; they cite a lack of consultation about where existing rural fire managers would fit in and their relationship to a possibly unqualified manager with little understanding of fire; and they question the need. 'In the Wellington region,' says Darragh, 'we've got [a memorandum of understanding] in place and that works with a simple document and a handshake. If there has to be change it has to be better than what we've got … At the moment it appears as if they've got an answer in search of a problem.'[85]

There is a further set of operational concerns. The National Rural Fire Authority has introduced performance assessment criteria, a 'minimum standards' approach used to audit the rural fire authorities. It considers these a valuable tool for identifying strengths and weaknesses. But those in the field describe the process as cumbersome, costly, time-consuming, unable to deliver

Fire-fighters, c. 1946. These men's clothes contrast dramatically with the specialised gear of today's rural fire-fighters.

Millman collection

clear directions for future development and so hands off that it fails to pick up on those authorities performing below par. Instead of adding efficiencies, the system 'is something that's been pulled around to work for rural fire and it is not an appropriate tool'.[86]

Nor are the safety and welfare standards that fire-fighters are now expected to reach viewed unequivocally. The shorts and singlet days are a thing of the past, but some practitioners, as we have seen, question how today's standards fit with the reality of having to use the people available, whether it be farmers with wet sacks or contractors with old sewage trucks who do not have the skills to be on the fire-line and are 'never wearing the right clothing'.[87] More stringent standards may mean losing volunteers. Might careful forward negotiations and contracts, or the stringent rules on supervision that the Forest Service used to corral in the untrained, obviate the problem? There are concerns, too, that new fitness standards may drive volunteers away. Such standards take time and commitment to achieve. Says Barnes, 'They're struggling out there. We've gone a long way in getting [the volunteers] on board, but how much farther do you go?' He also raises the difficult question of whether the competence gained through training via skills standards matches up to that gained through fire-fighting experience: 'Put some of the people with some of the highest unit standards in the country into a fire situation and they don't want to know.'[88]

That issue is compounded by the increasingly frequent imposition of burning restrictions.

Newcomers to New Zealand, whether of Polynesian or European descent, used fire as a land-management tool – and with dramatic results. Today a better balance has been achieved between the use of fire and environmental considerations. However, managed fires still burn annually over some 100,000 hectares – of the wheat stubble left after harvest, of tussock grasslands, gorse and scrub.[89] While many consider that one of the strengths of our rural fire regime is that it accommodates fire as a land-management tool rather than attempts to ban it, a number of factors now raise questions about this approach. Environmental care regulations, the cost of fire suppression, and transfers of marginal land to DOC estate have cut back the use of fire somewhat. In addition, the rate of dairy conversions and urban subdivision increased over the last decade, biofuel potential may encourage planting, and the demand for wood as a sustainable and renewable resource may grow. All these trends may help to heighten sensitivities about burning-off practices.[90]

DOC works at helping farmers to burn safely, but the wider debate over the role of conservation in managing the threat of fire has recently intensified as large tracts of pastoral leasehold land have been retired to public conservation land. One perspective is backed by a long tradition of using fire as a method of land management. Proponents point to Australia's managed burns, arguing that managed fire reduces fuels loads, promotes palatable regrowth, reduces 'woody' species and 'reduces the risk of devastating wildfire whose heat destroys the ecology of the region and makes regeneration unlikely'.[91]

From a different perspective, environmentalists argue that if the land were allowed to revert over time to the original species, the highly flammable tussock and *Dracophyllum* would be replaced by less flammable native species. In support of this argument they point to biodiversity loss in Northland, for example, where in the aftermath of big fires eucalyptus and acacia have invaded huge areas, and to the high country where 'mat-dominated plant communities' such as *Hieracium* easily replace tall tussock killed by fire.[92] A 2006 review of the economic impacts of New Zealand's public conservation lands recognises the 'non-market values' of ecosystems. Fire can detract from New Zealand's natural capital by increasing flood risk, reducing whitebait catches, degrading water quality, impoverishing tourist experience and damaging New Zealand's clean, green image.[93]

As the body in charge of conservation lands, DOC notes that escaped fire from burn-offs is a major cause of high-country wildfire. It appreciates the delicate balance of regenerating vegetation at a time when increasing human activity and incursion into these areas augment the risk, but argues that managing this by creating buffer zones between managed and regenerating areas is preferable to large burns.[94] Today a warming world and the need to minimise carbon release give a new perspective to the 'critical questions about controlled burning as a silviculture tool' first raised over three decades ago.[95] It is a perspective that DOC and the National Rural Fire Authority have yet to explore.

Vegetation fires generate enormous heat and energy. One metre of fire-line has the equivalent energy of 4000 one-kilowatt bar heaters stacked on top of one another. Fires that burn below 4000 kilowatts can be stopped. For those between 2000 and 4000 kilowatts heavy machinery and aircraft are essential for ploughing through firebreaks and delivering water mixed with fire suppressants or retardants onto the fire edge. Over 4000 kilowatts, fire becomes very difficult to control without rain or the wind shifts that drive it back on itself, depriving it of fuel.[96] The aftermath is always hot-spots, often underground and fed by air that can circulate through tortuous root systems, allowing the fire to flare up again and present new dangers.

The long argument about the flammability of green New Zealand bush has been settled. Podocarps, we now know, need extreme drought to burn, whereas beech forest goes up like pine forest. But as the early Polynesian and European settlers slashed and cut into the bush, driven by economics and emotion, the very conditions they worked to create as they cleared the hills would, in dry years, lead to fires that swept uncontrolled outside the confines of the planned burn and through vast tracts of countryside. The flames, the choking smoke, the terrible heat and the horror that early Maori must have known, and that were part of the European settler experience, were not products of journalistic imagination. They were the reality that families or isolated pockets of settlers faced as they banded together desperately trying to save their homes and properties and livestock. Was companionship heightened and were the communities of interest, so valuable for successful settlement, strengthened as they buried their valuables, battled alongside neighbours with wet sacks or branches and handed on the next bucket? Certainly good grass sprang from the devastated landscapes. And planning for the next year's fires would begin.

Today the landscape is transformed. The bush is now gone from arable country; those in the trade talk not about bush fire, but about wild and vegetation fire. Rural dwellers still use fire in working their land, but the changed attitude that arose from a perceived need to save the bush for future use, that demanded that burning be done not just at individual whim but within the larger community interest, has resulted in burning being placed within a system

Burkes Pass Volunteer Rural Fire Force with its new fire shed and rural fire appliance, 2009.

National Rural Fire Authority

where land managers and users are integrally involved in managing the risk of wildfire. When fires get out of control it is possible to muster an array of local, then regional, then national fire forces, backed by experience, training, scientific knowledge, equipped with machinery and gear and with a mobility that would have beggared any early settler's imagination. It is a system, says New Zealand Fire Service Commissioner Piers Reid, that has 'tended to avoid some of the high fixed-cost structures' incurred in the urban fire setting; it is also one that is 'absolutely reliant' on the rural and forestry sectors' goodwill and continued involvement.[97]

Critical as today's farmers and forest owners and managers can be of New Zealand's rural fire management, the system works because of nine decades of developing a systematic approach to combating rural fire and maximising forests as a valuable resource. The attempts to raise public consciousness of fire danger and change public practice, particularly from the 1920s to the 1960s, contribute to a fascinating aspect of the public education initiatives that marked the lives of two generations. The early Forest Service men battled mud, recalcitrant millers and reluctant landowners; they persuaded or prosecuted; they fought fires, often alongside those who had illegally lit them, in efforts to enforce a fire regime, develop relationships and induce the attitudinal change that would assist in local protection.

In later years Forest Service employees' presence, experience and equipment (assembled to combat fire within the exotic plantations that by then were to give New Zealand its future) made them the nation's rural fire-fighters. At the same time the training they offered and the equipment they wrote off to rural communities helped channel the impulses behind the elderly Cust women who decades ago armed themselves with buckets of water and biked out to extinguish some railway fires. The individuals or small bands joining forces to meet a threat, or the volunteers from permanent brigades became trained rural fire-fighting groups.

Rural dwellers have long participated, contributing their knowledge of the land and its typography, as well as their manpower. Social and legislative change, technological developments and the incursion of urban areas into the country allow women and urbanites to participate, and urban fire brigades are an important first-strike force. But it is the volunteer fire crews who make up 87 per cent of the fire-fighting force, and particularly the 3000 volunteers in the rural fire system, who are 'integral to [rural] fire suppression in New Zealand'.[98] Some are driven by a passion for fire and its management, or by an urge for achieving the self-discovery that comes with tough physical effort. For others the primary drivers are about contributing to their community's safety or protecting its natural assets. There are obvious social satisfactions that come from being part of an important group. But whatever the motivation, the volunteers give time and effort for little or no pay, largely out of the public eye and largely unsung. They, their predecessors, and their histories take their place alongside the Forest Service in the annals of New Zealand's history.

Endnotes

Abbreviations

ACL Auckland City Library
AJHR *Appendices to the Journals of the House of Representatives*
ANZ Archives New Zealand (all relevant files are housed in Wellington)
ATL Alexander Turnbull Library
NFU National Film Unit
NRFA National Rural Fire Authority
NZJF *New Zealand Journal of Forestry*
NZPD *New Zealand Parliamentary Debates*
OHC Oral History Centre, Alexander Turnbull Library
RFB Report on Fire Brigades
RFP Rural Firefighting Project, 2009–10, Oral History Centre, Alexander Turnbull Library
SFSAR State Forest Service Annual Report
WRFF Wellington Rural Fire Force

Preface

1 Department of Conservation, *State of New Zealand's Environment 1977*, Ministry for the Environment, GP Publications, Wellington, 1997, p. 8.30.

Chapter One: The burning of New Zealand

1 M. S. McGlone, 'The Polynesian settlement of New Zealand in relation to environmental and biotic changes', *NZ Journal of Ecology*, vol. 12 (supplement), 1989, p. 1.
2 M. S. McGlone, 'Interpreting the pollen record: Reconstructing the deforestation of New Zealand', 2009, www.landcareresearch.co.nz/research/ecosystems/past_env/pollen_interpretation.asp. McGlone found the forest had regenerated within about 200 years of the eruption.
3 For a brief outline of the arguments that climate change had a role in creating the treeless South Island east coast, and the Holloway/Cumberland debate, see for instance P. J. Molloy, 'Distribution of subfossil forest remains, eastern South Island, New Zealand', *NZ Journal of Botany*, vol. 1, 1963, p. 76; and McGlone, 'Polynesian settlement of New Zealand', p. 116. However, other ecological research findings (including Molloy's) increasingly suggest the destructive fires began with the Polynesian arrivals.
4 Information from Brian Molloy, March 2010.
5 Department of Conservation, *New Zealand Threat Classification List*, 2005: www.nhc.net.nz/index/extinct-new-zealand/extinct.htm (accessed 25 January 2012).
6 McGlone, 'Interpreting the pollen record', p. 2; R. J. Cameron, 'Maori impact upon the forests of New Zealand', *Bay of Plenty Journal of History*, vol. 9, no. 3, 1961, pp. 131–41.
7 C. O'Loughlin, 'Historical perspectives of indigenous forests', in *Native Trees for the Future: Potential, possibilities, problems of planting and managing New Zealand forest trees*, proceedings of the forum held at the Centre for Continuing Education and Department of Biological Sciences, University of Waikato, Hamilton, 8–10 October 1999, p. 1; Rowan Taylor and Ian Smith, *The State of New Zealand's Environment, 1997*, Ministry for the Environment, GP Publications, Wellington, 1997, 8.80.
8 Taylor and Smith, *State of New Zealand's Environment*, 8.30.
9 Narena Oliver, 'Moa', *NZBirds*, 2005: www.nzbirds.com birds/moa.html (accessed September 2008).
10 'A forceful impact', in Malcolm McKinnon (ed.), *New Zealand Historical Atlas: Visualising New Zealand*, David Bateman/Department of Internal Affairs, Auckland, 1997, plate 12.

11 The history of the debate is summarised in Molloy, 'Distribution of subfossil remains', pp. 75–76. McGlone ('Polynesian settlement of New Zealand', p. 126) suggests that the reluctance to attribute the burning to Maori arises from a reluctance to move 'from the proposition that the pre-industrial hunter-gatherers and neolithic agriculturalists lived in harmony with their environment'. This statement does not, in my view, attribute blame but simply points to a people who, like their later European counterparts, were attempting to wrest a living from a land generally covered in dense vegetation.

12 D. B. McWethy, C. Whitlock, J. M. Wilmshurst, M. S. McGlone and X. Li, 'Rapid deforestation of South Island, New Zealand, by early Polynesian fires', *The Holocene*, 19(6), 2009, pp. 883–87. Other studies used in working on this section are M. S. McGlone, 'The origin of the indigenous grasslands of southeastern South Island in relation to pre-human woody systems', *New Zealand Journal of Ecology*, vol. 25, no. 1, 2001, pp. 1–15; McGlone, 'Polynesian settlement of New Zealand'; John Odgen, Les Basher and Matt McGlone, 'Fire, forest regeneration and links with early human habitation: Evidence from New Zealand', *Annals of Botany*, vol. 81, 1998, pp. 687–96.

13 Information from Brian Molloy, March 2010.

14 J. Halkett, *The Native Forests of New Zealand*, GP Publications, Wellington, 1991, pp. 12–22; Dennys Guild and Murray Dudfield, 'A history of fire in the forest and rural landscape in New Zealand, Part 1: Pre-Maori and pre-European influences', *NZ Journal of Forestry*, vol 54, no. 1, pp. 34–38.

15 'Notes of a journey through a part of the Middle Island of New Zealand', *Nelson Examiner and New Zealand Chronicle*, July 1844, p. 83.

16 Cameron, 'Maori impact upon the forests of New Zealand', pp. 134–35.

17 Taylor and Smith, *State of New Zealand's Environment*, 8.20 and 8.30. However, the authors point out that the causes of erosion have to be examined in relation to when they occurred. The Lake Tutira erosion rates, for instance, only increased as European farmers replaced deeply rooted bracken with lightly rooted grasses; O'Loughlin, 'Historical perspectives of indigenous forests', p. 2.

18 J. C. Beaglehole, (ed.), *The Journals of Captain James Cook, Vol. 1: The Voyage of the Endeavour 1768–1771*, Cambridge University Press for the Hakluyt Society, Cambridge, 1955. See, for instance, entries on 11 October 1769 (p. 173) where he recounts seeing fires 'dayly', and on 17 October (p. 180).

19 Guild and Dudfield, 'A history of fire in the forest and rural landscape', pp. 6–7.

20 William Swainson to Michael Murphy, Chief Police Magistrate, 13 December 1842, *NZ Gazette and Wellington Spectator*, 17 December 1842, p. 2.

21 In relation to settlers' increasing prosperity, see Cameron, 'Maori impact upon the forests of New Zealand', p. 139; *Appendices to the Journals of the House of Representatives* (*AJHR*), 1869, D.–22, Correspondence relative to the present condition of the forests of New Zealand, Enclosure No. 9, Charles Sealey, Chief Provincial Surveyor comments on the decline of Maori burning in the Hawke's Bay.

22 'Notes of a journey through a part of the Middle Island of New Zealand', *Nelson Examiner and NZ Chronicle*, 27 July 1844, p. 83.

23 Ninety-five years later a State Forest Service conservator described the same process. Fires from burning off invariably 'spread into the forests to a greater or lesser extent, and even when they do not actually encroach far along the ground the flames scorch the standing bush to a depth of a chain or so, with the result that the scorched trees die and are usually superseded by bracken fern. The next time a fire occurs in the same locality, a further stretch of bush gets scorched, until in the course of time serious inroads are made into the forests if they are not totally eliminated.' F1 12/0, Pt. 5, Fire protection general 1939–40, Archives New Zealand (ANZ). R. D. Campbell, Auckland conservator to Director of Forestry, 2 June 1939, p. 2.

24 'Fires', *NZ Gazette and Wellington Spectator*, 10 January 1844, p. 2.

25 Hooker, quoted in Cameron, 'Maori impact upon the forests of New Zealand', p. 134.

26 Swainson to Murphy, *NZ Gazette and Wellington Spectator*, p. 2.

27 'Notes from a journal', *New Zealander*, 22 December 1847, p. 2.

28 Taylor and Smith, *State of New Zealand's Environment*, 8.80.
29 McGlone, 'Polynesian settlement of New Zealand, p. 126.
30 O'Loughlin, 'Historical perspectives of indigenous forests', p. 2.
31 'Deforestation', *Forestry Insights*, 2005: www.insights.co.nz/story_behind_d.aspx (accessed October 2008).
32 'The kauri harvest: Timber and gum in the north 1860s to 1920s', in McKinnon (ed.), *New Zealand Historical Atlas*, plate 48.
33 'Alarming fires', *Taranaki Herald*, 19 November 1859 (from the *Lyttelton Times*, 15 October 1859), p. 3.
34 Taylor and Smith, *State of New Zealand's Environment*, 8.30.
35 'From forest to pasture: The clearance of the lower North Island bush, 1870s to 1910s', in McKinnon (ed.), *New Zealand Historical Atlas*, plate 47.
36 Taylor and Smith, *State of New Zealand's Environment*, 8.30.
37 'Fire, flood and quake', in McKinnon (ed.), *New Zealand Historical Atlas*, plate 87.
38 'A visit to the bush fires', *Otago Witness*, 3 February 1872, p. 9.
39 'Through the stricken bush districts: Our special reporter's impressions', *Evening Post*, 17 January 1898, p. 6.
40 Journal of T. F. Cheeseman on board HMS *Buffalo*, August 1836–March 1841 (ATL), quoted in Michael Roche, *History of New Zealand Forestry*, New Zealand Forestry Corporation/GP Books, Wellington, 1990, p. 21.
41 'Alarming fires', *Taranaki Herald*, 19 November 1859, p. 3.
42 A. D. MacIntosh (ed.), *Marlborough: A provincial history*, Marlborough Provincial Historical Committee, Blenheim, 1940, p. 126.
43 Rollo Arnold, *New Zealand's Burning: The settler's world in the mid 1880s*, Victoria University Press, Wellington, 1994, p. 22.
44 *Marlborough Express Weekly Edition*, 23 January 1886, p. 10, quoted in Arnold, *New Zealand's Burning*, p. 22.
45 I used the National Library of New Zealand's excellent online resource *Papers Past*: http://paperspast.natlib.govt.nz/cgi-bin/paperspast (accessed March to April 2008). At time of writing digitisation had progressed to 1915.
46 For instance, commenting on a fire in the Puhipuhi block in 1885, Professor Kirk noted that the damage was 'not as much as I was led to believe', *AJHR*, 1886, vol. 1. C.–3, Native forests and the state of the timber trade, The Puhipuhi block, p. 21, Professor Kirk to the Minister of Lands, 16 November 1885. However, note that this fire, and others in the 1880s, preceded the disastrous 1887 fire ('the burn which almost finished the forest', D. E. Hutchings, *New Zealand Forestry*, 1919, quoted in Thomas E. Simpson, *Kauri to Radiata: Origin and expansion of the timber industry of New Zealand*, Hodder and Stoughton, Auckland, 1973, p. 95) which burned for several weeks and destroyed some 300 million superficial feet of timber (Guild and Dudfield, 'A history of fire in the forest and rural landscape', p. 10); in 1903 reports of a loss of 3000 sheep in a fire around Morere were later found to be exaggerated (*Taranaki Herald*, 10 January 1903, p. 2); and in 1908 the *Evening Post* (18 February 1908, p. 7) claimed there were 'exaggerated rumours' over the extent of the losses experienced. Reading the reports today, it is obvious that statements of the value of property destroyed need to be assessed sceptically, as the following three instances demonstrate. In 1872, when 'much of Carterton burnt', the cost of damage to homes, outhouses, a smithy and other buildings was said to be £539. On the other hand the *Taranaki Herald* on 5 February 1879 assessed the value of 200 bags of cocksfoot seed, pasture and a house as £200, and in the 1908 fires the *Otago Witness* (4 March 1908) valued one miller's holding of about 1820 hectares in the Makerua Swamp as £120,000 – well beyond the £7 per acre given for swamp land out of Dargaville.
47 *Taranaki Herald*, 17 February 1908, p. 7.
48 'A trip up the Hutt Valley', *Evening Post*, 14 January 1898, pp. 5–6.
49 *AJHR*, 1908, C.–1, Department of Lands, Extracts from the reports of Commissioners … pp. 19–38.
50 'The burning region', *Taranaki Herald*, 9 January 1886.

51 Hochsetter, quoted in *AJHR*, 1874, H.–5, Papers relating to state forests: 1 Parliamentary debates and resolutions, The forests of the Colony, p. 1.
52 'Deforestation', *Forestry Insights*: www.insights.co.nz/story_behind_d.aspx (accessed October 2008).
53 Review, New Zealand Institute, *Transactions and Proceedings of the New Zealand Institute*, vol. 1, 1868, *Otago Witness*, 17 July 1869, p. 9.
54 'Bush fires', *Evening Post*, 30 January 1872, p. 2.
55 *AJHR*, 1874, H.–5, Papers relating to state forests. 1. Parliamentary debates and resolutions, The forests of the colony, Mr O'Neill, p. 12.
56 'Reefton', *Evening Post*, 20 January 1876, p. 2.
57 'Great bus fire near Wellington', *Evening Post*, 2 April 1881, p. 2.
58 Guild and Dudfield, 'A history of fire in the forest and rural landscape'.
59 Francis Fry, 'How we fought the bush fires', Programme 75, Open Country Sound Recordings, OHT 5-0064, OHC.
60 Tom Brooking, 'Use it or lose it: Unravelling the land debate in late 19th-century New Zealand', *NZ Journal of History*, vol. 30, no. 2, October 1996, pp. 141–45. Richard Boast, *Buying the Land, Selling the Land: Governments and Maori land in the North Island 1865–1921*, Victoria University Press, Wellington, 2008, pp. 121–34, discusses these issues and their relationship to Liberal land purchase policies.
61 Boast, *Buying the Land*, p. 132.
62 R. Dalziel, *Julius Vogel: Business Politician*, Auckland University Press, Auckland, 1986, p. 105, see also pp. 80–81, 97–99; Preece, Diary, 15 April, MSS 249, Auckland Institute and Museum Library, quoted in Judith Binney, *Encircled Lands: Te Urewera 1820–1921*, Bridget Williams Books, Wellington, 2009, p. 213.
63 'New Zealand: Its past, present and future', Paper read at the Imperial Institute, 4 December 1893, London, 1879, p. 4, in Dalziel, *Julius Vogel*, p. 105. Arnold, *New Zealand's Burning*, pp. 24–26, also discusses the military goals of the public works initiative.
64 J. Cowan, 'Preece's Road', *Tales of the Maori Bush*, p. 34, quoted in Binney, *Encircled Lands*, p. 222.
65 *AJHR*, 1872, A.–1, 'Notes of a Journey of Sir George Bowen, April 1872, p. 86, quoted in Binney, *Encircled Lands*, p. 222. I have widened the ideas where Binney discusses Bowen's words in terms of the Urewera.
66 Boast, *Buying the Land*, p. 135.
67 Brooking, 'Use it or lose it', pp. 141–62.
68 On the settlement of the development of the lower North Island, Palmerston North and Hawke's Bay see Arnold, *New Zealand's Burning*, pp. 29–32 and McKinnon, 'From forest to pasture'; 'In 1875', *Evening Post*, 21 June, p. 20; 'Men and their deeds: Landmarks in history', *Evening Post*, 3 November 1910, p. 9.
69 'Out of the bush: The birth of a town', *Evening Post*, 21 June 1911, p. 11.
70 Roche, *History of New Zealand Forestry*, p. 137.
71 See *AJHR*, 1892 and successive years, C.–1, Crown Lands, reports on the Land Act 1982. The relevant sections of the act are in Parts III, V and VI (the last being the only part to contain conditions designed to prevent fires).
72 Boast, *Buying the Land*, p. 135. Drawing on E. C. R. Warr, 'Pakeha agriculture', in K. W. Thomson (ed), *The Legacy of Turi: An historical geography of Patea County*, Dunmore Press, Palmerston North, 1976, pp. 19–44, Boast states that until the 1890s, when dairying took off, bush farming was largely subsistence in nature.
73 On the incentives to settle and farmers' interests in sawmilling, see Arnold, *New Zealand's Burning*, pp. 28–29; McKinnon, 'From forest to pasture'. 'Remarks on clearing land in the valley of the Hutt', *NZ Gazette and Wellington Spectator*, vol. 5, no. 339, 10 April 1844, p. 2, indicates that even then there were opportunities to sell wood, especially kahikatea and other pines, to sawyers. Arnold, *New Zealand's Burning*, p. 26, discusses the changing perceptions of the cost of land and settling in the bush country.
74 'Clearing, draining and fencing for a garden', *Daily Southern Cross*, 1 September 1869, p. 5.

The title is slightly misleading – one section does deal specifically with clearing bush for farming.

75 See the Reserve Bank inflation calculator: www.rbnz.govt.nz/statistics/0135595.html. This figure and those following were calculated using the general category and fourth-quarter values. Amounts are rounded up.
76 'Remarks on clearing land in the valley of the Hutt', *NZ Gazette and Wellington Spectator*, 10 April 1844, p. 2. This early European settler's account of how the Maori he employed cleared and planted land describes their efficient technique for pushing inroads into dense bush.
77 Arnold, *New Zealand's Burning*, p. 26; 'Clearing, draining and fencing for a garden', p. 5; McKinnon, 'From forest to pasture: The clearance of North Island bush 1870–1910s', plate 47. The details of how the fires burned and of sowing seed come from Programme 117, Open Country Sound Recordings, 'The flame in the forest', OHT5-0104, OHA. The men were recalling a time when they were 12 and 13 years of age. Unfortunately, neither the programme itself nor the programme notes indicate exactly what period they were talking about. However, internal evidence relating to the effect of the fires in the Whanganui–Stratford area suggests the 1920s or thereabouts.
78 Arnold, *New Zealand's Burning*, p. 28.
79 Programme 117, Open Country Sound Recordings, 'Bushwackers and bushfires', OHT5-0104, OHA.
80 P. Walsh, 'The effect of deer on the New Zealand bush: A plea for protection of our forest reserves', *Transactions and Proceedings of the Royal Society of New Zealand 1868–1961*, vol. 25, 1892, p. 439. (Before 1933 the society was known as the New Zealand Institute.)
81 *AJHR*, 1907, C.–1, Appendix 1, Settlement of Crown land, Canterbury; Settlement of Crown land, Otago, Bush fires, p. 37ff.
82 *AJHR*, 1908, C.–1, Department of Lands, Bush fires, p. 8.
83 'Bush and grass fires round Masterton and Eketuhuna [*sic*]: A fire-swept district', *Evening Post*, 14 January 1908, p. 3.
84 *Wanganui Herald*, 21 January 1908, p. 7.
85 'Bush fires', *West Coast Times*, 19 February 1908, p. 4. This is a late report, giving accounts of the fires a month earlier.
86 *Taranaki Herald*, 17 February 1908, p. 7.
87 'Bush fires: Latest reports of damage', *Hawera and Normanby Star*, 17 February 1908, p. 5; 'Great alarm, Latest bush fire reports', *Evening Post*, 19 February 1908, p. 8; 'The bush fires, Auckland', *Otago Witness*, 26 February 1908, p. 31.
88 'The bush fires, Auckland', *Otago Witness*, 26 February 1908, p. 31. It is difficult to discover just when those fires started, and how accurate the *Otago Witness* retrospective report was. The Wilton Bush fire was reported on 18 February, the Khandallah fires on 22 February. The smoke from fires further north could have been drifting over the city for up to two weeks.
89 *Evening Post*, 18 February 1908, p. 7.
90 'Great alarm: Latest bush fire reports', *Evening Post*, 19 February 1908, p. 8.
91 'Fires', *Otago Witness*, 19 February 1908, p. 32.

Chapter Two: A good burn?

1 *Appendices to the Journals of the House of Representatives* (*AJHR*), *1908*, C.–1, Department of Lands, Reports from rangers on the settlement of Crown lands: Nelson and Marlborough, pp. 37 & 38.
2 *AJHR*, 1908, C.–1, Department of Lands, Annual report, Appendix IV, Report on the timber industry in the Auckland land district, by H. P. Kavanagh, Chief Timber Expert, pp. 80–81.
3 *Evening Post*, 22 February 1908, p. 9.
4 G. T. Bloomfield, *New Zealand: A handbook of historical statistics*, G. K. Hall and Co., Boston, 1984, Table VII.4.
5 *AJHR*, 1907, C.–4, Department of Lands, Summary of the sawmills of New Zealand, p. 4.
6 Malcolm McKinnon (ed.), *The New Zealand Historical Atlas: Visualising New Zealand*, David

Bateman/Department of Internal Affairs, Auckland, 1997, plate 64.

7 *Evening Post*, 25 February 1908, p. 2.

8 Letter to the editor by 'Fifty Years Here', *Evening Post*, 25 February 1915, p. 3.

9 Bloomfield, *A Handbook of Historical Statistics*, Table V.12, Livestock numbers 1851–1917, p. 181. Figures for 1906 and 1908 are not in the table.

10 'A visit to the bush fires', *Otago Witness*, 3 February 1872, p. 9; 'Pine Hill Bush fire', *Otago Witness*, 2 March 1897, p. 9; 'From our own correspondent', *Evening Post*, 30 December 1878, p. 2; *Evening Post*, 14 January 1897, p. 5; 'The calamitous bush fires', *Evening Post*, 19 January 1898, p. 6.

11 *Taranaki Herald*, 16 March 1867, p. 2.

12 'The calamitous bush fires', *Evening Post*, 18 January 1848, p. 5.

13 Rollo Arnold, *New Zealand's Burning: The settler's world in the mid 1880s*, Victoria University Press, Wellington, 1994, pp. 258–59.

14 Arnold, *New Zealand's Burning*, pp. 258–59. Arnold, in the preceding and following pages, has an interesting discussion on the local social networks that made it easier to organise effective help.

15 'Calamitous bush fires: Interrupted railway traffic', *Evening Post*, 17 January 1898, p. 6.

16 *North Otago Times*, 10 January 1879, p. 2. Arnold, *New Zealand's Burning*, p. 45, documents the special train to Waipawa.

17 'Damage to the Korokoro bush: Statement by the Premier', *Evening Post*, 22 February 1908, p. 6.

18 *AJHR*, 1908, C.–1 Department of Lands, Annual report, Bush fires, p. 6.

19 *Evening Post*, 22 February 1908, p. 6; *Evening Post*, 24 February 1908, p. 2; *Taranaki Herald*, 25 February 1907, p. 7. However, as noted, the official government reporting did not list these initiatives and I have not been able to establish the accuracy of the papers reporting on them.

20 *AJHR*, 1908, C.–1 Department of Lands, Annual report, Bush fires, p. 6.

21 'Assisting settlers: Supply of grass seed', *Evening Post*, 30 May 1908, p. 2. In total £13,223 was distributed among 319 settlers in the Wellington district (which included the Wairarapa); £7837 among 201 in Auckland; £1514 among 102 in Taranaki; £133 among eight in Hawke's Bay; £1023 among 44 in Nelson; and £1681 among 21 in Marlborough. One man in Southland got £17 5s 1d.

22 *AJHR*, 1908, C.–1, Crown Lands, Appendix 1, Settlement of Crown lands, p. 22. Only the Commissioner of Crown Lands in Auckland gives the figures for his province. The content of commissioners' reports was not standard, so no conclusions can be drawn from comment that appears in one report and not in another.

23 'Bush-fires in the King Country', *New Zealand Parliamentary Debates* (*NZPD*), 1908, vol. 143, pp. 669–70.

24 *Wanganui Herald*, 20 March 1908, p. 7.

25 A. D. McIntosh (ed.), *Marlborough: A provincial history*, Marlborough Provincial Historical Committee, Blenheim, 1940, pp. 161, 194, comments on the unintended results of Grey's 1853 cheap land regulations. These were intended to settle small-time farmers but pastoralists used them in their own favour, extending initial one-year licences to seven or 14 years and buying up portions or the whole of the holdings at minimum prices. Runs, too, could be auctioned only at their own request, and early franchises were weighed in their favour. *AJHR*, 1908, Department of Lands, Annual report, Table J: Summary of Settlements established upon estates, p. 10, indicates that settlement in Marlborough was primarily on the large scale needed for grazing and sheep farming.

26 See, for instance, *AJHR*, 1902, C.–1, Department of Lands, Annual report, p. vi; *AJHR*, 1904, Department of Lands, Annual report, pp. xix–xx.

27 'Minister for Lands in the South', *Evening Post*, 28 October 1908, p. 12.

28 *AJHR*, 1907, C.–1, Department of Lands, Settlement of Crown Lands, Otago, Bush fires, p. 46.

29 Arnold, *New Zealand's Burning*, p. 254.

30 'The Wairarapa fires', *Evening Post*, 13 January, p. 4.

31 'The gale', *Evening Post*, 10 February 1898, p. 6.
32 'The up-country bush fires, Hastwell relief fund', p. 6, 'The destruction still going on', p. 6, 'The Wairarapa fires', p. 4, *Evening Post*, 13 January 1897. Again the estimate of loss appears extravagant. At Hastwell the fund was closed down once timber for Mr McGregor's house had been ordered, and he was to get a new chimney. Clothing was to be distributed to 'the two sufferers'.
33 Arnold, *New Zealand's Burning*, pp. 253–55. In relation to viewing assistance as charity, see the letter he quotes on p. 267.
34 *Evening Post*, 25 February 1908, p. 8.
35 '1918 bush fires', *New Zealand Herald*, 1 December 1938, p. 11. Karen Hawke Grimwade, *100 Years of Fire Fighting in Ohakune 1908–2008: A history, to commemorate the centennial jubilee of the Ohakune Volunteer Fire Brigade*, The Brigade, Ohakune, 2008, p. 22, notes the estimated losses.
36 'The calamitous bush fires', *Evening Post*, 19 January 1898, p. 6.
37 J. D. Wickham, a travelling correspondent for the *Auckland Weekly News* during the disastrous fires in January 1887, quoted in Arnold, *New Zealand's Burning*, p. 249.
38 Francis Guy, 'How we fought the bush fires', Programme 75, Open Country Sound Recordings, OHT 5-0064, OHA. Francis was five at the time of the 1898 fires.
39 'The bush fires: Settlers' terrible hardships', *Otago Witness*, 26 February 1908, account dated 22 February, p. 31.
40 'Local and general', *Evening Post*, 27 May 1910, p. 6.
41 *AJHR*, 1908, C.–1, Department of Lands, Annual report, Reports on settlement of Crown lands, Hawke's Bay, pp. 24–25, and Taranaki, p. 27.
42 Judith A. Johnston, 'New Zealand bush: Early assessments of vegetation', *New Zealand Geographer*, April 1981.
43 Ibid., pp. 22–23.
44 In New Zealand a few suggested that fires were not beneficial. In 1869 the Rev. Buchanan warned that good grass did not always result from repeated burning. The finer species of grass, he argued, were shallow-rooted and so destroyed by fire, sun or frost, while the deep-rooted tussock and speargrass were preserved. Review: The New Zealand Institute, *Transactions and Proceedings of the New Zealand Institute, Vol. 1, 1868, Otago Witness*, 17 July 1869, p. 9.
45 Peter Stanbury (ed.), *Bushfires: Their effect on Australian life and landscape*, Macleay Museum, University of Sydney, Sydney, 1981, p. 33.
46 *AJHR*, 1904, C.–1, Department of Lands, Appendix 1, Administration.
47 'Minister for Lands in the south', *Evening Post*, 28 October 1908, p. 12.
48 *Nelson Examiner and NZ Chronicle*, 7 January 1843, p. 174. The long and fascinating article on breaking in the land behind Nelson describes the practice and discusses its pros and cons.
49 Arnold, *New Zealand's Burning*, p. 238; S. Pyne, *Fire in America: A cultural history of wildland and rural fire*, Princeton University Press, Princeton, 1982, p. 133, comments on the effects of enclosure etc.
50 F1, 10/2/46, Pt. 1, Reports of fire control officer, 1949–56, 'New Zealand Forest Service: The fire protection of rural lands in New Zealand', undated, [1951], Archives New Zealand (ANZ).
51 T. Foster, *Bushfire in Australian History: History, prevention, control*, A. H. & A. W. Reed, Sydney, 1976, p. 12. As late as the 1950s, an official report also suggests that the New Zealand settlers' lack of forestry traditions 'was bound to mean the destruction of forest on a large scale' – William Wright, F1, 10/2/46, Pt. 1, Reports of fire control officer 1949–56, 'New Zealand Forest Service: The fire protection of rural lands in New Zealand', undated (1951), p. 1, ANZ.
52 Pyne, *Fire in America*, p. 135.
53 Foster, *Bushfire in Australian History*, pp. 13–14. The quote on the Peshtigo fire is on p. 14; Pyne, *Fire in America*, p. 92.
54 *West Coast Times*, 14 February 1898, p. 4; *Hawera & Normanby Star*, 22 January 1899, p. 4; *Otago Witness*, 7 February 1906, p. 9.

55 Arnold, *New Zealand's Burning*, p. 238. Arnold quotes Stephen Pyne, *Burning Bush: A fire history of Australia*, New York, 1991, p. 261, in which Pyne suggests that burning was integral to the making of both Australian and North American frontier economies, and 'was regarded as a folk right'.
56 John Sheenan, *NZPD*, 1874, vol. 16, p. 351.
57 William Gibbs, *NZPD*, 1874, vol. 16, p. 358.
58 *AJHR*, 1904, C.–1, Department of Lands, Annual report, p. xx.
59 Christchurch *Press*, 21 March 1928, F. V. Bevan Brown, Letters to the Editor, p. 11. The newspapers carry many such statements.
60 *AJHR*, 1890, C.–1, Crown Lands Department, Annual report, State forests, p. 5. The Secretary of Crown Lands considered it pointless to reserve small forest areas close to land being cleared: unable to be protected from the annual fires, those lands would themselves become a danger. A similar view was expressed the previous year (p. 7) and in the department's annual reports in 1894 in regard to whether the Auckland Land Board should sell its forests given the inability to stop fires (p. vi); 1896, Appendix 1 (p. 2); 1898, on the risk of fires from settlement (p. vi).
61 Arnold, *New Zealand's Burning*, p. 238.
62 'Out of the bush: The birth of a town', *Evening Post*, 21 June 1911, p. 11.
63 'Legislative approach', *Otago Witness*, 16 April 1864, p. 13.
64 *AJHR*, 1938, C.–3, State Forest Service Annual Report (SFSAR), p. 3.
65 Michael Roche, *History of New Zealand Forestry*, New Zealand Forestry Corporation/GP Books, Wellington, 1990, pp. 22–23.
66 J. Macklow, 'To readers and correspondents', *Daily Southern Cross*, 17 February 1872, p. 2.
67 Bush fire ordinance 1863, *Canterbury Provincial Council Ordinances, 1853–64*, p. 555. This provided a penalty (unspecified) for firing bush, scrub or grass over certain months of the year (effectively a fire ban); in the event of a fire, the owner or occupier was deemed to be guilty unless proved otherwise; sawpit refuse was to be destroyed by 1 November each year (at the beginning of the dry period); offences were to be determined by confessions or by witnesses under oath; penalties could not be recovered after six months after the offence. The move was disallowed by the governor; *New Zealand Gazette*, 17 January 1864, p. 1.
68 Bush fire ordinance 1864, *Otago Provincial Council Ordinances*, p. 941, provided for a penalty of up to £25 for lighting fires near crops or imprisonment with or without hard labour, and permitted burning if owners of adjoining properties notified 24 hours beforehand. Firebreaks of 5 yards (4.56 metres) had to be made from any fence dividing properties and, in the event of fire, any owner not complying had to re-erect the fence at his own cost. The governor signalled he was pleased 'to leave the same to their operation': Bush fire ordinance, *NZ Gazette*, 11 January 1865, p. 12.
69 As reported in 'Bush Fires Act 1866 (Draft)', *Daily Southern Cross*, 23 November 1866, p. 5. The bill provided for (limited) penalties for damage to another's property with cases to be heard before a resident magistrate or two JPs; no negligence could be claimed if a firebreak 20 metres wide had been developed close to or adjoining neighbouring property and notice had been given; all cultivated land within 200 metres of fern fired was deemed to be immediately adjoining; owners of land 200 metres from burn-offs of more than 0.4 hectares of bush or 2 hectares of fern could claim fires had been negligently or unlawfully lit if due notice had not been given; fires could not be lit three days either side of notified date and time of burning; and all land, other than timber or light bush, was held to be fern land within the meaning of the act. On 26 January 1867 the *Daily Southern Cross* (p. 4) reported that at its second reading the bill was removed from the order paper because of lack of time. I found no further reference to it.
70 *AJHR*, 1869, D.–22, Correspondence relative to the present condition of the forests of New Zealand, Returns: Enclosure no. 5, Otago.
71 'The forests of New Zealand', *Otago Witness*, 31 October 1868, p. 10. This was reprinted verbatim in the *Otago Daily Times* a month later.
72 'The new Governor of New Zealand', *West Coast Times*, 29 October 1867, p. 2 describes the provisions of the South Australian act. From November to May anyone smoking in the open

air within 9 metres of any 'stable, or of any rick or field of grass, hay, corn, straw, or stubble, unless within a town; or [using] a pipe without a metal cover, or having in his possession other than safety matches [is] to pay, on conviction no less than 10 shillings and no more than £5, or, in default of payment, liable to imprisonment no longer than one month, with or without hard labour'.

73 *AJHR*, 1874, H.–5, Papers relating the state forests. 1. Parliamentary debates and resolutions, p. 11. See also Hochsetter, quoted by Mr Potts in *AJHR*, 1874, H.–5, Papers relating to state forests. 1. Parliamentary debates and resolution, The forests of the colony, p.1 – though Hochsetter viewed the destruction solely through a conservation lens.

74 Simon Nathan and Mary Varnham (eds), *The Amazing World of James Hector*, Awa Science, Wellington, 2008, p. 4.

75 *AJHR*, 1874, H.–5, Papers relating to state forests, 3. Provincial reports: Analysis of reports received in 1869, Dr Hector to the Hon. Colonial Secretary, pp. 14–16.

76 *NZPD*, 1874, pp. 467–68. The bill prescribed rudimentary firebreaks, times to burn and quite stringent penalties for burning others' crops, grass or 'wood land' and was criticised as impractical and an impediment to settlement (information from Michael Roche).

77 *Evening Post*, 17 July 1874, (untitled article), p. 2.

78 New Zealand Forests Act 1874, s. 11.

79 Roche, *History of New Zealand Forestry*, p. 93.

80 See *AJHR*, 1880, 'Suggestions on forests in New Zealand', Mr A. Lecoy to the Hon. Minister of Lands. Lecoy's broader interest in forestry is also evident in a further paper to the Wellington Philosophical Society on 26 July 1879 – see *Transactions and Proceedings of the Royal Society of New Zealand 1868–1961:* http://rsnz.natlib.govt.nz/volume/rsnz_12/rsnz_12_00_000560.html (accessed November 2008).

81 *AJHR*, 1886, C.–3D, State Forests Department, Report on progress by T. Kirk, pp. 1 and 2.

82 *Regulations or the New Zealand Forests and Forest Reserves Subject to the Provisions of 'The New Zealand State Forests Act 1885' and Forests Growing on any Crown land*, ATL.

83 *AJHR*, 1886, C.–3D, State Forests Department, Report on progress by T. Kirk.

84 Roche, *History of New Zealand Forestry*, p. 95.

85 Ibid., p. 136; *AJHR*, 1888, C.–1, Crown Lands Department, Annual report, Forests and Agriculture Branch.

86 *AJHR*, 1889, C.–1, Crown Lands Department, Annual report.

87 J. H. Young, cited in J. P. Grossman, *The Evils of Deforestation*, Brett Printing and Publishing Co., Auckland, 1909, p. 14.

88 *AJHR*, 1874, H.–5, Papers relating to state forests, 4: Extent and rate of destruction of forest lands: profits of planting, rough memorandum by T. Calcutt, of the cost of planting and subsequent management of 100 acres of suitable land in trees, pp. 36–37.

89 *AJHR*, 1874, H.–5. Papers relating to state forests. 3: Provincial reports, Analysis of report received in 1869, Dr Hector to the Colonial Secretary, Parts 2 and 3.

90 Roche, *History of New Zealand Forestry*, p. 89.

91 *AJHR*, 1898, C.–1, Department of Lands, Annual report, p. vi, comments on the growing extraction rate.

92 In relation to estimates of the diminishing resource, see, for instance, Department of Lands, Annual report 1907; 1908, C.–1B, Department of Lands, State nurseries and plantations, pp. 1 and 4; *AHJR*, 1909, C.–4, Forestry in New Zealand, William Kensington to Minister of Lands, p. 3. Kirk's 'Forestry in New Zealand' (*AJHR*, 1909, C.–4) reported on the amount of bush remaining, how much should be retained for climatic, scenery and soil-protection purposes, and other related matters. The quote is from *AJHR*, 1909, C.–4, Forestry in New Zealand, Preface, p. 3.

93 *AJHR*, 1918, Session II, C.–3, Department of Lands, Annual report, Part 1, Native forests, p. 37.

94 *AJHR*, 1877, C.–3, Crown Lands, Report by the conservator of forests, p. 48.

95 *AJHR*, 1909, C.–4, Forestry in New Zealand. The terms in which Kirk discussed the climatic and protection values of the bush were rather more sophisticated than those of his predecessors.

96 'A visit to the bush fires', *Otago Witness*, 3 February 1872, p. 9.
97 'A great gift in danger', *Evening Post*, 28 February 1898, p. 4.
98 *AJHR*, 1886, C.–3D, State Forests Department, Report on progress, p. 9.
99 F1, 45/217/5 Historical records, 1899–1092, ANZ. Report for April 1902 by chief forester to surveyor-general, 2 May 1902.
100 M. Roche, 'Edward Phillips Turner: The development of a "forest sense" in New Zealand 1890s to 1930s', Proceedings of the 6th National Australian Forest History Conference, Augusta, WA, 12–17 September 2004, p. 2.
101 *AJHR*, 1911, C.–6, Department of Lands, Scenery Preservation, Appendix B, Report by the Inspector of Scenic Reserves, p. 6.
102 *AJHR*, 1919, C.–3, Department of Forestry, Part 1, Indigenous forests: Legislation. The state holdings comprised: (a) land controlled by the Commissioner of State Forests as state forest proclaimed under the State Forests Act 1908 and certain forest reserves under the Land Act before 1885; (b) land not controlled by the commissioner: forest on national endowment lands, forests acquired under the Land for Settlements Act, forest reserves or reserves for growing and preserving timber or under the Land Act or other acts, scenic reserves and national parks. Subsequent statements of the extent of state forests, which give lesser acreages, refer to exotic and indigenous forests under Forest Service control.
103 *AJHR*, 1905, Department of Lands, Annual report, Appendix VI, Report by Henry J. Mathews, chief forester, p. 75.
104 Richard Monk, *Our Forests and How to Conserve Them: A paper on the New Zealand State Forests Act 1885. Read before the Waitemata County Council 5 February 1886*. Printed at the *Star* office, Auckland. For details on Monk himself, see 'Monk family meets to celebrate 150 yrs', *Rodney and Waitemata Times*, 29 April 1980.
105 *NZPD*, 1891, vol. 105, p. 556.
106 *AJHR*, 1902, C.–1, Department of Lands, Report upon forests of kauri and other timbers in the Auckland, Poverty Bay and Westland Districts, pp. x–ix.
107 Monk, *Our Forests and How to Conserve Them*, p. 7. Monk wanted legislation to cover: Crown and privately owned forests; rangers not necessarily under government employ; district regulations providing for what were effectively fire bans, except under specific conditions; adequate enforcement for those guilty of careless use or not preventing fires through suspension of licences or imprisonment; and enforcement for those guilty of unintentionally lighting fires, with measures ranging from expressions of displeasure to complete censure.
108 *AJHR*, 1897, vol. 1, Session II, C.–8, Department of Lands, George Perrin, Report upon the conservation of New Zealand's forests, p. 37. Perrin wanted: the best lands to be designated as timber lands and made as unalienable as possible; fire bans in some districts over a very limited period; lighting fires or leaving any fire to burn or letting it spread to be prisonable offences; prohibitions on smoking, shooting and using any lighting device other than safety matches; extra assistance to the officers in charge to burn off and clear land near settlements over the summer months; the ability for officers to call on local men to help extinguish fires, with refusal punishable by imprisonment or loss of licence or both; specific provisions for issuing gum-diggers' licences; heavy fines for pastoral tenants for burning mountain tops or bush; and that all pastoral lands should be declared as fire districts, with fines for shepherds and possible fines or forfeiture of licence for pastoral tenants.
109 *AJHR*, 1908, Department of Lands, Annual report, State nurseries and plantations, duties of the Forestry Branch, p. 10.
110 *AJHR*, 1909, C.–4, Forestry in New Zealand, Preface, p. 3.
111 F1, 45/271/5, Historical records 1899–1902, chief forester to surveyor-general, 4 January 1900 re Rotorua nursery; F1, 45/272/5, Historical records 1903–03, Report for December 1902, chief forester to surveyor-general, 10 January 1903; re Waiotapu plantation, ibid., Report for November, 24 December 1903; re Puhipuhi plantation, F1, 45/271/5 Historical records 1906–06, October report, 10 November 1906. All at ANZ.
112 F1, 45/272/5, Historical records 1903–05, Report for July, 12 August 1904, ANZ.
113 F1, 45/271/5, Historical records 1906–06, Under-secretary Marchant to H. J. Mathews, chief

forester, ANZ.
114 F1, 45/271/5 Historical records 1906–06, ANZ. October report, 1 November 1906, Waiotapu plantation, pp. 17–18, ANZ.
115 *AJHR*, 1911, C–1B, Department of Lands, State afforestation in New Zealand, p. 18. Re the risk from grass breaks see Appendix B, Report on tree-growing operations in South Island, p. 54.
116 *AJHR*, 1911, C.–1B, Department of Lands, State afforestation in New Zealand, p. 18.
117 *AJHR*, 1913, C.–12, Report of the Royal Commission of Forestry, p. x.
118 *AJHR*, 1913, C.–1B, Department of Lands, Annual report, Summary of operations; and Report on afforestation operations in the North Island, p. 5.
119 *AJHR*, 1914, C.–1B, Department of Lands, State nurseries and plantations, Report on afforestation operations in the South Island, Fire-resisting trees.
120 *AJHR*, 1917, C.–3, Department of Lands, State nurseries and plantations, Report on afforestation operations in the South Island, p. 18.
121 *AJHR*, 1919, C.–3, Department of Forestry, Appendix A: Report upon afforestation operations in the North Island; and Report on afforestation operations in the South Island.
122 *AJHR*, 1913, C.–1B, Department of Lands, State nurseries and plantations, Report on afforestation operations in the North Island, p. 5.
123 *AJHR*, 1914, C.–1B, Department of Lands, State nurseries and plantations, p. 2; Report on afforestation operations in the North Island, Protecting the plantations from fire; Report on afforestation in the South Island, Fire-preventative measures, p. 25. The exact information on what was where conflicts slightly between the main report and Robinson's (South Island) report, and the number of depots at Dusky Hills is unclear. Superintending nurseryman H. A. Goudie's 1913 report and recommendations (*AJHR*, 1913, C.–1B, Report on the afforestation operations in the North Island, p. 7) on what was needed may well have had some bearing on these developments.
124 F1 45/171/1 Historical records 1907–07, chief forester to under-secretary of lands, 13 February 1907, ANZ (emphasis in the original). Phillips Turner, Inspector of Scenic Reserves, made a similar statement in 1911 (*AJHR*, 1911, C.–6, Department of Lands: Scenery Preservation, Appendix II, p. 5).
125 In 1913 fire damaged trees at Puhipuhi and in the 1916 fire at Hanmer Springs destroyed about 12 hectares of 14-year-old trees.
126 *AJHR*, 1917, C.–3, Department of Lands, State forestry, Part 1, State nurseries and plantations, Report on afforestation operations in the South Island, p. 17.
127 *AJHR*, 1915, C.–1B, State nurseries and plantations, Appendix 1, Report on forestry conference in Scotland and tour through America and Europe, p. 49.
128 The first example of staff contributions I found was H. A Goudie's afforestation operations report in 1913, when he drew on his reading of *Forestry in New England* by Hawley and Hawes to observe shortcomings in New Zealand fire-fighting equipment and systems (*AJHR*, 1913, C.–1B, Department of Lands, State nurseries and plantations, Report into the afforestation operations of the North Island, p. 7). In terms of the wider public appreciation of current ideas, timber merchant Alexander McColl's representations on Waipoua Forest stressed the need for forestry training and practical forest management as a way of minimising the number of forest fires (*AJHR*, 1913, C.–12, Report of the Royal Commission on Forestry, p. 56).
129 *Rules and Objects of the New Zealand Forestry League (Inc)*. For its achievements, see *History of the New Zealand Forestry League: A remarkable record*, Wright and Carman, Wellington, 1935? For an idea of its message and form, see the *Forest Magazine of New Zealand: Official journal of the New Zealand Forestry League*, which described itself as '[a] monthly magazine about Forestry, Tramping, and Outdoors'.
130 *AJHR*, 1915, C.–1B, State nurseries and plantations, Appendix 1, Report on forestry conference in Scotland and tour through America and Europe, p. 49.
131 *AJHR*, 1921, Session II, C.–3, State Forest Service, p. 7.
132 *AJHR*, 1916, C.–6, Department of Lands, Scenery preservation, p. 2.

133 *AJHR*, 1908, C.–4, Department of Lands, Annual report, Scenery preservation, p. 9.
134 *AJHR*, 1910 and 1911, C.–6, Department of Lands, Annual report, Appendix B, Report by the Inspector of Scenic Reserves, pp. 6 and 5 respectively.
135 *AJHR*, 1910, C.–6, Department of Lands, Annual report, Appendix B, Report by the Inspector of Scenic Reserves, p. 7.
136 *AJHR*, 1919, C.–3, Department of Forestry, Forest fires, p. 6.

Chapter Three: 'A practical, constructive, and well-ordered' fire policy

1 Alex R. Entrican, 'Looking back, MacIntosh Ellis had great influence on New Zealand forest policy', excerpt from 'Influence of forestry on forest policy and forest products trade in Australia and New Zealand', MacMillan Lecture, Vancouver, January 1963, *New Zealand Journal of Forestry* (*NZJF*), August 1996, p. 46.
2 Peter McKelvey, 'Macintosh Ellis in France', *NZJF*, August 1989, pp. 15–18.
3 Michael Roche, 'Leon MacIntosh Ellis', *Dictionary of New Zealand Biography:* www.teara.govt.nz/en/biographies (accessed 23 February 2012).
4 Michael Roche, personal communication, October 2008.
5 McKelvey, 'Macintosh Ellis in France', p. 16.
6 *Appendices to the Journals of the House of Representatives* (*AJHR*), 1920, C.–3A, Department of Forestry, Report on forest conditions in New Zealand, and the proposals for a New Zealand forest policy, pp. 3, 7, 8 and 20.
7 Sir Francis Dillon Bell's 1919 Forests Amendment Act should also be noted in terms of forest conservation. Prior to that act, the Department of Lands and Survey effectively could decide to take Crown land from its reserved status and sell it for farms, while foresters had to argue for its retention as state forest. After the amendment act the situation was turned on its head, with Lands and Survey having to justify its wish to change the land use.
8 Roche, 'Leon Macintosh Ellis'.
9 *AJHR*, 1925, C.–3, State Forest Service Annual Report (SFSAR), p. 17.
10 *AJHR*, 1920, C.–3A, SFSAR.
11 *AJHR*, 1920, C.–3, SFSAR, pp. 6 and 10.
12 *AJHR*, 1921, Session II, C.–3, SFSAR, p. 7.
13 Christine Kenneally, 'The inferno', *New Yorker*, October 2009, p. 50.
14 *AJHR*, 1922, C.–3, SRSAR, p. 8.
15 Roche, 'Leon MacIntosh Ellis'.
16 *AJHR*, 1922, C.–3, SRSAR, p. 8.
17 Stephen Pyne, *Fire in America: A cultural history of wildland and rural fire*, Princeton University Press, Princeton, 1982, pp. 224–25.
18 Ibid., pp. 190, 218, 260.
19 Ted Foster, *Bushfire: History, prevention, control*, A. H. & A. W. Reed, Sydney, 1976, pp. 61–62.
20 F1, 12/0, Pt.1, Annual fire reports and fire protection 1920–24, Archives New Zealand (ANZ). Memorandum to all conservators, 14 January 1924, copy of 'Forest policies and the problems of wood operations' in *American Lumberman*, November 1923.
21 Pyne, *Fire in America*, p. 168.
22 Ibid., pp. 167–68; *AJHR*, 1919, C.–3, Department of Forestry, p. 6.
23 *New Zealand Parliamentary Debates* (*NZPD*), 1921, vol. 192, pp. 490–91.
24 F1, 1/5/- Pt.1 Forest legislation general 1921–22, ANZ. William Carey, forest ranger, British Columbia, to Director of Forestry, 4 November 1921.
25 Foster, *Bushfire*, p. 39.
26 R. H. Luke and A. G. McArthur, *Bushfires in Australia*, Australian Government Publishing Service, Canberra, 1978, p. 18.
27 H. C. Morgan, 'Fighting an Australian bush-fire', *Chambers Journal*, June 1937, pp. 445–58.
28 *NZPD*, 1921, vol. 192, pp. 490–91.
29 F1, 1/5/-, Pt. 1, Forest legislation general 1921–22, ANZ. C. E. Poole, Forests Department, Perth, 25 May 1921.
30 Luke and McArthur, *Bushfires in Australia*, p. 18.

31 F1, 12/0, Pt. 1, Annual fire reports and fire protection 1920–24, ANZ. Memorandum to all conservators, 14 January 1924, copy of 'Forest policies and the problems of wood operations', in *American Lumberman*, November 1923.
32 *AJHR*, 1921, Session II, C.–3, SFSAR, p. 7.
33 F1, 1/5/-, Pt. 1, Forest legislation general 1921–22, ANZ. Director of Forestry to Crown Law draftsman, 24 May 1921.
34 Ibid. On the Australian input, see correspondence from C. E. Poole, Forests Department, Perth, 25 May 1921; on the British Columbian, see William Carey, forest ranger, British Columbia, to Director of Forestry, 4 November 1921; on the industry comments, see New Zealand Council of Agriculture to Commissioner of State Forests, 19 August 1921, urging the passage of the bill to safeguard the forests; amendments suggested by Dominion Federated Sawmillers Association, 'Forests Bill, 1921', 8 November 1921 and change to the supplementary order paper, Legislative Council, 6 January 1922.
35 *AJHR*, 1921, Session II, C.–3, SFSAR, p. 13.
36 F1, 1/5/-, Pt. 2, Forest legislation general 1921–22, ANZ. There is a discrepancy here that I have not been able to resolve, but it suggests that Ellis was talking to his men about the legislation before it was passed. Information from British Columbia on its fire districts and how they were operated was received on 4 November 1921, and Part 2 of the Forest Legislation file (F1, 1/15/- Forest legislation general, ANZ) indicates that, after progressive amendments, the fire district provisions were inserted, for the first time, into the Forests Bill. However, as early as May 1921 rangers in the Ohakune district were discussing setting up fire districts within their part of the conservancy (F1, 12/3, Pt. 1, Fire prevention Wellington 1921–25, ANZ).
37 F1, Box 288, 12/0, Pt. 2, Fire protection general 1922–29, ANZ.
38 A. V. Galbraith, *Forest Fires: Cause and effect*, H. J. Green, Government Printer, Melbourne, 1926, pp. 2–3.
39 Two good examples of how he encouraged his men to take independent decisions can be seen in letters to the conservator in Southland, D. Macpherson: 'I have to state that provided that you keep within your allotments and the authorised rates of pay there is no necessity for you to refer to me for authority to engage fire patrol men,' he wrote on 26 November 1921. Again, on 17 February 1925 he wrote on stopping logging machinery to obviate the threat they caused. 'The Conservator of Forests has full authority to control this menace in forests under his control …' (see F1, 12/7, Pt. 1, Fire reports Southland 1920–26, ANZ).
40 *AJHR*, 1923, C.–3, SFSAR, p. 9.
41 *AJHR*, 1926, C.–3, SFSAR, p. 14.
42 F1, 12/3, Pt. 1, Fire prevention Wellington 1921–25, ANZ. The earliest reference I found to fire patrols was from Wellington conservator S. A. C. Darby to rangers, 27 May 1921, when he suggested setting up patrols. The circular is also interesting because it suggests the staff were coming to terms with new ideas, perhaps presented by Ellis as he sought comment and direction on the new act he was developing. Darby's obituary, *NZJF*, 30(1), 1985, pp. 16–17) suggests frequent interchanges between the two men. The file also holds elaborated suggestions by A. Hansson, chief inspector, on how the patrols might be implemented: Some notes on fire protection by A. Hansson, chief inspector, undated [c. 1921].
43 F1, 12/0, Pt. 1, Annual fire reports and fire protection 1920–24, ANZ. Secretary of Forestry to Public Service Commissioner, 'Fire prevention (state forest), undated, May (c. August/September 1922). Hansson's 'Notes on fire protection' (ibid.), which I used to elaborate on the summary given in this document, also carried rates of pay slightly higher than in the 1922 document. I have used the later pay rates, as the patrolmen were more embedded in the system by then and the later rates are more likely to represent continued practice.
44 F1, 12/3, Pt. 1, Fire prevention Wellington 1921–25, ANZ. Ranger R. Norman Uren to senior ranger, 2 October 1921.
45 *AJHR*, 1923, C.–3, SFSAR, p. 9–10; F1, 45/11, Pt, 1920–22, ANZ. 'Fire season 1922–23', 18 October 1922.
46 F1, 12/5, Box 301, Fire reports Westland 1921–44, ANZ. 'Forest prevention', letter from McGavock to millers, 10 November 1922; McGavock to Director of Forestry, 'Forest fire

prevention', 7 December. Phillips Turner's positive response was dated 18 December 1922.
47 F1, 12/3, Pt. 1, Fire prevention Wellington 1921–25, ANZ. See comment on note 36 above.
48 F1, 12/0, Pt. 2, Fire protection general 1925–29, ANZ. 'Prevention and control of plantation fires by means of the fire district principle' [undated, c. 1926].
49 F1, 1/5/-, Pt. 4, Forest legislation 1925–26, ANZ. Proposals for amendments to the Act, 21 May 1925.
50 *AJHR*, 1927, C.–3, SFSAR, p. 14.
51 F1, 1/15/-, Pt. 4, Forest legislation 1925–26, ANZ. Secretary of Forestry to Commissioner of State Forests, 29 May 1925. The district could be constituted or a closed fire season declared without written notices being provided.
52 F1, 12/0, Pt. 2, Fire protection general, ANZ. 'Prevention and control of plantation fires by means of the fire district principle' [undated but c. 1926].
53 F1, 1/15/-, Pt. 4, Forest legislation 1925–26, ANZ. Secretary of Forestry to Commissioner of State Forests, 29 May 1925.
54 *AJHR*, 1925, C.–3, SFSAR, p. 15.
55 F1, 12/0/1, Fire protection of flax areas 1927–27, ANZ.
56 *AJHR*, 1948, C.–3, SFSAR, p. 25.
57 *Daily Telegraph*, 22 February 1946. In an editorial, 'The menace of fire', the writer commented on recurring loss, a high price for 'the persistence of the theory that what is everybody's business is nobody's'.
58 *AJHR*, 1921, C.–3, SFSAR, p. 7.
59 *AJHR*, 1926, C.–3, SFSAR, p. 1.
60 See, for instance, New Zealand Forestry League, 'Forestry facts: Something to think about', *Forest Magazine of New Zealand*, vol. 1, no. 1, February 1922, in which Ellis promoted the need for 'practical forestry … The continuance of the laissez-faire-talk policy will certainly mean that, in a few years, your National forest estate will be as extinct as the Moa!' (p. 6).
61 F1, 1/15/-, Pt. 3, Forest legislation 1922–24, ANZ. E. Phillips Turner to Commissioner of State Forests, 30 June 1922.
62 F1, 12/9, Fire-prevention signs, leaflets etc 1922–21, ANZ.
63 F1, 12/0, Pt. 1, Annual fire report and fire protection, ANZ. In September 1922 a number of designs and possible wordings were submitted, aiming to appeal to a number of audiences: the general public, sportsmen, timber workers. Another suggestion, focusing more on the punitive aspects, read: Fires: A burning match or cigarette carelessly thrown down may by starting a fire in this forest, mean Imprisonment or a fine of £50 to you. Don't run the risk. Fire raising is a destructive and expensive form of amusement.
64 In 1920 signs were posted at the main routes into forest reserves (F1, 45/11, Pt. 1, Circular letters to conservators 1920–22, ANZ. Memorandum to forest rangers, 15 November 1920). In the 1923–24 fire season 2350 calico signs were distributed (*AJHR*, 1924, C.–3, SFSAR, p. 17). However, in early 1927 Ellis wrote to Hansson complaining that although he had travelled through most of the conservancies he had not seen a notice (F1, 12/0, Pt. 2, Fire protection general 1925–29, ANZ. Director of Forestry to chief inspector, 9 February 1927).
65 F1, 12/9, Fire-prevention signs, leaflets etc 1922–31, ANZ. Ellis to conservator, Whakarewarewa, 22 August 1923.
66 Ibid. Ellis to conservator, Whakarewarewa, 21 November 1922. Ellis's Rotorua conservator did not agree: worried about the 'tremendous' fire loss in forests in the Rotorua–Taupo area due to perceived Maori carelessness with fire, he wanted penalties emphasised, not appeals for assistance (ibid., conservator, Whakarewarewa, to Director of Forestry, 29 August 1923).
67 F1, 12/0, Pt. 1, Annual fire report and fire protection, ANZ. Arnold Hansson to Director of Forestry, 2 September 1924.
68 F1, 12/9, Fire-prevention signs, leaflets etc 1922–31, ANZ. See correspondence 19 January 1925 to October 1925; and 26 November 1926, 4 December 1926.
69 F1, 45/11, Circular letters to conservators, 'Forestry propaganda', 27 May 1924.
70 *AJHR*, 1927, C.–3, SFSAR, p. 29.
71 Ellis and Burnand, *Ellis and Burnand: Fifty years of service in the timber industry of N.Z.*,

Hamilton, 1953; Thomas E. Simpson, *Kauri to Radiata: Origin and expansion of the timber industry of New Zealand*, Hodder and Stoughton, Auckland, 1973, p. 75.

72 McKelvey, 'Macintosh Ellis in France', p. 15.

73 *AJHR*, 1924, C.–3, SFSAR, p. 7.

74 *AJHR*, 1923, C.–3, SFSAR, pp. 9–10.

75 F1, 12/0, Pt. 5, Fire protection general 1939–40, ANZ. Conservator, Westland, to Director of Forestry, 26 May 1939, p. 5.

76 Anon., 'Obituary, Arnold Hansson', *NZJF*, vol. 1, no. 27, 1982, pp. 19–21 for further details on the man and his work.

77 Anon., 'Obituary, Norman James Dolamore', *NZJF*, vol. 7, no. 3, 1956, p. 9.

78 F1, 12/3, Pt. 1, Fire prevention Wellington 1921–25, ANZ. 'Some notes on protection' by A. Hansson, chief inspector, undated [c. 1921].

79 F1, 12/0, Pt. 2, Fire protection general 1925–29, ANZ. 'Prevention and control of plantation fires by means of the fire district principle', undated [c. 1926]. The comment on liability comes from *AJHR*, 1923, C.–3, SFSAR, pp. 9–10.

80 See, for instance, F1, 12/5, Box 301, Fire reports Westland 1921–44, ANZ. Forest Service, Fire report, 27 May 1922.

81 F1, 12/7, Pt. 1, Fire reports Southland 1920–26, ANZ. Director of Forestry to conservator, 3 February 1925.

82 F1, 45/11, Pt. 1, Circular letters to conservators 1920–21, ANZ. 'Fire Season 1922–23', 18 October 1922.

83 F1, 12/0, Pt. 1, Annual fire reports and fire protection 1920–24, ANZ. Director of Forestry to all conservators, 21 October 1922.

84 Ibid. Director of Forestry to all conservators, 10 January 1924. In this memo Ellis deplored that fires reported in daily papers had not been reported to head office, and ended sharply: if conservators couldn't report the full particulars with their present state patrolmen 'your protective system is weak and needs revision'.

85 F1, 13/7, Forest rangers' handbook 1921–24, ANZ.

86 F1, 12/0, Pt. 1, Annual fire reports and fire protection 1920–24, ANZ. Memorandum, all conservators re fire protection proposals, 1922–23, 19 August 1922

87 F1, 12/0, Pt. 1, Annual fire reports and fire protection 1920–24, ANZ. Summary of annual fire reports 1923–24.

88 W. B. Osborne, 'The lookout system', in *The Western Fire Fighters' Manual* (6th edn), Western Forestry and Conservation Association, Seattle, 1934 (Millman collection), p. 1.

89 *AJHR*, 1922, C.–3, SFSAR, p. 19, and Annex III, Report on forestation operations in the South Island.

90 F1, 12/3, Pt. 1, Fire prevention Wellington 1921–25, ANZ. Memorandum to Director of Forestry, 8 July 1921.

91 F1, 12/0, Pt. 1, Annual fire reports and fire protection 1920–24, ANZ. Summary of annual fire reports 1923–24.

92 F1, Box 301, 12/5, Fire reports Westland 1921–44, ANZ. Westland region, Summary of annual protective operations 1921–22, undated [May 1922].

93 F1, 12/0, Pt. 1, Annual fire reports and fire protection 1920–24, ANZ. Inspector of Forests to all conservators, 6 August 1924.

94 F1, 12/3, Pt. 1, Fire prevention Wellington 1921–25, ANZ. S. A. C. Darby to rangers, 27 May 1921.

95 Ibid. R. Norman Uren to senior ranger, undated.

96 Ibid. Memorandum to Director of State Forest Services, 8 July 1921.

97 Ibid. Director to ranger Dolamore, Ohakune, 21 October 1921.

98 Ibid. Uren to senior ranger 2 October 1921.

99 Ibid. Uren to senior ranger, 14 February 1922.

100 F1, 12/7, Pt. 1, Fire reports Southland 1920–26, ANZ. Director of Forestry to conservator, Invercargill, 10 November 1922. In this memo Ellis stressed the importance of patrolmen supervising burning to lessen the spread of fires.

101 F1, 12/3 Pt. 1, Wellington fire prevention 1921–25, ANZ. Uren to senior ranger, Wellington, 4 June 1922.
102 F1, 12/7, Pt. 1, Fire reports, Southland 1920–26, ANZ. Conservator to Director of Forestry, 6 July 1921. The sources for the rest of this section are from this file and are referenced only where material is directly quoted.
103 Ibid. Director to conservator, 26 November 1921.
104 *AJHR*, 1922, C.–3, SFSAR, 'Forest protection', 'Forest fires', p. 8.
105 *AJHR*, 1923, C.–3, SFSAR, 'Forest protection', 'Forest fires' p. 9.
106 *AJHR*, 1920, C.–3A Department of Forestry Report on *Forest Conditions in New Zealand, and the proposals for a New Zealand Forest Policy*, p. 21. Exactly what lay behind this plan is not clear. In his first fire plan Macpherson had identified the spark arrester question as needing attention. However, the Forest Service annual fire statistics on reported fires for 1924/25 and 1925/26 attribute the majority of fires to land-clearing operations.
107 F1, 12/0, Pt. 1, Annual fire reports and fire protection 1920–24, ANZ. A. E. Entrican to conservators, 8 August 1921.
108 F1, 12/7, Pt. 1, Fire reports Southland 1920–26, ANZ. Conservator to Director of Forestry, 16 April 1923.
109 Ibid., More & Sons to conservator, Invercargill, 20 April 1924.
110 This fire seems to have been reported only some five months later in Macpherson's account of Hansson's visit to his conservancy – see below and F1, 12/7, Pt. 1, Fire reports Southland 1920–26, ANZ. Conservator, Invercargill, to Director of Forestry, 26 February 1925.
111 F1, 12/7, Pt. 1, Fire reports Southland 1920–26, ANZ. Conservator, Invercargill, circular to millers, 3 February 1925.
112 F1, 12/7, Pt. 1, Fire reports Southland 1920–26, ANZ. More & Sons to conservator, Invercargill, 9 February 1925.
113 Ibid. Conservator, Invercargill, to Director of Forestry, 26 February 1925.
114 F1, 13/7, Forest rangers' handbook 1921–24, ANZ.
115 Ibid.
116 *NZ Perpetual Forests Ltd 1924–26: Reports covering operation and progress.* The first quote is taken from the 1926 report of Professor Owen Jones, forestry administrator, p. 28, the second from the report of Professor Corbin, Professor of Forestry, University of Auckland, p. 128. The reference to 250 employees is in the 1927 document.
117 F1, 12/2, Rotorua: Fire protection 1928–32, ANZ.
118 *AJHR*, 1925, C.–3, SFSAR, p. 6.
119 *AJHR*, 1926, C.–3, SFSAR, p. 3.
120 *AJHR*, 1955, C.-3, SFSAR, p. 8.
121 A. N. Cooper, 'Forest fires in New Zealand: History and status, 1880–1980', National Rural Fire Authority (NRFA) files.
122 Ibid.
123 These figures are taken from the totals given in the forest protection tables in the annual reports 1922–26.
124 Francis Pound in *The Invention of New Zealand Art and National Identity 1930–1970* (Auckland University Press, 2009, pp. 170–71) discusses the parallels that by the late 1920s some of the country's writers and painters drew between the still-standing fire-blackened trunks in otherwise stark landscapes and the battlefields of World War I. (Pound does not elaborate on whether the soldiers saw this likeness.) In his complex argument about the rise of a 'nationalist' art, Pound suggests that such images continued to be used during World War II. That such images resonated with a viewing public suggests something of the continued scale of burning off.
125 *AJHR*, 1922, C.–3, SFSAR, Annex III, Report on forestation operations in the South Island, 'Fire protection', pp. 20–21; *AJHR*, 1923, C.–3, SFSAR, pp. 9–10; F1, 12/0, Pt. 2, Fire protection general 1925–29, ANZ. 'Prevention and control of plantation fires by means of the fire district principle', undated [c. 1926].

Chapter Four: Consolidation – and change: 1928–47

1 'Plan to combat forest fires', Christchurch *Press*, 22 October 1941.
2 *Appendices to the Journals of the House of Representatives* (*AJHR*), 1930, C.–3, State Forest Service Annual Report (SFSAR), p. 5.
3 F1, 12/1, Pt. 4, Rotorua fire protection 1939–44, Archives New Zealand (ANZ). Girling-Butcher to Minister of Internal Affairs, Fire protection: Rotorua thermal area, 16 May 1944.
4 See, for instance, *AJHR*, 1938, C.–3, SFSAR. McGavock argued (p. 3) 'that as a long-range policy the use of indigenous species and the adaptation of the virgin forest composition both for wood-production and for protection purposes should be regarded as the essence of New Zealand forestry'.
5 For the figures on the constitution of state forests see *AJHR*, 1947, C.–3, SFSAR, p. 18. In relation to 'sustainable logging', *AJHR*, 1944, C.–3, SFSAR, p. 3 sets out the perceived benefits of such an approach in terms of ensuring a constantly regenerating bush rather than one that rots away and decays unused.
6 G. T. Bloomfield, *New Zealand: A handbook of historical statistics*, G. K. Hall and Co., Boston, 1984, Table V.1, Estimated value of production by sector 1900/1–1960/1, pp. 601–02.
7 Michael Roche, *History of New Zealand Forestry*, New Zealand Forestry Corporation/GP Books, Wellington, 1990, pp. 224–25.
8 *AJHR*, 1931, C.–3, SFSAR, p. 5.
9 *AJHR*, 1947, C.–3, SFSAR, p. 4.
10 Roche, *History of New Zealand Forestry*, Table 4.5 (p. 231) and Table 4.5 (p. 241).
11 *AJHR*, 1935, C.–3, SFSAR, p. 3.
12 *AJHR*, 1943, C.–3, SFSAR, p. 3.
13 There are no Forest Service reports on these fires, so it is difficult to assess how much the Service used such fires prior to the major burn-offs from the 1960s, or how often they broke away. However, one 1950s circular to conservators' comments on the lack of knowledge of burning as a silviculture tool, concerns with fires getting out of control (although the forests were younger and therefore more manageable), and the lack of records on earlier burning off. It is clear that the Service did use fire to clear land.
14 F1, 12/4, Pt. 3, Fire reports etc Nelson–Marlborough 1930–37, ANZ. 'Fires in plantation areas', 5 August 1932.
15 N. Taylor, *New Zealand People at War: The home front*, vol. II, Historical Publications Branch, Department of Internal Affairs, Wellington, 1986, pp. 752–55.
16 F1, 12/1, Forest fires general Auckland 1925–36, ANZ. Ranger Uren to conservator, Auckland, 30 June 1931.
17 *AJHR*, 1931, C.–3, SFSAR, p. 5.
18 'Obituary, A. D. McGavock', *New Zealand Journal of Forestry* (*NZJF*), vol. 8, no. 1, 1959, p. 168
19 F1, 45/11, Pt. 6, Circulars to conservators 1930–31, ANZ. 'Fire prevention 1931', 7 January 1931.
20 A. N. Cooper, 'Forest fires in New Zealand: History and status 1880–1980', National Rural Fire Authority, undated.
21 F1, 12/0, Pt. 3, Fire protection, general, ANZ. Commissioner of Forests to Federated Sawmillers' Association, 17 December 1937, a figure given in confidence.
22 *AJHR*, 1944, C.–3, SFSAR, pp. 7–8. AD1 270/5/5, Pt. 1, Preservation of native bush/Forest and bush fires/Prevention of/Rotorua thermal district 1938–46, ANZ provides information of the saga of continuing fires especially around Taupo over those years.
23 F1, W3219, Box 108, 12/1/15/1, Pt. 1, Fire control and prevention/Taupo fires/Correspondence, ANZ. Forest Products Ltd to Director of Forestry, 7 March 1946ff. The fires were barely out before the companies began approaching Entrican, asking if the Forest Service, as a big user of wood, wanted some of the damaged timber. On 14 March Entrican requested the Rotorua mill manager to reduce the input of saw logs from the Whakarewarewa forest in favour of fire-damaged timber.
24 *AJHR*, 1946, C.–3, SFSAR, p. 15.

25 Alex R. Entrican, 'Looking back: MacIntosh Ellis had a great influence on New Zealand forest policy', *NZJF*, August 1966, p. 46; John Halkett, Peter Berg and Brian Mackrell, *Tree People: Forest Service memoirs*, New Zealand Forestry Corporation/GP Print, Wellington, undated (c. 1987–88), p. 24.
26 Halkett et al, *Tree People*, p. 58.
27 'A gruff, no-nonsense man' – Halkett et al, *Tree People*, p. 58; 'by no means automatically pleased' – Taylor, *New Zealand People at War*, vol. 1, p. 277, referenced in the *Press*, 15 April 1943, p. 4.
28 Pers. comm., Michael Roche, 2009; Halkett et al, *Tree People*, p. 59.
29 For McGavock's employment history see *Tree People*, p. 24; V. T. F., Review of *The First Fifty Years of New Zealand's Forest Service* by F. Allsop, *NZJF*, vol. 19, no. 1, 1974, pp. 146–48.
30 F1, 12/0, Pt. 7, Fire protection general 1942–46, ANZ. Director of Forestry to commissioner of works, 18 January 1944. Entrican also makes an oblique reference to these concerns in *AJHR*, 1944, C.–3, SFSAR, p. 8.
31 F1, 12/2, Rotorua fire protection 1932–38, ANZ. Rotorua fires, 7 October 1938. The sheet onto which the report is pasted does not record the originating paper.
32 Unless otherwise signalled, the material on these files comes from F1, 12/1, Fire prevention general 1936–38, ANZ. The extent of kauri forests and their flammability made fires in this conservancy more likely and potentially more destructive than elsewhere. The number and location of fires indicates the level of threat the Forest Service faced.
33 *AJHR*, 1938, C.–3, SFSAR, p. 19; *AJHR*, 1940, C.–3, SFSAR, Appendix V1, Schedule of fires in state forests 1939–40, p. 35; *AJHR*, 1941, SFSAR, Appendix V, Fires outside state forests … and Appendix V1, Fires in state forests 1940–41, p. 44.
34 *AJHR*, 1946, C.–3, SFSAR, p. 15.
35 F1, 45/11, Pt. 19, Circulars to conservators 1939–40, ANZ. Reporting and recording of fires, 2 December 1940.
36 F1, 12/1, Fire prevention general 1936–38, ANZ. Arnold Taylor, solicitor, to Minister in Charge, State Forests Department, 15 July 1938.
37 Ibid. H. Moore to Minister of Lands, 14 June 1938.
38 Ibid. See letters: Farm appraiser to manager, State Advances Corporation, 16 July 1938. This was obviously forwarded to McGavock, who replied on 21 July 1938 pointing out that his staff had a duty to grant permits and that weather dictated the need for restrictions.
39 F1, 12/0, Pt. 5, Fire protection general 1939–40, ANZ. Conservator, Palmerston North, Protection of forests from fire, 17 May 1939.
40 For further examples of the courts' responses, see *AJHR*, 1935, C.–3, SFSAR, p. 9. McGavock, railing against the extraordinary instance a man who had pleaded guilty to lighting a fire yet was not convicted on the grounds of his age, that he had no previous convictions, and had not received a personal warning against lighting fires, wrote, 'The law becomes almost impossible of enforcement.' In relation to Forest Service staff attitudes to miscreants see F1, 45/11, Pt. 11, Circular letters to conservators 1942–44, ANZ. Law enforcement policy, 3 February 1943. This circular, on the increasing number of prosecutions and the importance of Forest Service staff not giving undue weight to extenuating evidence, related not only to fires, though fires and the removal of forest property were the most serious offences.
41 F1, Box 299, 12/3/7, Fires outside state forests in Southland region 1941–47, ANZ. Re evidence to bring prosecutions successfully, undated (c. August 1945).
42 H. T. Gisborne, *Measuring Fire Weather and Forest Inflammability*, US Department of Agriculture, circular no. 398, Washington DC (Millman collection), 1936; W. B. Osborne, 'The lookout system', in *The Western Fire Fighters' Manual* (6th edn), Western Forestry and Conservation Association, Seattle (Millman collection), 1934. The Forest Service accessioned Gisborne's book in 1939, the year that Entrican began importing and building new weather stations.
43 R. H. Luke and A. G. McArthur, *Bushfires in Australia*, Australian Government Publishing Service, Canberra, 1978, pp. 169–72 and 197–89.
44 F1, 12/0, Pt. 2, Fire protection general 1925–29, ANZ. B. J. Dunsheath Ltd, October 1927.

45 F1, Box 338, 14/7/9, Fire-fighting equipment 1930–36, ANZ. Leaflet from Watson, Jack & Co., filed 3 November 1931.
46 Ibid. See correspondence, Girling-Butcher to Ashburton County Council, 22 September 1939, and Colonial Motor Company to Conservator of State Forests, 14 February 1934
47 In relation to the tools held in the conservancies see F1, Box 303, 12/6/2, Pt. 1, Fire-fighting and control equipment 1939–47, Returns on fire-fighting equipment, correspondence 19 July 1939ff. The Auckland conservator, for one, raised questions about the adequacy of hand tools, while acknowledging his dependence on them. In 1944 R. Girling-Butcher commented on their adequacy in relation to Rotorua district (F1, 12/1, Pt. 4, Rotorua fire protection 1939–44, ANZ. Girling-Butcher to Minister of Internal Affairs, 'Fire-protection Rotorua thermal area', 16 May 1944).
48 *AJHR*, 1946, H.–12, Report on Fire Brigades (RFB), p. 3.
49 F1, 45/11, Pt. 14, Circular letters to conservators 1948–48, ANZ. Distribution of fire tools, 9 September 1947.
50 F1, Box 303, 12/6/2, Pt. 2, Fire-fighting & control equipment general 1948–58, ANZ. Correspondence, September 1948.
51 F1, 12/2, Pt. 3, Rotorua fire protection 1932–38, ANZ. Entrican to W. T. Morrison, 14 October 1938; Ibid., correspondence to Director of Forestry, re distribution of motor vehicles, 18 October 1938; F1, 12/2, Pt. 2, Rotorua fire protection 1928–32, L. H. Bailey to conservator, Whakarewarewa Plantation, fires on 7 October 1938.
52 F1, 45/11, Pt. 10, Circulars to conservators 1939–47, ANZ. Suppression of fires, hire of transport, 12 September 1939.
53 F1, 12/1/6/, Pt. 1, Fire control Canterbury 1944–58, ANZ. Conservators' conference, Canterbury conservancy, Fire organisation 1944–45.
54 F1, 12/0, Pt. 7, Fire protection general 1942–46, ANZ. L. H. Bailey to Director of Forestry, 1 February 1943ff.
55 F1, 1.12.2/1, Transport: Fire-fighting, ANZ. See correspondence 9 January 1939ff.
56 F1, 12/1, Pt. 4, Rotorua fire protection 1939–44, ANZ. Girling-Butcher to Minister of Internal Affairs, Fire protection Rotorua thermal area 16 May 1944.
57 *AJHR*, 1945, C.–3, SFSAR, p. 8. My description also uses material from F1, 12/6/1, Pt. 1, Fire engines and water tankers 1944–48, ANZ. Report on tests of fire engine, Whangamata, on 25 October 1945 and 'Fighting against forest fires', *Evening Post*, 20 November 1944.
58 *AJHR*, 1946, C.–3, SFSAR, p. 17.
59 F1, 45/11, Pt. 14, Circular letters to conservators 1948–48, ANZ. Mechanised fire equipment, 21 September 1948. The equipment specified for the Quad engines was: 10 lengths of 3.8cm canvas hose, 36.5 or 30.5m; one length of 3.8cm canvas hose, 7.6m; 4.6m of 1.9cm rubber hose; one 5.3cm rubber hose, approximately 7.6m long; one suction strainer with guard ring; three hose packsacks; two 3.8cm branch pipes; two adaptors, 1.9–3.8cm; two different-sized nozzles and two shut-off nozzles; a breaching piece with control valve 3.8cm; one spanner (suction hose); six water bottles; two knapsack pumps; one first aid kit; two electric torches; two pumice or similar torches; two canvas buckets (if available from local sources); 9.1 litres of diesel; rations (type and quantity optional); one grease gun for the pump and one for the vehicle; one fire extinguisher (0.9 litres carbon tetrachloride); also hand tools – the number to vary with local conditions, but here it was specified as six shovels and slashers, two axes and two mattocks or grubbers. See also *AJHR*, 1948, C.–3, SFSAR, p. 24.
60 F1, 12/1/2, Fire control Rotorua 1944–48, ANZ. Fire equipment officer to Director of Forestry, Fire protection, Rotorua, 4 September, 1947; Ibid., Fire protection, Rotoehu, 2 September 1947. Instructions to order a range of equipment through head office were issued in an attempt to standardise the equipment used nationally in September 1947 – F1, 45/11, Pt. 14, Circular letters to conservators 1949–48, ANZ. Purchase of fire-fighting equipment, 11 September 1947.
61 F1, 45/11, Pt. 14, Circular letters to conservators 1948–48, ANZ. Circular memorandum no. 356, Mechanised fire equipment, 24 August 1947; Ibid., Fire equipment operation and maintenance course, 28 May 1948.

62 Ibid. Circular memorandum no. 356, Mechanised fire equipment, 24 August 1947 and 21 September 1948. The Marine Department, especially in the early days of the electrification of the coastal lighthouses, had to issue similar elementary instructions to ensure the diesel engines used to power the lights were properly equipped and maintained. See Helen Beaglehole, *Lighting the Coast: A history of New Zealand's coastal lighthouse system*, p. 255, Canterbury University Press, Christchurch, 2006.

63 F1, 12/6/1, Pt. 1, Fire engines and water tankers 1944–48, ANZ. Director to all conservators, 13 September 1946; F1, 45/11, Pt. 13, Circular letters to conservators 1946–46, 13 September 1946. It is worth noting that only three months earlier the general position vis-à-vis truck availability and the impossibility of getting 'even some' vehicles approved by Cabinet for normal operations had meant a contrary instruction had been issued: no vehicles were to be set aside for fire-fighting (F1, Box 648, 1.12.2/1, Letters to conservators, Transport for fire-fighting, fire season 1946–47, 13 June 1946, ANZ). However, the vehicle shortage was rectified by hiring from the army and from local bodies: Entrican to Quartermaster-General, Army HQ, Hire of Army vehicles …, 29 August 1946 and Quartermaster-General to Entrican, 5 September 1946. On local body loans, see, for instance, F1, Box 303, 12/6/2, Pt. 1, Fire-fighting equipment and control general 1939–47, Correspondence between Palmerston North conservancy and Waipapa Borough Council, June and August 1947.

64 F1, 12/1/24, Pt. 1, Fire-prevention training of personnel 1948–53, ANZ. Teaching notes, Auckland Conservancy fire-fighting course, 29 September 1950.

65 F1, 45/11, Pt. 13, Circular letters to conservators 1946–46, ANZ. Fire-protection courses, 22 October 1946; F1, 45/11, Pt. 14, Circular letters to conservators 1947–48, Fire-fighting, 30 April 1947. Harold Morgan ('Fighting an Australian bush-fire', *Chambers Journal*, June 1937, pp. 445–58) describes but does not name the technique. Luke and McArthur (*Bushfires in Australia*, pp. 211–13) describe the methods used by the fire-line.

66 F1, 12/1/II A Anti-fire publicity campaigns 1946–49, ANZ. Correspondence 25 September 1948, 22 October, 26 October 1948.

67 Roy Knight, Interview, RFP.

68 F1, 12/1, Pt. 4, Rotorua fire protection 1939–44, ANZ. R. Girling-Butcher to Minister of Internal Affairs, Fire protection, Rotorua thermal area, 16 May 1944. This memo contains recommendations for fire protection for the local camps and settlements.

69 *AJHR*, 1940, C.–3, SFSAR, p. 16; F1, 12/2, Pt. 4, Rotorua fire protection 1939–44, ANZ. Dunn to conservator, Safety of employees, Kaingaroa Forest, 17 October 1939; Ibid., chief inspector to Director of Forestry, Fire-safety precautions, Kaingaroa, 30 October 1939. On traffic safety in the case of fire in the Central Plateau forests see ibid., Road Board to Entrican, June 1940 and Entrican's reply on diverting traffic, 24 July 1940; and *AJHR*, 1942, SFSAR, p. 11 re extending provisions of s. 63.

70 F1, 12/0, Pt. 7, Fire protection general 1942–46, ANZ. Fire plan 1942–43, Golden Downs Exotic Forest, 23 September 1943.

71 *AJHR*, 1927, C.–3, SFSAR, Experimental treatment of firebreaks, p. 25.

72 *AJHR*, 1940, C.–3, SFSAR, pp. 14–15. The value of prisoners-of-war in this activity was also noted.

73 *AJHR*, 1941, C.–3, SFSAR p. 20; F1, 12/0, Pt. 7, Fire protection general 1942–46, ANZ. L. H. Bailey to Director of Forestry, 1 February 1943.

74 *AJHR*, 1940, C.–3, SFSAR, p. 20 and *AJHR*, 1948, C.–3, SFSAR, p. 25.

75 *AJHR*, 1940, C.–3, SFSAR, p. 15.

76 F1, 12/2/IIB, Anti-fire publicity campaigns 1946–49, ANZ. Director of Forestry, Reporting of forest fires/ co-operation of telephone exchanges, 7 January 1947.

77 F1, 12/0, Pt. 1, Annual fire reports and fire protection 1920–24, ANZ. T. H. Bedford, Westport, to secretary, Forest Service, 28 September 1921.

78 *AJHR*, 1922, C.–3, SFSAR, p. 19; *AJHR*, 1941, C.–3, SFSAR, p. 23.

79 *AJHR*, 1946, C.–3, SFSAR, pp. 17–18 and 25.

80 F1, 45/11 Pt. 5, Circulars to conservators 1929–30, ANZ. Plantation maps, 6 November 1929.

81 F1, 12/0, Pt. 3, Fire protection general 1929–38, ANZ. McGavock to Director of Forestry,

Tasmanian Forestry Department, 22 April 1931.

82 *AJHR*, 1942, C.–3, SFSAR, p. 6.

83 An additional tower at Eyrewell, postponed in 1930, was still delayed by shortages of building materials in 1947 (F1, 12/7/1/6, Fire prevention, buildings, towers, Canterbury, ANZ); and construction of fire-lookout cabins for Kaingaroa begun by June 1945 was only completed in 1952 (F1, W3129, Box 113, 12/7/1/2, Pt. 1 Fire prevention and control/Fire-prevention/Buildings and lookout towers/Rotorua conservancy 1941–62, ANZ).

84 *AJHR*, 1948, C.–3, SFSAR, p. 25.

85 F1, W3129, Box 113, 12/7/2/1, Fire prevention and control/lookout station equipment Auckland conservancy 1938–53, ANZ; F1, W3129, Box 113, 12/7/2, Fire prevention and control/lookout station equipment Canterbury conservancy 1938–60, ANZ; F1, W3219, Box 122, 12/7/2/4, Fire prevention and control/lookout station equipment Nelson conservancy 1938–54, ANZ.

86 'Patrol of forests: Aeroplanes used', *Evening Post*, 25 May 1929; 'Aeroplanes used to guard forests', *New Zealand Herald*, 5 January 1930; F1, 12/0, Pt. 2, Fire protection general 1925–30, ANZ.

87 F1, 12/0, Pt. 3, Fire protection general 1929–38, ANZ. Canadian Knight & Whippet Motor Co. to the Commissioner of Forests, 27 August 1930.

88 Osborne, 'The lookout system', p. 5.

89 *AJHR*, 1941, C.–3, SFSAR, p. 16.

90 For the 1944 fires and forester Pollock's recommendations on introducing plane spotting as a permanent feature see F1, 12/2, Pt. 4, Rotorua fire protection 1939–44, ANZ, assistant forester Pollock to conservator, 17 February 1944. The developments in conjunction with the armed forces over 1944–45 are noted *AJHR*, 1945, C.–3, SFSAR, pp. 3 and 8. Comments on the air patrols in the 1946 Taupo fires are in *AJHR*, 1946, C.–3, SFSAR, p. 17. There was also a degree of public interest, with the president of Forest and Bird suggesting their value in observing and reporting fire – F1, 12/0, Pt. 7, Fire protection 1925–, president of Forest and Bird to Director of Forestry, 27 April, 1943.

91 F1, 12/1/II, B, Anti-fire publicity campaigns 1946–49, ANZ. Text of the minister's radio talk, undated (end of 1947).

92 *AJHR*, 1948, C.–3, SFSAR, p. 24.

93 F1, Box 647, 1.12, Pt. 4, Fire-fighting equipment 1939–43, ANZ.

94 F1, 45/11, Pt. 3, Circulars to conservators August 1925–May 1926, ANZ. Instructions for using psychometers, 21 January 1926. On the weather stations being in both exotic and indigenous forest, see *AJHR*, 1942, C.–3, SFSAR, p. 6. A psychrometer is an instrument that measures an area's relative humidity. It uses a wet and a dry bulb. When the water on the wet bulb has evaporated the temperature of each bulb is taken and the difference is plotted to establish the area's temperature and its relative humidity. A small difference between the bulbs indicates a low evaporation rate and a consequent high relative humidity. A large temperature difference indicates faster evaporation and a low relative humidity. http://simple.wikipedia.org/wiki/Psychrometer (last accessed August 2012).

95 *AJHR*, 1940, C.–3, SFSAR, p. 15; *AJHR*, 1942, C.–3, SFSAR, p. 6; *AJHR*, 1943, C.–3, SFSAR, p. 6; *AJHR*, 1944, C.–3, SFSAR, p. 7.

96 *AJHR*, 1939, C.–3, SFSAR, p. 17.

97 In relation to the ban see *AJHR*, 1946, C.–3, SFSAR, p. 19. On the value of staff training, enthusiasm and interest see ibid., p. 16; F1, Box 647, 1.12, Pt. 4, Fire-fighting equipment 1939–43, ANZ, correspondence 2 June 1939ff in relation to the presence of keen Mr Kenderdine and the value of the instruments.

98 *AJHR*, 1946, C.–3, SFSAR, p. 16.

99 *AJHR*, 1947, C.–3, SFSAR, p. 24.

100 *AJHR*, 1932, C.–3, SFSAR, p. 5; *AJHR*, 1942, C.–3, SFSAR, p. 6; *AJHR*, 1946, C.–3, SFSAR, p. 18.

101 F1, 45/11, Pt. 9, Circulars to conservators 1936–38, ANZ. Fire season 1928–39, 17 October 1938. This states the circular was reiterating instructions already in force.

102 F1, 45/11, Pt. 11, Circular letters to conservators 1942–44, ANZ. Fire-protection measures, instructions to all field personnel, 23 January 1943.
103 'Obituary, A. D. McGavock', *NZJF*, vol. 8, no. 1, 1959, p. 168.
104 *AJHR*, 1935, C.–3, SFSAR, Recreation in state forests, p. 17.
105 F1, 12/0/3A, Fire-prevention stickers etc 1929–38, ANZ. File note, 2 November 1938.
106 F1, 12/0, Pt. 7, Fire protection general 1942–46, ANZ. R. F. Hammatt, US Forest Service, to Director of Forestry, 29 October 1942.
107 F1, 12/0/3B, Fire-prevention poster & signs 1928–45, ANZ. 'Hell and high timber'.
108 F1, 12/0/3A, Fire-prevention stickers etc 1929–38, ANZ. A note from that remarkable public servant, Joe Heenan, under-secretary of Internal Affairs, to McGavock on 12 October 1938 first signalled this initiative. Heenan offered his department's assistance in minimising fire risks. Except where otherwise indicated, the material in the following paragraph is taken from this file.
109 Ibid. Director of Forestry to conservators, 2 March 1939. See also F1, 12/0, Pt. 4, Fire protection general 1938–39, ANZ. F. Mahy to Director of Forestry, 17 August 1938ff in relation to a motor-camp proprietor's initiative for sticking them to his camp information folders.
110 F1, 12/0, Pt. 4, Fire protection general 1938–39, ANZ. Senior draughtsman to Director of Forestry, 21 August 1939.
111 Ibid. Director of Forestry to various firms, undated (November 1947); F1, 12/2/11A, Anti-fire publicity campaigns 1946–49, ANZ. K. R. Mitchell, executive secretary, Junior Chamber of Commerce, Wellington, 2 December 1947.
112 Ibid. Director-general, P&T, to Director of Forestry, 19 December 1938. An attached Tasmanian stamp franked with 'Prevent Forest Fires' suggests where the idea came from.
113 Ibid. Correspondence, November 1941.
114 F1, 12/1/2/1, Pt. 1, Fire-prevention stickers and slogans 1940–50, ANZ. Director of Forestry to managing director, National Tobacco Co. Ltd, 6 November 1941.
115 F1, 12/0/3B, Fire-prevention poster & signs 1928–45, ANZ. Entrican to director-general, P&T Department, 6 November 1941; to Sanderson, president of Forest and Bird, 27 February 1942 re distribution through Forest and Bird; Entrican to all conservators, 12 October 1943; Railways response: general manager, Railways, 22 October 1943.
116 F1, 12/0/3A, Fire-prevention stickers etc 1929–38, ANZ. J. Heenan to Director of Forestry, 3 November 1938. However, calico signs were still being used into the 1950s (F1, Box 555d 45/11, Pt. 15, Circular letters to conservators 1949–51).
117 F1, 12/0/3B, Fire-prevention poster & signs 1928–45, ANZ. Entrican, 5 November 1940.
118 Ibid. Director of Forestry to all conservators, 19 September 1945ff.
119 F1, 12/1/12/3, Fire-prevention posters & signs 1944–49, ANZ. Entrican to all conservators, 28 June 1946 and 11 August 1947.
120 Ibid. Entrican to all conservators, 28 June 1946.
121 Ibid. Entrican to all conservators, 15 October 1947.
122 *AJHR*, 1938, C.–3, SFSAR, p, 20. The idea was first mooted by the Sawmillers' Association, which advocated the appointments because some AA men were already honorary rangers and were closely co-operating with forestry interests (F1, 12/0, Pt. 3, Fire protection general, ANZ. Correspondence, Sawmillers' Association, 17 November 1937).
123 F1, 45/11, Pt. 9, Circulars to conservators 1936–38, ANZ. To all conservators, AA patrolmen as honorary rangers, 22 December 1937; Fire season 1938–39, 17 October 1938. This circular stresses the positive relationship head office wanted its conservators to develop with the AA men, involving them in training in prevention and ensuring good avenues of communication. The initiative was first reported in the department's annual report in *AJHR*, 1938, C.–3, p. 20.
124 F1, 12/0, Pt. 4, Fire protection general 1938–39, ANZ. McGavock, correspondence, 1 July 1938.
125 See, for instance, F1, 12/0, Pt. 5, Fire protection general 1939–40, ANZ. The conservator in Palmerston North thought the amount of forest they oversaw was minimal (conservator to Director of Forestry, Protection of forests from fire, 17 May 1939). On the other hand, the

Rotorua conservator thought appointing the mobile AA men and acclimatisation rangers was better than indiscriminately appointing locals (correspondence, 17 July 1939). The Auckland conservator also put in a positive word (correspondence, 2 June 1939).

126 'Plan to combat summer fires', *Press*, 22 October 1941.

127 F1, 12/1/111B, Anti-fire publicity campaigns 1946–49, ANZ. Commissioner on radio, undated (end of 1947).

128 F1, W3291, Box 112, 12/2/3, Pt. 1, Fire control and prevention/fires in state forests/ Wellington conservancy 1938–64, ANZ. A. H. Cockayne, Director of Agriculture, to Director of Forestry, 15 November 1939.

129 Ibid. Conservator to Director of Forestry, 4 February 1941.

130 Ibid. Fire, Owhango box factory, 13 February 1941, ranger Reveirs to Palmerston North conservator; and Entrican, 20 February 1941.

131 Thomas E. Simpson, *Kauri to Radiata: Origin and expansion of the timber industry of New Zealand*, Hodder and Stoughton, Auckland, 1973, pp. 283 and 332.

132 F1, 12/2, Pt. 4, Rotorua fire protection 1939–44, ANZ. Assistant forester Pollock to conservator, Forest fires, spotting by aeroplane, 17 February 1944, entries 1–12 January 1944.

133 AD1 270/5/5, Pt. 1, Accidents and disasters/Fires/Bush/Assistance by army, ANZ. Army commandant, Fire-fighting at Taupo, Diary of events, 19 February 1946, p. 1.

134 F1, 12/3/2/1, Pt. 1, Fires in Kaingaroa, Rotorua district 1946–48, ANZ; Editorial, 'Fire danger', *Press*, 20 December 1948 p. 6.

135 F1, 12/1/11A, Anti-fire publicity campaigns 1946–49, ANZ. Palmerston North conservator to Director of Forestry, 29 March 1946.

136 AD1 270/5/5 Pt. 1, Preservation of native bush/Forest and bush fires/Prevention of/ Rotorua thermal district 1938–46, ANZ. The papers of the Taupo Town Board meeting 9 November 1938 record that V. T. Fail, the man in charge of the Afforestation Proprietary forests, spoke of the concerns with bush fires, difficulty of detection and prosecution, and the value of the minister's radio talks on fire prevention.

137 The availability of the school film is in F1, 12/1/11A, Anti-fire publicity campaigns 1946–49, ANZ, correspondence, 28 November 1947; the planned publicity is in F1, 12/1/11, B, Anti-fire publicity campaigns 1946–49, ANZ, undated (1947). For the requests from overseas groups see F1, 12/1/11A, Anti-fire publicity campaigns 1946–49, ANZ, secretary, Save the Trees Campaign, Brisbane, to Director of Forestry, 20 May 1948; and F1, 12/1/11B, Anti-fire publicity campaigns 1946–49, ANZ, National Veld Trust, 30 June 1947.

138 F1, 12/1/11B, Anti-fire publicity campaigns 1946–49, ANZ. Text of commissioner on radio, undated (end of 1947).

139 F1, 12/0, Pt. 4, Fire protection/general 1938–39, ANZ. Review of service's organisation and management of fire control, 28 April 1939; *AJHR*, C.–3, SFSAR, p. 6.

140 *AJHR*, 1939, C.–3, SFSAR, p. 2.

141 *AJHR*, 1940, C.–3, SFSAR, p. 5.

142 F1, Box 157, 1/5/-, Pt. 9, Forest legislation general 1930–45, ANZ. In a memo to the Crown solicitor on 30 May 1940 and a later undated memo to the Attorney-General, Entrican sought to amend the act in a number of ways. In relation to fire he wanted to be able to prohibit entry into state forests (and possibly fire districts); to close roads through state forests and fire districts during fires; to insist that rights-owners and operators, sawmill owners and forest owners in fire districts maintain adequate patrols, lookouts and manpower to fight fires, and that they have adequate fire-fighting equipment. He wanted, too, to compel those with milling or ownership rights in the forests to provide adequate fire shelters for employees and endangered settlers, and fire-fighting plans that would safeguard any evacuation. Entrican also wanted an 8km impress provision, the ability to insist on immediate assistance and the power to recover fire-suppression costs. As will be seen, his regulations eventually covered all of these points except for planning for travellers (he subsequently thought it better if roads were closed in emergencies).

143 F1, Box 647, 1.12, Pt. 4, Fire-fighting equipment 1939–43, ANZ. Entrican to Director of Forestry, Northern Rocky Mountain Forest & Range Experimental Station, 2 June 1939.

144 F1, 12/3, Pt. 3, Fire prevention general Wellington 1929–34, ANZ. Palmerston North conservator to Director of Forestry, 14 October 1929. The issue was not restricted to the Central Plateau. In 1931 the Canterbury conservator communicated with McGavock after requests to the Railways district traffic manager to restrict stoking while running through the plantation danger zone. Eight days and five additional fire reports later, and with no comment from the traffic manager, the conservator wrote, 'The matter has been taken up so often with the Railways Department that further action seems useless from here. Do you desire Ministerial action to be invoked?' It was not (F1, 12/6, Fire reports Canterbury–Otago 1929–38, December 1931). On the other hand, the Rotorua conservator reported that the Public Works officers controlling the men working on the railway in the vicinity of the Whakarewarewa plantation assisted in every way (F1, 12/2, Rotorua fire protection 1928–32, conservator to Phillips Turner, 14 November 1928).

145 F1, 12/3, Pt. 3, Fire prevention general Wellington 1929–34, ANZ. Palmerston North conservator to Director of Forestry, 20 August 1929ff.

146 F1, 12/3, Pt. 4, Fires Wellington region 1935–39, ANZ. Macpherson to McGavock, December 1937.

147 F1, 12/0, Pt. 5, Fire protection general 1939–40, ANZ. Palmerston North conservator to Director of Forestry, Protection of forests from fire, 17 May 1939.

148 F1, 12/0, Pt. 1, Annual fire reports and fire protection 1920–24, ANZ. Entrican to conservators, 8 August 1921.

149 F1, W3129, Box 107, 12/1/10, Pt. 1, Co-ordination by governmental departments, ANZ. File note on discussion with G. Mackley, 26 September 1939.

150 F1, W3219, Box 112, 12/2/3, Pt. 1, Fire control and prevention/Fires in state forests Wellington conservancy 1938–64, ANZ.

151 F1, 12/0/6, Anti-fire publicity 1939–46, Entrican to general manager, New Zealand Railways, 20 November 1940.

152 F1, Box 647, 1.12, Pt. 4, Fire-fighting equipment 1939–43, ANZ. J. S. Reid, forest officer, to head office, 12 December 1939.

153 In 1939 Perpetual Forests Ltd wanted those controlling a fire district to be able to burn firebreaks on adjoining properties, and tighter trespass provisions (F1, 12/0, Pt. 5, Fire protection general 1939–40, ANZ. Perpetual Forests Ltd to Director of Forestry, 24 May 1939; Ibid., New Zealand Forest Products, 14 November 1939; Ibid., New Zealand Farmers' Union, 19 December 1939, on issue of balance between forestry and land clearance; and again, on 8 May 1940, over permits to burn logs and stumps and the Forest Service ability to withhold permission until wet weather set in).

154 The main provisions are taken from Entrican's description in *AJHR*, 1941, C.–3, SFSAR, pp. 17–18. Within that general text, comment on specific provisions is from F1, Box 289, 12/0/7, Forest fire regulations 1940–43, Commissioner of State Forests to Executive Council, Forest (Fire prevention) Regulations, undated.

155 The length of the fire season and the regional variation in danger periods were issues a number of conservators addressed in their responses to Entrican's memo – and probably explains the length of the fire season. In Southland, for instance, the greatest hazard was during August and September, when runholders burned the tussock and plantation staff burned firebreaks and carried out other protective burning. In the Hokitika conservancy, where the fire season was January to March, the graziers burned the high country for feed for young sheep and the pakihi rushes for feed for young cattle; hunters burned for amusement and so that, once on the flats, they could see where they were; prospectors burned to facilitate mining licences. Meanwhile bracken, which dried out over the winter, presented a spring fire hazard. Their season, the conservator thought, should run from 1 September to 31 March (F1, 12/0, Pt. 5, Fire protection general 1939–40, ANZ. See replies 19 May and 26 May 1939).

156 F1, Box 157, 1/5/-, Pt. 9, Forest legislation general 1930–45, ANZ. Memo, 25 September 1940.

157 F1, 12/0, Pt. 4, Fire protection, general, 1938–39, ANZ. Review of Forest Service's organisation and management of fire control, 28 April 1939; F1, Box 289, 12/0/7, Forest

Fire Regulations 1940–43, Commissioner of State Forests to Executive Council, Forest (Fire prevention) Regulations (undated).

158 F1, 12/0/6, Anti-fire publicity 1939–46, ANZ. Entrican to all conservators, Fire prevention and control 1941–42 season, undated (about July 1941), and following correspondence from conservators.

159 In 1931 the conservator at Hanmer Springs, concerned about sparks from the local railways, sought to have stoking restricted while the train was running through the plantation danger zone (F1, 12/6 Fire reports Canterbury–Otago 1929–38, ANZ, Correspondence, December 1931). In 1941 the Canterbury conservator reported the Christchurch locomotive engineer's helpfulness in ensuring that Railways trains hauling through Balmoral Forest would use only hard coal (F1, 12/0/6, Anti-fire publicity 1939–46, ANZ, conservator, Christchurch, 15 September 1941). However, in 1946, in one of a number of articles on fires from trains, the *Press* reported that Canterbury Plains residents had sent a deputation to meet with the Minister of Railways to discuss the dangers from trains. Railways' response to a deputation from the Canterbury Rural Fire Prevention Committee – that the men were too busy to follow the trains – indicated little shift in its position (*Press*: 'Mid-Canterbury fires', 2 March 1946, p. 6; 'Fire hazard: Danger of sparks from engines', 19 March 1946; letter on engine sparks and soft coal, 5 March 1946; 'Semple to be approached by delegation from Canterbury Progress League', 4 April 1946, p. 4; 'Fire hazard from engine sparks', 16 September 1946). At Golden Downs in 1947 in times of fire hazard one staff member was responsible at all times for watching for Railways locomotive fires, patrolling the lines after incoming trains and keeping track of the railway linesmen so he could call on them (F1, 12/1/4 Control Nelson 1937–53, Golden Downs Forest fire plan 1944–45, ANZ).

160 *AJHR*, 1946, C.–3, SFSAR, p. 15.

161 Jack Barber, Interview, RFP.

162 On the ongoing nature of this problem see F1, W3129, Box 2, 1/1/4/6, Pt. 1, Forest and Rural Fires Act Booklet 1955, ANZ. W. Girling-Butcher to L. G. Ardell, county clerk, Maniototo County Council, 13 August 1974. Girling-Butcher was replying to a request 'for strong action to have N. Z. Railways bound by the Forest & Rural Fires Act'.

163 F1, 12/0/6, Anti-fire publicity 1939–46, ANZ. Entrican to MacIntosh Ellis, 4 December 1940.

164 F1, 12/2, Pt. 4, Rotorua fire protection 1939–44, ANZ. Taupo Road Board to Director of Forestry, 11 April 1941 and 18 February 1942.

165 *AJHR*, 1942, C.–3, SFSAR, p. 11. In 1946 the Commissioner of Transport told all traffic officers that in cases of extreme hazard they were to co-operate with the Forest Service in controlling traffic, arranging diversions and impressing on motorists the need for care (F1, W3129, Box 107, 12/1/10, Pt. 1, ANZ. Commissioner of Transport to all traffic officers, 1 March 1946).

166 F1, Box 289, 12/0/7, Forest Fire Regulations 1940–43, ANZ.

167 *AJHR*, 1943, C.–3, SFSAR, p. 6. The legislation was the Forest (Fire prevention) Regulations 1940, Amendment no. 1 (1943/31).

168 *AJHR*, 1943, C.–3, SFSAR, p. 3

169 F1, 12/2, Rotorua fire protection 1928–32, ANZ. Taupo County Board to Commissioner of State Forests, Fire menace, 30 July 1938; Ibid., commissioner to board, 10 August 1938; Ibid., W. Parry, Minister of Internal Affairs, to commissioner, 26 August 1938.

170 F1, 12/2, Pt. 4, Rotorua fire protection 1939–44, ANZ. Taupo Road Board to Commissioner of State Forests, 11 March 1929 ff, 30 March 1939 and reply from the commissioner, undated.

171 Ibid. Commissioner of State Forests to Minister of Internal Affairs, 2 May 1939; Ibid., Taupo Road Board to Minister of Internal Affairs, 17 December 1940; Taupo Road Board to chief inspector, Forest Service, 17 January 1941. The history of local concern and attempts to deal with the situation can also be traced in AD1 270/5/5, Pt. 1, Preservation of native bush/Forest and bush fires/Prevention of Rotorua thermal district, 1938–46, ANZ.

172 F1, 12/2, Pt. 4, Rotorua fire protection 1939–44, ANZ. Girling-Butcher to Minister of

Internal Affairs, Fire-protection: Rotorua thermal area, 16 May 1944, p. 5. Similar calls to action are in ibid., conservator to Director of Forestry, 4 April 1944, forwarding a report from Detective Sergeant White; Ibid., Boardman to Director of Forestry, 20 June 1944. Ibid., Smith to Director of Forestry, 32 June 1944, comments on Girling-Butcher's proposals, suggesting that the Forest Service could be overloaded in protecting private interests and lose what all within the Service considered as its vital 'tactical command'.

173 *AJHR*, 1946, H.–12, FBR, pp. 3–4.

174 'Bush fires in Northland', *Press*, 9 February 1946, p. 6. Quotes have been specifically attributed. Because it would have been cumbersome to note other material from each particular account, I note that following papers were used: *Bay of Plenty Times*, 11 February and 20 February 1946. *Dominion*, 11 February, pp. 6, 8; 12 February, p. 8; 15 February, p. 8; 18 February, p. 8; 19 February, p. 8; 20 February, p. 8; 25 February, p. 6; 27 February, p. 6. *Press*, 9 February, pp. 6, 10; 25 February, p. 4; 27 February, p. 6. *Waikato Times*, 11 February, p. 4; 12 February, p. 4; 13 February, p. 4; 22 February, p. 4. Also valuable is F1, W3219, Box 108, 12/1/15/1, Pt. 1, Fire control and prevention/ Taupo fires/Correspondence, ANZ. Log compiled by Courtney Biggs, undated, beginning Sunday 10 February.

175 'Bush fires in Northland', *Press*, 9 February 1946, p. 6; 'North Island power: Reduced demand essential', *Press*, 27 February 1946, p. 6.

176 'Bush fires in Northland', *Press*, 9 February 1946, p. 6.

177 'Hawkes Bay faces terrific loss', *Dominion*, 11 February 1946, p. 6.

178 *AJHR*, 1946, H.–12, RFB, p. 4.

179 'Huge areas of forest ablaze as Taupo fires leap out of control', *Dominion*, 11 February 1946, p. 8.

180 This account draws on the reports on the Saturday fires: from 'Fire extends along 50-mile front', *Dominion*, 11 February 1946, p. 8; and 'Fires sweep Taupo area', *Waikato Times*, 11 February 1946. The *Dominion* has referred variously to the 'Australasian Afforestation' area and the 'Perpetual Afforestation' area. No New Zealand-registered company of either of these names appears to have existed. Michael Roche has provided knowledgeable pointers. The Australasian Forestry Bondholders Trust Company 1925 was set up to ensure that Australian bondholders' money in New Zealand Perpetual Forests was properly used. That firm had holdings around Putaruru and towards Tokoroa. In 1935 NZ Perpetual Forests was incorporated into New Zealand Forest Products (now Carter Holt Harvey). The Afforestation Proprietary Company ('Proprietary' signals that it, too, was an Australian firm) took over some 1012-hectare blocks of Kaingaroa Forest in 1932. Working in relation to those comments and to the accounts of where the fires were at the times mentioned and the various firms' holdings, I have changed the names of the firms that the *Dominion* used in its accounts to what I think is correct. A lot of Australians must have been hoping the whole lot did not go up in flames.

181 'Fire-fighters' efforts', *Dominion*, 11 February 1946, p. 8. Re the Oruanui fire see 'Unchecked growth brings danger', *Weekly News*, 27 February 1956, p. 20.

182 'Fires sweep Taupo area: Valuable timber lost', *Waikato Times*, 11 February 1946. However – and newspaper hyperbole again comes into question – the Forest Service annual report stated that much of that scorched timber could be salvaged over the next two or three years without significant loss (*AJHR*, 1946, C.–3, SFSAR, p. 15).

183 *AJHR*, 1946, H.–12, RFB, p. 4.

184 'Taupo Fires', VHS video, R.V. 812 (ex. C.D. Neg), ANZ. Different newspapers differ on how many members of the armed forces were initially involved. The *Dominion* ('Fire extends along fifty-mile front', 12 February, p. 8) initially refers to 'hundreds of soldiers from Waioruru', then, more factually, states there were 100 army and 20 navy personnel; the *Waikato Times* ('Fires sweep Taupo area', 11 February) states 68 and 22 respectively.

185 'Seen as national calamity', *Dominion*, 12 February 1946, p. 8.

186 The film is still viewable – see note 183. The report is in 'Airmen's close call', *Waikato Times*, 12 February 1946, p. 4.

187 AD1 270/5/5, Pt. 1, Accidents and Disasters/Fires/Bush/Assistance by army, ANZ.

Commandant, 19 February 1946, Fire-fighting at Taupo, Diary of events.
188 Ibid. Telegram, 19 February 1946.
189 Ibid. Commandant, 19 February 1946, Fire-fighting at Taupo, Diary of events.
190 'Fire-fighters given mobile headquarters', *Taranaki Daily News*, 30 March 1948.
191 *AJHR*, 1946, H.–12, FBR, p. 5.
192 Ibid.
193 Ibid., p. 4.
194 F1, W3219, Box 108, 12/1/15/1, Pt. 1, Fire control and prevention/Taupo fires/ Correspondence, ANZ. Conservator N. J. Dolamore to Entrican, Employment of labour for fire-fighting etc, 1 March 1946. Dolamore reported that the Public Works Department was concerned at what its assistance cost, and he himself did not know on whom the responsibility for paying 'the very large wage bills incurred by Forest Products will fall'. He went on to report that a large number of men, including 100 Public Works men, were still doing a lot of work in the Mokai district to prevent 'the very extensive fire' there from getting into Forest Products' forest, and if the Australian-owned Afforestation Proprietary Company was not prepared to pay its full share of the costs, withdrawing Forest Service operations to better protect the Kaingaroa forest was an option.
195 Ibid. Telegram, 12 April 1946 ff.
196 *AJHR*, 1946, H.–12, RFB, p. 3. In paragraph 9 R. Girling-Butcher writes, 'Small fires have, of course, occurred from carelessness in burning off …'
197 *AJHR*, 1946, C.–3, SFSAR, p. 15.
198 Editorial, 'Prevention of forest fire', *Gisborne Herald*, 15 February 1946.
199 B. C. Cathcart, *Water on! The first hundred years of the United Fires Brigades' Association of New Zealand*, United Fire Brigades' Association, Te Awamutu, 1979, entry for 1945–46 year.
200 *AJHR*, 1946, H.–12, RFB, p. 4.
201 F1, 1/1/4, Box 151, Pt. 1, Forest and Rural Fires Bill Feb 1946–November 1946, ANZ. Girling-Butcher to Director of Forestry, Fire Service co-operation in forestry fire protection, 31 May 1946.
202 'Prevention of fires', *Dominion*, 25 February 1946, p. 6.
203 AD1 270/5/5, Pt. 2, Preservation of native bush/Forest and bush fires/Prevention of/Rotorua thermal district 1946–58, ANZ.
204 AD1 270/5/5, Pt. 2, Accidents and Disasters/Fires/Bush/Assistance by army, ANZ.
205 F1, W3219, Box 108 12/1/15/1, Pt. 1, Fire control and prevention/Taupo fires/ Correspondence, ANZ. Impressions of fire hazard etc, 11 June 1946. The quoted material is on pp. 2 & 6. Other recommendations included: ensuring that water, the most effective way of suppressing fire, was available by having the right types of engines and pumps with trained operators, as well as permanent dams and improved water-carriers; ensuring that pruning and planting patterns would enable engines and pumps to move easily; badges to allow ease of recognition; trained observers and radio operators on aerial patrols, as well as more detailed maps; a dedicated and trained victualling officer; a means of hasty recruitment (all Public Works teams within range, he suggested, could have a reserve of fire-fighters, and a form of national fire service, at least for high-hazard areas); siting lookouts stations on boundaries to have them looking in so that their visibility was not totally obscured by smoke.
206 *AJHR*, 1946, C.–3, p. 19.
207 Ibid., p. 18.
208 F1, 1/1/4, Forest and Rural Fires Bill, February–November1946, ANZ. R. Girling-Butcher to Director of Forestry, Fire Service co-operation in forestry fire protection, 31 May 1946, set out the bones of what that role might be; conservators disagreed only in the matter of detail.

Chapter Five: 'With hearts ahigh and courage aglow': Fire brigade history to 1977
1 D. G. Trass, 'Preamble', *Nuhaka Fire Service: The first 30 years, 1960–1990*, The Brigade, Nuhaka, 1990.
2 A. B. Scanlan, *100 Years of Firefighting: The New Plymouth Fire Brigade 1866–1966*, New Plymouth Fire Board, 1966.

3 Finding aids, Fire, vol. 1, Archives New Zealand (ANZ), pp. 1–2.
4 Rollo Arnold, *New Zealand's Burning: The settler's world in the mid 1880's*, Victoria University Press, Wellington, 1994, p. 244.
5 B. C. Cathcart, *Water On! The first hundred years of the United Fire Brigades' Association of New Zealand*, Te Awamutu, United Fire Brigades' Association, 1979, p. 4.
6 'News of the week', *Otago Witness*, 27 August 1864, p. 13.
7 J. K. Molloy, *Port Chalmers Volunteer Fire Brigade*, Port Chalmers Volunteer Fire Brigade, Port Chalmers; W. I. Bradley, *Lyttelton Volunteer Fire Brigade Centennial 1873–1973*, The Brigade, Lyttelton, 1973.
8 Cathcart, *Water On!*, p. 4.
9 Arnold, *New Zealand's Burning*, pp. 242–43.
10 Geraldine Volunteer Fire Brigade, *History of the Geraldine Volunteer Fire Brigade*, pp. 3–4.
11 Arnold, *New Zealand's Burning*, pp. 236–37.
12 G. R. Kear, *A Century of Service: A history of the Palmerston North Fire Brigade, 1883–1983*, Dunmore Press, Palmerston North, 1983, pp. 5–17.
13 Clyde Volunteer Fire Brigade, *75th Jubilee Programme and History, 1906–1981*, written and prepared by R. J. and V. I. Davidson, The Brigade, 1981.
14 John J. McCormick, *Oamaru Volunteer Fire Brigade: Looking Back 1879–2004.*
15 *Feilding Star*, 9 January 1886, quoted in Arnold, *New Zealand's Burning*, p. 239. Arnold also describes the Feilding council's actions, pp. 239–40.
16 Arnold, *New Zealand's Burning*, pp. 248–49. Arnold quotes on p. 249 the *Wairarapa Times*, 13 January 1886. For the use of periwinkle see B. S. Kingsbury, *The Cust Volunteer Fire Brigade 1948–2000: A record of the brigade's formation and activities over its first 52 years*, Cust Volunteer Fire Brigade, 2000, p. 7.
17 Arnold (*New Zealand's Burning*, p. 248) points out that even in the devastating Victorian fires of 6 February 1851, when about a quarter of the state was swept by fire and wool exports were cut by half, only 10 people perished; London's Great Fire of 1666 destroyed most of the city's civic buildings and burned some 300,000 homes, but the only loss of life was apparently caused by looting.
18 Arnold, *New Zealand's Burning*, p. 241. On the cost of the fire engine see 'Our Woodville letter', *Evening Post*, 11 January 1889, p. 4.
19 Arnold, *New Zealand's Burning*, p. 248.
20 Cathcart, *Water On!*, p. 31.
21 *New Zealand Parliamentary Debates* (*NZPD*), 1907, vol. 139, pp. 462, 447.
22 *Appendices to the Journals of the House of Representatives* (*AJHR*), 1909, H.–6A, Report by the Inspector of Fire Brigades (p. 3) first instances this concern, which lasted well into the 1930s.
23 'Cities and boroughs and their estimated populations', *New Zealand Official Year Book 1907*, pp. 129–31. The fire districts were in Wellington (the council in February 1908 asked the governor to declare that the city should no longer be a fire district), Auckland, Christchurch, Dunedin, Whangarei, Gisborne, New Plymouth, Hawera, Feilding, Dannevirke, Palmerston North, Masterton, Petone, Waimate, Alexandra, Oamaru, Maori Hill, Lawrence, Milton, Greymouth, Hokitika – see *AJHR*, 1909, H.–6A, Fire brigades of the Dominion, p. 1.
24 *AJHR*, 1909, H.–6A, Report by Inspector of Fire Brigades, p. 4.
25 *AJHR*, 1922, H.–6A, Report on Fire Brigades (RFB), p. 2.
26 The 1929 figure is in *AJHR*, 1929, H.–12, RFB, p. 3; the English comparison is in *AJHR*, 1928, H.–12, RFB, p. 3.
27 *AJHR*, 1909, H.–12, Report by the Inspector of Fire Brigades, p. 1.
28 *AJHR*, 1930, H.–12, RFB, p. 3. The problem is first mentioned in *AJHR*, 1928, H.–12, RFB, p. 4.
29 *AJHR*, 1935, H.–12, RFB, p. 3; see also *AJHR*, 1936, H.–12, RFB, pp. 1–2 on incendiarism, the problems of over-insurance and the more stringent risk inspection needed.
30 *AJHR*, 1933, H.–12, RFB, p. 3.
31 *AJHR*, 1934, H.–12, RFB, p. 5; AJHR, 1936, H.–12, RFB, p. 2.
32 *AJHR*, 1915, H.–6A, RFB, pp. 9, 10, 12, 19, 20, 21.

33 *AJHR*, 1917, H.–6A, RFB, p. 5.
34 *AJHR*, 1918, H.–6A, RFB, p. 12.
35 Ibid., p. 8.
36 *AJHR*, 1916, H.–6A, RFB, pp. 2–3.
37 Ibid., p. 2.
38 *AJHR*, 1925, H.–6A, RFB, p. 2.
39 Karen Hawke Grimwade, *100 Years of Fire Fighting in Ohakune 1908–2008: A history, to commemorate the centennial jubilee of the Ohakune Volunteer Fire Brigade*, The Brigade, Ohakune, 2008, pp. 8–9, 20.
40 *AJHR*, 1926, H.–6A, RFB.
41 *AJHR*, 1930, H.–12, RFB, p. 3.
42 *AJHR*, 1933, H.–12, RFB, p. 2.
43 Fire Brigades Amendment Act, 1908, *New Zealand Statutes*, 1908, p. 159ff.
44 *AJHR*, 1920, H.–6A, RFB, p. 2. The problem continued: see *AJHR*, 1930 H.–12, p. 5.
45 *NZPD*, 1926, vol. 209, p. 432: the consultative process with every fire board, after an earlier failure, ensured the legislation's passage.
46 The importance of these provisions were commented on in the debate – see *NZPD*, 1926, vol. 209, pp. 431.
47 *AJHR*, 1932, H.–12, RFB, pp. 4–5. The provisions were apparently used fairly extensively, with many fire boards contracting under the amendment act to provide protection to areas with reticulated water, to individual or private buildings, and to private property outside the district.
48 *AJHR*, 1935, H.–12, RFB, p. 2.
49 *AJHR*, 1933, H.–12, RFB, p. 3; 1935, H.–12, RFB, p. 2 for comparisons with the US and Canada.
50 *AJHR*, 1937, H.–12, RFB, p. 2.
51 *AJHR*, 1936 H.–12, RFB, pp. 2–3.
52 *AJHR*, 1937, H.–12, RFB, p. 2. Similar concerns had also been voiced in *AJHR*, 1935, H.–12, RFB, pp. 5–6.
53 *AJHR*, 1937, H.–12, RFB, p. 3.
54 *AJHR*, 1934, H.–12, RFB, p. 6.
55 *AJHR*, 1937, H.–12, RFB, pp. 3 and 5.
56 N. Taylor, *The New Zealand People at War: The home front*, vol. II, Historical Publications Branch, Department of Internal Affairs, Wellington, 1986, pp. 487–88; Finding aids/Fire vol. 1, ANZ.
57 *AJHR*, 1945, H.–12, RFB, pp. 8, 10–11; in relation to the Cabinet decision see p. 5.
58 *AJHR*, 1945, H.–12, RFB, pp. 13–15 and Schedules 2, 3 and 4.
59 F1, Box 151, 1/1/4, Forest and Rural Fires Bill, Pt. 3, ANZ. Inspector of Fire Brigades to Minister of Internal Affairs, 'Fire Service Bill', 22 August 1946. The purpose behind the Fire Service reorganisation that Girling-Butcher wanted was to ensure that 75 per cent of county towns had a pumping appliance, a minimum of 304 metres of hose and a working crew available for 24 hours a day. As most brigadesmen in those centres were volunteers, proposals were being developed for permanent staff and equipment to come in from larger centres. Means of funding the Fire Service were also developed. Girling-Butcher's comments on the minister's willingness to drop the Fire Service legislation and simply enact the rural fires legislation suggest an unwillingness to recognise that reorganisation and financial input was necessary for efficient brigade functioning. For comment on the experience, equipment and funding of the wartime Emergency Fire Service, see *AJHR*, 1945, H.–12, RFB, pp. 5–7.
60 Cathcart, *Water On!*, 1943–44 year, p. 61. The association distinguished between a permanent fire brigade, which meant a 'fire brigade established and maintained by the Board, the services of whose members are wholly at the disposal of the Board', and 'Volunteer Firemen' – those 'enrolled in a Volunteer Fire Brigade, by the decision of its members, whether or not such enrolment is subject to the approval of the controlling authority, and who give their services at drills, practices, fires, and in the fire station, either with or without remuneration, but who

rely for their living on some other vocation'.

61 *AJHR*, 1947, H.–12, RFB, p. 4.

62 Ibid., pp. 6–7.

63 After lengthy discussions with the union, the act defined a volunteer fireman as someone paid less than £52 annually. This amount was set so that any union fees an individual might accrue to the brigade would not impact on them.

64 *NZPD*, 1949, vol. 285, pp. 629, 632 for clauses dealing with rural protection and brigades' reinforcement role.

65 F1, Box 153, 1/1/10, Pt. 2, Fire Service Bill 1948–50, ANZ. Entrican to Commissioner of State Forests, 1 July 1948. The quoted material is included in this minute; Ibid., minute, Boardman, 26 July 1949.

66 F1, Box 153, 1/1/10, Pt. 2, Fire Service Bill 1948–50, ANZ. Girling-Butcher to Minister of Internal Affairs, 24 November 1950.

67 'Co-ordinating summer fire control', *Southland Times*, 6 July 1971.

68 183 brigades were entirely voluntary; only 20 employed one or more permanent officers. There were 500 permanent fire-fighters and 3600 members of the United Fire Brigades' Association. To bring the volunteers into the system would have demanded 6500 men and cost some £4 million – *NZPD*, 1949, vol. 285, pp. 668, 671.

69 *An Encyclopaedia of New Zealand 1966*: www.teara.govt.nz/en/1966/fire-services/4 (accessed September 2009). For the 1974 figure see Cathcart, *Water On!*, 1974–75 year.

70 This simplified account does not mention the Fire Services Act 1972, as the result of which a new Fire Services Council – to all intent and purposes the old Fire Service Council – took over. The structures under which the new body worked were substantially the same as the first.

71 *AJHR*, 1953, H.–12 Report of the Fire Service Council, pp. 5–7. There were three classifications:

- Class F+ would generally apply to most secondary fire districts already constituted, and to new ones where risk demanded a fire engine but where the local authority could not meet the cost and where there was no, or only substandard, reticulated water. Such brigades were to have no fewer than 10 volunteer members, a fire engine, plus a trailer or portable petrol-driven pump (meeting certain specifications), an electric fire siren, and station premises with space for socialising; steps were to be taken to make any natural or static water supply available. The Fire Service Council estimated the annual expenditure should normally be below £300.
- Class F was applicable to small towns outside an established brigade's protection range, where fire risk did not justify a fully motorised brigade but where isolation demanded a separate district. The brigade would consist of no fewer than six volunteer members and generally not more than 10, and was to operate as a fire party. It was to be equipped with a light trailer pump or hand-cart pump; the local authority was to provide somewhere to store equipment (not necessarily a separate fire station) and to make natural or static water supplies available. Expenditure was not to exceed £100.
- Class F– classifications applied where new fire districts were not justified because they were within an urban area (or areas) already protected by an existing brigade but where the density of risk or distance meant a local fire party was needed to give 'first aid' (not in the medical sense) before the brigade arrived. A volunteer fire party was to have no fewer than six men and not generally more than 10. It was to be attached to the district brigade under an appropriate agreement, be equipped (by arrangement with the urban fire authority) with a light trailer pump or trailer hand-cart pump, have somewhere to store it, and have satisfactory arrangements for fire calls and for notifying the district brigade; natural and static water supplies were to be made available. The fire party came under the district brigade but the territorial local authority was to pay the local proportion of its cost to the United Fire Brigades' Association, in addition to any general agreed out-of-district fire protection. Fire parties were to be maintained at no more than £75.

72 I am indebted to Gavin Wallace for this succinct definition.

73 *AJHR*, 1959, H.–12, Report of the Fire Service Council, p. 7.

74 D. W. Blewett and Mark Hutton (compilers), *Alexandra Volunteer Fire Brigade 1903–2003:*

100 Years, Otago Daily Times Print, Alexandra, 2003, pp. 19, 21.

75 *Titirangi Volunteer Fire Brigade Jubilee 1949–1999*, The Brigade, Titirangi, 1999, p. 13.
76 Arnold Carr, OHInt-0641/03, OHC.
77 *Titirangi Volunteer Fire Brigade Jubilee 1949–1999*, p. 21.
78 Penny Walker, *Sound the Siren: A history of the Wainuiomata Volunteer Fire Brigade 1944–1994*, The Brigade, Wainuiomata, 1994, pp. 97–98.
79 Mervyn Allaway, OHInt-05342/2, OHC.
80 Eric Tikey, OHC-010136, OHC.
81 Glenn Richards, *The History of Firefighting at Rolleston: Written to celebrate 25 years as a fire brigade*, Rolleston Volunteer Fire Brigade, Rolleston, 1994.
82 F1, W3219, Box 114, 12/9/3/20, Fire control and prevention/Fire districts/Wellington Conservancy/Eastern Bays 1937–56, ANZ, illustrates well the extent to which the Forest Service was working alongside local volunteers, bringing in men and equipment.
83 *Titirangi Volunteer Fire Brigade Jubilee 1949–1999*, p. 19.
84 F1, W3219, Box 113, 12/2/6/a, Fire control and prevention, ANZ. Mt White fire investigation, Recommendations by the Combined Fire Committees for improvements in fire prevention and control, Wellington, 1973, p. 12.
85 Penny Walker, *Sound the Siren*, pp. 94–99.
86 G. C. Hensley, *A Review of Rural Fire Services in New Zealand: Report of a committee chaired by G. C. Hensley, Coordinator Domestic and External Security*, Rural Fires Review Committee, Wellington, August 1989, pp. 5, 9–11.
87 *AJHR*, H.–12, Fire Service Council 1969 and 1972.
88 'City firemen to learn rural methods', *Evening Post*, 14 September 1970.
89 B. C. Cathcart, *Water on!*, pp. 90–92.
90 *AJHR*, H.–12, RFB, p.12
91 Hensley, *A Review of Rural Fire Services*, p. 11. Hensley also describe (p. 8) the Forest Service role in rural fires as 'underpin[ning] the whole rural fire system' that the successive Forest and Rural Fires Acts established.

Chapter Six: The nation's rural fire-fighters 1948–87

1 The National Film Unit (NFU) was initially set up to provide information on the country's war achievements. In 1946 its *Weekly Review* became part of the Prime Minister's information section, making documentaries on New Zealand's problems and undertakings. In 1950 it became part of the Department of Tourist and Publicity, then faded as a result of accusations of political bias. It was replaced in 1952 by *Pictorial Parade*, which ran until 1971. *New Zealand Mirror* was one of the unit's longest-running newsreel series.
2 *Appendices to the Journals of the House of Representatives* (*AJHR*), 1949, C.–3, p. 7. The mill, as that report goes on to explain, was established to ensure that the 'small-diameter, rough, knotty logs', which had eventuated because the trees had originally been planted too widely and then had not pruned, could be harvested economically.
3 The quoted material and the references to the shots are taken from *No Random Harvest*, NFU, Pictorial Parade, 1954, R.V. 225, Archives New Zealand (ANZ). However, Forest Service propaganda was not limited to one film. Similar messages (often using the same footage) can be seen in such films as *Nobody Ordered Fire*, NFU, Pictorial Parade no. 173, R.V. 373, 1965, ANZ; and 'Protection of wealth: Forest Service's fire fighters', *New Zealand Mirror*, R.V. 138, 1953, ANZ. These films also incorporated fire messages, which are covered later in this chapter.
4 G. T. Bloomfield, *New Zealand: A handbook of historical statistics*, G. K. Hall and Co., Boston, 1984, table V.1, p. 160.
5 John Halkett, Peter Berg and Brian Mackrell, *Tree People: Forest Service memoirs*, New Zealand Forestry Corporation, GP Print, Wellington, undated (c. 1987–88), pp. 93–94.
6 Mike Hockey, Interview, RFP. At that stage Hockey was officer-in-charge of Wharerata exotic forest between Wairoa and Gisborne.
7 Michael Roche, 'Exotic forestry, Harvesting the forests, 1950s–1980s', *Te Ara: The Encyclopedia*

of New Zealand: www.teara.govt.nz/en/exotic-forestry/4 (accessed March 2009); Morrie Geenty, Interview, RFP.

8 Mike Hockey, Interview, RFP.

9 'Guarding against fires', *Hauraki Plans Gazette*, 15 January 1951.

10 *AJHR*, 1947, C.–3, p. 8; F1, 10/2/46, Pt. 1, Reports/Fire officer 1949–56, ANZ. 'New Zealand Forest Service, Fire protection of rural lands in New Zealand, undated, (1951); AANI, W3219, Box 1, 1/1/4/0, Pt. 3, Revision Forest and Rural Fires Act 1958–74, ANZ.

11 These parties were county councils, forest owners, farmers, the sawmilling industry, fire underwriters, relevant government departments and representatives of the fire boards and fire brigades.

12 *AJHR*, 1948, C.–3, pp. 26–29.

13 NZ statutes; AADY, W3564, Box 2, 1/3, Pt. 1, Legislation proposals, ANZ. Director of Forestry to Commissioner of State Forests, 4 November 1948, ANZ.

14 AANI, W3219, Box 1, 1/1/4/0, Pt. 3, Revision of Forest and Rural Fires Act, 1968–74, ANZ. Forest fires, undated (c. 1968–69). This appears to be the text of a publicity booklet, some of which appeared in other more specific publications such as on fire-fighting methods.

15 F1, Box 152, 1/1/4, Pt. 2, Proposed rural fire regulations 1950–51, ANZ. Bodies involved included the New Zealand Catchment Association, Dominion Federated Sawmillers' Association, Public Service Association, Forest Owners' Association, Farmers' Union, Soil Conservation Council, Ministry of Works, New Zealand Workers' Union, Timber Workers' Union and the Tokoroa Fire Committee. It was obviously a time to air differing viewpoints: that the Forest Service was preventing burning of land adjacent to state forests, land was being forced out of use, and would eventually constitute a threat itself. With land highly valued and the government well disposed towards farmers, it seemed obvious to some that the Service should allow well-farmed land to be located close to forest as the best possible form of protection, and some considered that some Forest Service officers' attitude was unsatisfactory.

16 *AJHR*, 1953, c.–3, pp.13–14.

17 A. N. Cooper, 'Forest fires in New Zealand, history and status 1880–1980', p. 4.

18 *AJHR*, 1955, C.–3, p. 49; *AJHR*, 1957, p. 62; F1, W3219, Box 113, 12/2/6A, Fire control and prevention, ANZ. Mt White fire investigation, report, 'Points of agreement on fire hazard and allied problems: Lincoln Conference 1969'. The paper notes that land-clearance fires in rural areas was only legitimate when other means of achieving the same results were not feasible, but stresses various safeguards (topdressing ridges, creating firebreaks and access tracks, and grazing) were essential.

19 *AJHR*, 1960, C.–3, Appendix 5, Fires, p. 73.

20 Jiani Wu, William Kaliyati and Kerl Sanderson, *The Economic Cost of Wildfires: Report to the New Zealand Fire Service Commission*, BERL Economics, Wellington, 2009, pp. 7, 9.

21 Forest and Rural Fires Association of New Zealand: www.ruralfirehistory.org.nz/4th.htm (accessed 19 March 2010). The site lists the major fires and provides brief information on some of the burns.

22 *Dominion*, 25 January 1954. 'Fire season begins', *Dannevirke Evening News*, 1 October 1954.

23 *AJHR*, 1953, C.–3, p. 35. That year, for example, 18 fires in the Westland conservancy destroyed 1767 hectares of cut-over forest and 40 hectares of millable rimu; in Southland four fires burned 225 hectares in total. Elsewhere in state forests, 27 fires burned over 657 hectares of scrub and bracken country.

24 *AJHR*, 1962, C.–3, p. 28.

25 AANI, W3219, Box 61, 12/3/1, Pt. 7, Fires outside state forests in Auckland fire districts 1961–63, ANZ. V. Singham, rural fire officer, to district ranger, NZFS, 11 October 1962.

26 Ibid. See Report, Karaka Creek, 24 February 1961, for the comment on night-time operations; F1, Box 300, 12/4/4, Pt. 1, Fires outside state forests and fire districts/Nelson conservancies 1947–58, ANZ. Fire report, acting conservator R. W. G. Janson, 17 December 1954. A month or so before a fire at Waiwhero threatened the Maorilands Block. Getting a tractor in to cut the necessary access to attack the fire would have meant going through

a gully with vegetation above head height (ibid., Report on fire at Waiwhero, 9 December 1954).

27 In 1949 Mr Horrell, a Waimea county fire officer, was killed when a gorse and scrub fire got out of control at Waiwhero ('Trapped by flames', *Nelson Mail*, 14 November 1949); Forest Service employee H. W. C Clift was killed in February 1950 in the Bay of Islands area (F1, Box 300, 12/4/1, Pt. 1, Fires outside state forest and state controlled fire districts/Auckland region 1946–53, ANZ. Report, 4 February 1950). Other deaths that occurred during controlled burns in both state and privately owned forests are mentioned in the text.

28 F1, Box 301, 12/4/7, Pt. 1, Fires outside state forests and fire districts/Southland conservancies 1942–57, ANZ. Report, 5 January 1953.

29 Ibid. Fire no. 11C, Scenic Reserve, Eglinton Valley, 8–20 March 1953; Entrican to Conservator of Forests, P. A. Reveirs, 5 May 1953, Extract from diary notes of W. F. Wright, paras 2, 4 (comments on Swale), 6 (comments on Linton). John Ward helpfully identified Tom (A. T.) Swale as the Swale concerned.

30 Ibid. Log of fire no. 12c, Notornis Reserve, 19 March 1953; Extract from diary notes, F. W. Wright, Fire, Notornis Reserve, 14 March 1953.

31 For the value of the army and the way its rations were enjoyed see Roy Knight, Interview, RFP.

32 F1, W3219, Box 113, 12/2/6/a, Fire control and prevention, ANZ. Mt White fire investigation report, Fire in Mt Cook National Park, 28–30 March 1970; Ibid., Report, Tussock fire, Mt Kyeburn area, 23 January 1973; Ibid., Mt White fire, 23 February 1972; 'New plans to fight hill fires', *Rotorua Post*, 4 December 1973.

33 Mike Hockey, Interview, RFP, describes the conservancy as 'a huge area. You get a sense of it if you draw a straight line from Tauranga through Taupo and across to Wairoa. Besides Kaingaroa and the private Fletcher Tasman forests it included the Whirinaki, Urewera, Waioeka and Raukumara indigenous forests.'

34 Significant fires/incidents 1947–87, www.ruralfirehistory.org.nz/4th.htm.

35 *AJHR*, 1956, C.–3, p. 63.

36 Jack Barber, Interview, RFP; F1, 10/2/46, Pt. 1, Reports/Fire officer 1949–55, ANZ. Annual report, Fire control section, undated (1956).

37 Roy Knight, Interview, RFP. The Woodsman Training School was designed 'to provide skilled forest-tradesmen, rather than leaders'. However, within five years Entrican recognised that the skilled craftsmen he had hoped to produce were in fact often able to do rather more. 'Family character plus late development of intellect appears to be the explanation,' he wrote in words that would receive tougher scrutiny now than then (J. Ward and N. Cooper, *Seventy Years of Forestry: Golden Downs Forest, Nelson, 1927–1997*, Forest History Trust, Richmond, pp. 92–93).

38 *AJHR*, 1973, C.–3, p. 21.

39 ABDT, W3092, Box 12/2/995(b), Fire Mohaka Forest 1973–80, ANZ. Fire log, Mohaka forest, 3–4 November 1973; Ibid., Girling-Butcher, Mohaka fire, 3–4 November 1973, 11 January 1974; ' "Burning-off" sparked Mohaka Forest blaze', *Daily Telegraph*, 5 November 1973. *AJHR*, 1974, C.–3, Fire control, p. 21, states the area destroyed.

40 *Report of the Commission of Inquiry in the Hanmer Fire* in ABFK, Box 7494, W4948, 50/16/9, Pt. 1, ANZ. Organisation and administration: Defence Fire Service: Forest and Rural Fires Act 1955, draft revision, 1976–76; AANI, W3219, Box 60, 12/2/B/1, Pt. 1, ANZ. Hanmer Forest Fire, Commission of Inquiry; Ibid., unidentified paper, 26 March 1976 for the estimated loss; 'Forest chief: Lives at risk', *Christchurch Star*, 23 March 1976. Volunteers from the general public had rushed in – undirected and unasked – and had almost got caught in the fire. When challenged on his failure to ask for volunteers, an exasperated conservator Levy pointed out: 'If there is a fire in the city, the fire brigade does not use volunteers from the street. They have to be trained.' 'Fire-fighters upset by attack' and 'Choppers save the day', Christchurch *Press*, 24 March 1976.

41 Carter Holt was formed in 1971 from the merger between Robert Holt & Sons and Carter Merchants. For comment on the fire see Morrie Geenty, Interview, RFP.

42 ABGE, 7498, Box 48, 12/1/3, Fires other than Marlborough–Nelson fire district, ANZ. Forest Service internal memo, Peter Maplesden, Hira fire, 5–7 February 1981, p. 1.
43 Ibid.
44 Ibid., p. 2.
45 Ibid., p. 5; Ross Hamilton, Interview with author.
46 F1, 10/2/46, Pt. 1, Reports/fire officer 1949–56, ANZ. Job description: regrading, W. F. Wright, 16 January 1951.
47 'Guarding against fires', *Hauraki Plains Gazette*, 15 January 1951.
48 John Ward, Interview, RFP. In terms of subsequent official acknowledgment of this position, see the recommendations on the Mt White fire. These specifically state that 'counties could not be expected to have the resources to fight large fires over long periods, and they must be able to call on the Forest Service to assist them in their operations, or to take control of fire-fighting operations': F1, W3219, Box 113, 12/2/6/a, Fire control and prevention, ANZ. Mt White fire investigation, Recommendations by the Combined Fire Committees for improvements in fire prevention and control, Wellington, 1973, p. 12.
49 Morrie Geenty, Interview, RFP. Geenty explains that it is impossible for everyone to be on the same system if using their system in normal operations. Big forest companies have radio for harvesting or for dispersing trucks on the same band so a uniform system could block urgent calls. Using designated channels is impractical because multiple organisations attend and it would be difficult to standardise equipment across them. Today the large, combined Rotorua fire district gets around the problem by having units of each organisation at fire headquarters so each one can relay messages through its own systems. This arrangement has proved essential – as has having a separate system for aircraft to prevent them getting conflicting instructions (it has happened!).
50 *AJHR*, 1949, C.–3, p. 21. In terms of later comments and slight shifts in emphasis in developing appreciations of the land's conservation value, then on the need for urban recreation, see *AJHR*, C.–3, 1961, pp. 35–36; and *AJHR*, C.–3, 1962, p. 30.
51 F1, 10/2/46, Pt. 1, Reports/Fire officer 1949–56, ANZ. New Zealand Forest Service, Fire protection of rural lands in New Zealand, undated (1951), Appendix C.
52 On the bridges see Halkett et al., *Tree People*, p. 123.
53 For Entrican's first suggestion see AADY, W3564, Box 27, 42/1, Pt. 2, Recreational uses of state forests/Permits to enter, shoot, picnic etc, ANZ. Memo to all conservators, Access to state forests for recreational purposes, 8 February 1950. The Forests Amendment Act 1965 allowed sectors of exotic forest with local recreational value to be opened to the public without a permit (*AJHR*, 1966, C.–3, p. 6). For some reason 'Forest Service has faith in public', *Dominion*, 18 September 1970, dates the opening of the forests as 1969.
54 Jack Barber, Interview, RFP, and John Ward comment on the hard bottom line. John Barnes, Interview, RFP, suggests that some forestry companies used lookouts longer than the Forest Service, and comments on the historical lookout at Eyrewell, 'where you climb about 30 metres to get into a sort of a box on top of a stand'. John Ward (pers. comm., March 2010) gives information on the five lookouts in the Waimea rural fire district. One is Richmond Hill, on the Barnicoat Range overlooking Richmond, and is manned all summer from about October to April. Mike Oliver, who at time of writing mans the lookout from October to April, has an important role and a wide range of duties: reporting fires, first port of call for after-hours 111 calls and (in relation to rural fire) instigating the initial attack, then monitoring the fire and monitoring daily weather readings for Nelson and Marlborough. He also provides fire permit management (i.e. identifies when controlled burns may be lit), provides local radio stations with information on fire hazards and prevention measures over the holiday seasons, and traps ferrets and stoats. The other lookouts are Inwoods in Golden Downs Forest, the lookout in the Whangapeka area, and the Greenhill lookout, which overlooks the lower Motueka and Ngatimoti areas. The last group is only used when the fire danger is extreme.
55 Jack Barber, Interview, RFP, explains that climatic conditions in Otago and Southland made lookouts a rarity, in contrast to Canterbury and Hawke's Bay where it could be 'hot and dry

for weeks, if not months, and lookouts were part of the scene'.

56 Jack Barber, Interview, RFP.

57 *Nelson Evening Mail*, 6 April 1955 (the ANZ copy – F1, 13/11/6, Pt. 2, Newspaper clippings – was clipped without its headline).

58 'The joys of country quietness', *Weekly News*, 1 April 1964.

59 F1, Box 283, 10/2/46, Pt. 1, Reports/Fire officer 1949–56, ANZ. New Zealand Forest Service, Fire protection of rural lands in New Zealand, undated (1951). Appendix C dates the campaign as beginning in 1951. However, the 1959 Forest Service annual report, p. 98 mentions the planning for the campaign taking place over the two previous years.

60 F1, 12/1/12/7, Pt. 3, Fire prevention screen advertisements 1952–54, ANZ. Entrican to J. Bardon, Dormer-Beck Advertising, 1 October 1954.

61 AATE, 889, W3321, Box 87, 74/4/7/2, Fire prevention under Forest and Rural Fires Act 1947/Rural fire districts 1952–74, ANZ. In what appears to have been a short-lived initiative, a Forest and Rural Fire Publicity Committee was set up on 30 September 1952 with members from the Dominion Forest Owners' Federation, Dominion Sawmillers' Association, Federated Mountain Clubs of New Zealand, Forest and Bird Protection Society, New Zealand Counties' Association, North Island Motor Union and South Island Motor Union and the Forest Service. I found no mention of this group beyond its first meeting on 3 December 1952.

62 AANI, W3219, Box 79, 13/82/8, Pt. 2, Radio, TV coverage 1969–80, ANZ.

63 AANI, W3219, Box 78, 13/82/4, Pt. 10, Exhibitions, displays, roadside hoardings and signs 1979–81, ANZ. Memo, 26 October 1976.

64 *Forestry Fire Season*, NFU, Weekly review no. 428, 1949, ANZ. Ref. R.V. 138. John Ward related the anecdote.

65 *No Random Harvest*, NFU, Pictorial Parade 1954, ANZ. Ref. R.V. 225.

66 *AJHR*, C.–3, 1959, p. 97.

67 *Nobody Ordered Fire*, NFU, Pictorial Parade no. 173, 1965, ANZ. Ref. R.V. 373.

68 F1, Box 299, 12/1/7, Pt. 2, Fires outside state forest/Southland 1948–50, ANZ. A. D. McKinnon, Fires no. 1(b) 2 August 1949; for Tapawera School see F1, Box 300, 12/3/3, Pt. 1, Fires outside state forests and fire districts/Nelson conservancies 1947–58, ANZ, conservator, Nelson, Courtney Biggs, to Director-General of Forests, 17 June 1949.

69 John Barnes, Interview, RFP.

70 Jack Barber, Interview, RFP.

71 *AJHR*, 1982, C.–3, p. 10.

72 John Barnes, Interview, RFP; *AJHR*, 1963, C.–3, p. 27; *AJHR*, 1980, C.–3, p. 9; *AJHR*, 1981, C.–3, p. 10.

73 Don Geddes, Interview, RFP.

74 *AJHR*, C.–3, 1982, p. 20; New Zealand Forest Service, *Remote Weather Stations for Forestry Use*, undated, in Reports, National Rural Fire Authority (NRFA).

75 Interviews: John Barnes, Don Geddes and John Ward, RFP.

76 Kerry Hilliard, Interview, RFP.

77 *AJHR*, 1955, C.–3, p. 46.

78 For the information on the numbers of stations and the need to replace the equipment, see *AJHR*, 1954, C–3, p. 46; *AJHR*, 1956, C.–3, p. 64. F1, 10/2/46, Pt. 1, Reports/Fire officer 1949–56 (ANZ), New Zealand Forest Service, Fire protection of rural lands in New Zealand, undated (1951), Appendix A, Radio communications, gives the technical details. John Ward's comments are from his RFP interview.

79 *AJHR*, 1957, C.–3, p. 3.

80 Ibid., p. 59.

81 *AJHR*, 1960, C.–3. p. 50; *AJHR*, 1963, C.–3, p. 27.

82 *AJHR*, 1972, C.–3, p. 24.

83 Mike Hockey, Interview, RFP.

84 *AJHR*, 1971, C.–3, pp. 34–5; *AJHR*, 1973, c.–3, pp. 32–33.

85 F1, Box 283, 10/2/46, Pt. 1, Reports/Fire officer 1949–56, W. F. Wright to

inspector-in-charge, Vehicle for Fire Control Section, 14 August 1953. In this memo he quotes Bailey's 1948 communication.

86 F1, W3219, Box 108, 12/1/10, Pt. 2, Fire prevention and control/Prevention, co-ordination by government departments 1953–62, Entrican to director-general, P&T, 27 October 1954.

87 *AJHR*, 1974, C.–3, p. 34.

88 Jack Barber, Interview, RFP.

89 John Ward, Interview, RFP.

90 F1, 10/2/46, Pt. 1, Reports/Fire officer 1949–56, ANZ. New Zealand Forest Service, Fire protection of rural lands in New Zealand, undated, 1951, Appendix B, Forest fire-fighting equipment in New Zealand.

91 *AJHR*, 1958, C.–3, Fire-fighting equipment, p. 70.

92 F1, W3129, Box 106, 12/1, Pt. 4, Fire control and prevention general 1958–62, ANZ. Discussion on HO instructions, 16 September, 1958, A. L. G. Taylor, Development of equipment, 17 September 1958. John Ward, Interview, RFP.

93 *AJHR*, 1959, C.–3, p. 95.

94 Jack Barber, Interview, RFP. For the reference to Calder's workshop see Ward and Cooper, *Seventy Years of Forestry*, p. 117.

95 *AJHR*, 1984, C.–3, Fire control, p. 11. A Forest Service publication, *Fire Engine Operators Manual*, 1983, also indicates that dual-purpose (but slow and under-powered) Bedfords were in use in the 1970s, but references to these machines are few.

96 *Fire Engine Operators Manual.*

97 On their selection see F1, Box 283, 10/2/46, Pt. 1, Reports/Fire officer 1949–56, ANZ. New Zealand Forest Service, Fire protection of rural lands in New Zealand, undated (1951), Appendix B, Forest fire-fighting equipment in New Zealand; Ibid., Diary notes, Nelson conservancy, 12–15 January 1952.

98 Roy Knight, Interview, RFP.

99 Warren Reekie, Interview, RFP.

100 F1, Box 283, 10/2/46, Pt. 1, Reports/Fire officer 1949–56, ANZ. Annual report, Fire control section, undated (1956).

101 *AJHR*, 1958, C.–3, Fire-fighting equipment, p. 70. John Barnes, Interview, RFP, comments on their origins and noise.

102 Interviews: Roy Knight and John Barnes, RFP. Forest and Rural Fires Association of New Zealand (www.ruralfirehistory.org.nz/pumps.htm) also gives an excellent detailed and technical account of the pumps used in the past and present.

103 AJHR, 1959, C.–3, p. 94; AANI, W3219, Box 62, 12/6/2, Pt. 7, Fire-fighting and control equipment/General info 1976–79, ANZ. NZ Forest Service 250 gal. dams, undated (c. July 1976).

104 John Barnes, pers. comm., April 2010.

105 Ian Millman, pers. comm., 9 April 2010.

106 Hydroblenders are first mentioned in AANI, W3219, Box 61, 12/3/1, Pt. 5, Fire outside state forests but inside fire districts/Auckland region 1955–58, ANZ. See also AANI, W3219, Box 61, 12/3/1, Pt. 7, Fires outside state forests in Auckland fire districts 1961–63, ANZ. Mike Hockey (Interview, RFP) describes their action. AANI, W3219, 12/6/4, Pt. 5, Chemical fire equipment and extinguishers 1977, ANZ. Hydroblenders, urban and rural areas, undated (January 1978).

107 For further information on Fire Trol see www. firetrolholdings.com/faqs; for Phos Check see www. phos-chek.com/products/retardant.php (both sites accessed June 2010).

108 AANI, W3219, Box 57, 12/1/24, Pt. 2, Fire prevention/Training of personnel 1954–78, ANZ. Urban fire officers' course for rural fire-fighting, Helicopters in fire-fighting, February 1977. See also Forest Service, *Helicopters in Forest Fire Operations*, p. 16.

109 John Barnes, Interview, RFP.

110 D. A. Campbell, 'Aerial fire fighting', *NZ Science Review*, December 1959, describes the trials. See also Forest Service comment in *AJHR*, 1957, p. 62; *AJHR*, 1959, p. 94.

111 See, for instance, Fire patrol/Canterbury Aero Club 1953–85, ACCQ, 828, W5616, Box 61,

12/1/14/6, ANZ. Contract accompanying memorandum, 12 February 1954, Forest fire aerial patrols. The surveillance these clubs provided can be judged from the land the Canterbury Aero Club flew over in Canterbury and Westland – 'the area bounded generally towards the north by a line drawn between the mouth of the Grey River and the mouth of the Clarence River; towards the south-east by the sea; towards the south-west by the Waitaki River, Lake Ohau and the Hopkins River to the Main Divide and thence by a line to Bruce Bay; and towards the north-west by the sea'. For this service it was paid a retainer of £10 per annum as well as an hourly rate for flying in fire emergencies.

112 'The spot against red alert', *Rotorua Post*, 4 March 1970.
113 *AJHR*, 1961, C.–3, pp. 37–38.
114 Mike Hockey, Interview, RFP.
115 *AJHR*, 1975, C.–3, p. 17.
116 *AJHR*, 1970, C.–3, p. 17.
117 *AJHR*, 1972, C.–3, p. 24 for other agencies' use of the buckets. The Titirangi Volunteer Fire Brigade first used one in 1974.
118 AANI, W3219, Box 57, 12/1/24, Pt. 2, Fire prevention/Training of personnel 1954–78, ANZ. Urban fire officers' course for rural fire fighting, Helicopters in fire fighting, February 1977. For their essential role in major fires see *AJHR*, 1982, C.–3, p. 10. Attending at an Australian fire control officers' conference helped New Zealand members conclude that helicopters were the only way of stopping serious fires such as the one at Hira Forest.
119 'Giant fire bucket', *Manawatu Standard*, 12 October 1971.
120 Mike Hockey, Interview, RFP.
121 *AJHR*, 1972, C.–3, p. 24.
122 F1, W3219, Box 113, 12/2/6A, Fire control and prevention/Mt White fire investigation, ANZ. A. F. A. Ferguson, forest ranger, Use of helicopters at Mt White fire.
123 AANI, W3219, Box 60, 12/2/B/2, Pt. 1, Hanmer forest fire/Commission of enquiry, ANZ.
124 F1, W3219, Box 113, 12/2/6A, Fire control and prevention/Mt White fire investigation, ANZ. A. F. A. Ferguson, forest ranger, Use of helicopters at Mt White fire.
125 AANI, W3219, 12/6/4, Pt. 5, Chemical fire equipment and extinguishers 1977, ANZ. Girling-Butcher to Frank M. Feffer, 24 July 1979.
126 AANI, W3219, Box 62, 12/6/2, Pt. 7, Fire-fighting and control equipment/General info 1976–79, ANZ. Quarterly report to 31 January, 2 March 1979, p. 2.
127 For monsoon buckets see *AJHR*, 1972, C.–3, p. 24; *AJHR*, 1977, C.–3, p. 25. For the infra-red devices see *AJHR*, 1978, C.–3, p. 12. Morrie Geenty (Interview, RFP) remembers the hand method. One of the Hanmer Commission of Inquiry's (1976) main recommendations was on introducingthe devices, which suggests that after the initial trial the devices were not immediately put to wider use.
128 F1, 10/2/46, Box 283, Pt. 1, Reports/Fire officer 1949–56, ANZ. Diary notes of fire control officer, duty tour of North Island stations 16 May – 23 June 1949, instances some of the discrepancies in tool allocation among conservancies. For rationalisation of this practice see *Fire*, New Zealand Forest Ranger Certificate, undated (early 1960s), Forest fire risk grading formula, section 2(a), Millman Collection, p. 3 ff.
129 Peter Amner, 'New Zealand Forest Service fire control staff 1947 until 1 April 1987', unpublished, ATL, p. 1.
130 Roy Knight, Interview, RFP. For John Ward's comment see RFP.
131 *AJHR*, 1959, C.–3, p. 92.
132 Doug Ashford, Interview, RFP.
133 Jack Barber, Interview, RFP. For other recollections of Girling-Butcher see Ian Millman and John Barnes, Interviews, RFP.
134 *AJHR*, 1953, C.–3, pp. 36–37; *AJHR*, 1969, C.–3, p. 34.
135 Subscriptions to international journals continued the practice noted in the previous chapter. For attendance at Australian courses see *AJHR*, 1982, C.–3, p. 10. The Australian State Forest Fire Control Officers' Conference that year, for instance, provided valuable insights into fire control in Australia and on their aerial fire-fighting systems.

136 The US booklets are: US Department of Agriculture, *Forest Service: Forest fire fighting fundamentals*; US Department of Agriculture, *Water vs. Fire*, which may have been the same as *Fighting Fire with Water*, of which 400 copies were ordered in 1954. Forest Service publications were: *Safety and Survival in Forest Fires*, Information Series no. 68, 1974 (reprinted 1984), which drew on an Australian CSIRO publication, *Bushfire Sense*, 1961; and *Basics of Fire Fighting*, 1980.
137 Jack Barber, Interview, RFP.
138 Ibid.
139 F1, 12/1/24, Pt. 1, Fire prevention training of personnel 1948–53, ANZ; AANI, W3219, Box 57, 1/1/24, Pt. 2, Fire prevention, training of personnel 1954–78, ANZ. For county fire officers' involvement see F1, Box 555, 44/11, Pt. 15, Circular letters to conservators 1949–51, ANZ. Memo no. 428, Fire-fighting instructions, 4 May 1950.
140 The history of the assistance given to Wellington's eastern bays illustrates these points. See F1, W3219, Box 114, 12/9/3/20, Fire control and prevention/Fire districts/Wellington conservancy/Eastern bays 1937–56, ANZ. On the willingness to bear the costs etc see ibid., Fire control officer, memo to Entrican, 20 March 1954.
141 A. N. Cooper's paper, 'Training using simulated forest fires', *Commonwealth Forestry Review* vol. 62 (2), 1986, pp. 131–39 discusses the decision after the 1976 Hanmer Forest fire to use the plantation fires as simulated fire exercises, and sets out the format used.
142 Such methods had significantly predated 1949, but other references to controlled burns are in relation to the Whakatane Board Mills (F1, 10/2/46, Pt. 1, Reports/Fire officer 1949–56, ANZ. Diary notes of fire control officer, duty tour of North Island stations, 27 March – 4 April 1949), while John Barber (Additional papers, OHC) participated in a large burn-off of felled manuka and similar scrub in Gwavas Forest in 1951.
143 Two men 'selected for physical fitness and adaptability' were sent to South Australia where the technique was widely practised: *AJHR*, 1959, C.–3, p. 94; AANI, W3219, Box 57, 12/1/24, Pt. 2, Fire prevention/Training of personnel, ANZ. Entrican to Minister of Forests, Training of Forest Service officers in the use of slash burning, 15 February 1957. Entrican here comments on the conservators' reluctance. The comment on the men's capabilities is from ibid., secretary, State Services Commission, 27 February 1957.
144 Jack Barber, Interview, Additional papers, OHC.
145 F1, W3219, Box 110, 12/1/29/3, Pt. 3, Fire control and prevention/Burning off in state forests/Wellington conservancy 1968–70, ANZ. Forest Service scientist Ashley Cunningham talks of the poor Kaweka land – the 162 hectares of bare erosion surface within the Tutaekuri, more in the Ngaruroro, and the tiny remaining pockets of the original podocarp forest in the Blowhard and Waiwhare areas. Today the Blowhard Bush preserves the main area of that vegetation.
146 *AJHR*, 1974, C.–3, p. 21; Girling-Butcher to Director-General of Forests, A. F. Thomson, Ventilation index, 22 November 1973, in Reports, Loose papers, NRFA; Girling-Butcher had recently attended a smoke symposium in the US. If guidelines to avoid these inconveniences were drawn up, as he suggested, there is little evidence that they were implemented, with the exception of the Meteorological Office initiative.
147 Don Geddes, Interview, RFP.
148 John Barnes, Interview, RFP.
149 Don Geddes, Interview, RFP.
150 Charlie Ivory, pers. comm.
151 ABDT, W3092, Box 12, 12/3, Controlled burns/Instructions/General 1963–82, ANZ. L. J. Ellis, Controlled burning, 14 April 1968.
152 Ibid. Memo, Control burning, 26 May 1981.
153 Warren (Nobby) Reekie, Interview, RFP.
154 'Smoke hangs low over city', *Rotorua Post*, 1 March 1972; 'Smoke from Mamaku Forest fire' *Thames Star*, 29 February 1972. For concern expressed about destruction of bush and wildlife see the *NZ Herald*: 'Bush burn involves many things', 4 March 1972; Letters to editor, 'Forest burn-off', 6 March 1972; '10 years of smoke haze likely', 30 March 1972. In relation

to complacency see ABDT, W3091, Box 12, 12/3, Controlled burns/Instructions/General 1963–82, ANZ. J. B. Everett, Controlled burns – wildlife, 16 March 1981. Everett pointed out that the Forest Service did not want the bad publicity from neglecting birdlife to 'grow out of all proportion to the value of well managed and achieved controlled burns'.

155 F1, W3219, Box 111, 12/1/29/7, Pt. 3, Fire control and prevention/Burning off in state forests/Southland conservancy 1969–76, ANZ. See also Don Geddes, Interview, RFP.

156 Jack Barber, Interview, RFP. The classes of burns are identified in the training materials on *Fire New Zealand Forest Ranger Certificate*, Circular to all conservators, no. 673, undated, p. 41, Millman Collection.

157 Different parts of the country, or the coast as opposed to inland, required different practices. The dry southwest winds, for instance, were good for burning at Rotorua, but in Otago that wind was too cold for the fire to light and a warmish day was better (Jack Barber, Interview, RFP). Mike Hockey instances other local variations: 'If it was an established wind, you could factor it into when you burned. For instance, on the East Coast at Wharerata the sea breeze would always come in by about 10am. It was very stable and from a constant direction. Further inland, the nor'wester would be established by 1pm. So you knew the local conditions, and timed the burns accordingly.'

158 This account is taken from ABDT, W3092, Box 12, 12/3, Controlled burns/Instructions/General 1963–82, ANZ. P. Amer, senior fire control officer, AFID burning, 14 March 1978. Most of the oral accounts also detail the methods and their variations. Additionally different conditions in a particular area could require special burning techniques. For instance, John Barnes, Interview, RFP: 'Some land-clearing burns were done in wind-rows. Canterbury only burned in wind-rows, that drew into each other to create effect. At Kaingaroa we used line burning mostly – burning in strips. For scrub burns there was the ring-burning technique – a strip across top to clear an area was burned, so that when the main fire arrived it came to a burnt-out break. With that safe area at top, you lit along bottom, then up the sides, so with luck the fire drew in together and headed up towards the area first burnt out. It worked most of the time.'

159 Interviews: Don Geddes, Warren (Nobby) Reekie and John Ward, RFP.

160 John Ward, Interview, RFP.

161 Interviews: Mike Hockey, John Barnes and John Ward, RFP; ABDT, W3092, Box 12, 12/3, Controlled burns/Instructions/General 1963–82, ANZ.

162 Jack Barber, quoted in John Ward, *A History of Ashley Forest including Mt Thomas, Okuku and Omihi Forests*, Kowhai Archives Society, Rangiora, 2010, p. 106.

163 Bill Studholme, pers. comm., May 2010.

164 ABDT, W3092, Box 12, 12/3, Controlled burns/Instructions/General 1963–82, ANZ. See papers September 1977 ff.

165 Roy Knight, Interview, RFP.

166 Interviews: John Ward and John Barnes, RFP. Barnes recalled the pilot in Kaingaroa who went right across out of the block and across to another block and lit up that patch. He 'just forgot to turn the AFID off, so there was a bit of excitement there while we had to put these other fires out'.

167 Mike Hockey, Interview, RFP.

168 ABDT, W3092, Box 12, 12/3, Controlled burns/Instructions/General 1963–82, ANZ. Memo, Controlled burning, 5 March 1980; Ibid., Memo, Control burning, 26 May 1981, which comments on the previous lack of attention to safety.

169 Jack Barber, Interview, RFP.

170 F1, Box 283, 10/2/46, Pt. 1, Reports/Fire officer 1949–56, ANZ. Fire protection of rural lands in New Zealand, undated (1951), ANZ.

171 For more detailed information on other companies' initiatives see the interviews with Don Geddes and Morrie Geenty, in which they discuss in-depth fire initiatives taken by Fletchers and, after 1972, by P. F. Olsen. Unless otherwise indicated, the following material is taken almost verbatim from the notes compiled in a phone interview with Rod Farrow, Interview, RFP.

172 Rod Farrow, Interview, RFP. The committee was made up of representatives from Forest Products, Cashmore farms (who were farming in the Cashmore hills behind Kinleith mill on land that had been logged in the 1920s and 1930s) and other forest owners (Taupo Totara Timber Co. Ltd, Bartholomew Timber Co., Hutt Timber & Hardware, Putaruru Timber Yards (PTY), Pacific Forests – all of which were subsequently taken over by Forest Products – and Ellis and Burnand, which became part of Fletchers, later Fletcher Challenge). A second fire district, the Tahurakuri fire district on the Rotorua–Taupo road, had also been established then. This covered the 12,000 hectares of Forest Products land that had been burned in the Taupo fires. Forest Products was the majority landholder in both fire districts, and in about the late 1950s both fire districts were merged into the Tokoroa fire district.

173 Don Geddes, Interview, RFP.

174 F1, W3219, Box 106, 12/1, Pt. 5, Fire control and prevention/General 1962–69. ANZ.

175 Ian Millman, Interview, RFP, also discusses the regional committee and its operations.

176 Don Geddes, Interview, RFP.

177 Ibid.

178 *AJHR*, 1956, p. 68; *AJHR*, 1971, C.–3, pp. 41–42. The comment on the need to appoint only active, experienced and dedicated forest users suggests the scheme did not always work to its full potential.

179 John Barrington, pers. comm., 2009.

180 *AJHR*, 1956, C.–3, p. 64.

181 *AJHR*, 1957, C.–3, p. 61; 'Fire-fighters in Sounds need constant vigilance', *Wairarapa Times-Age*, 8 November 1965.

182 AJHR, 1959, C.–3, p. 94; AANI, W3219, Box 62, 12/6/2, Pt. 7, Fire-fighting and control equipment/General info 1976–79, ANZ, Report, 13 December 1976.

183 Murray Dudfield, Interview, RFP.

184 John Barnes, Interview, RFP.

185 F1, Box 298, 12/3/1, Pts 1, 2, 3 & 4, Fires outside state forests but inside fire districts/ Auckland region 1948–63, ANZ. In relation to equipment, one Forest Service manager to head office on the need for a fire tanker to replace the hose layer that could be maintained and operated by the Kaitaia Volunteer Fire Brigade. 'If a suitable vehicle is available, irrespective of whether it is supplied by this Service of the Fire Services Council, it would prove invaluable to the three rural fire authorities in the Mangonui County and the Kaitaia Urban area. It would also save long travelling with the Kaikohe Tanker.' (F1, Box 298, 12/3/1, Pt. 3, Fires outside state forests but inside fire districts, Auckland region 1948–63.) On the willingness to use schoolboys, see R. G. Lawn, Forest Fire no. 3B2, Kaikohe Fire Zone, 1961, AANI, W3219, Box 61, 12/3/1, Pt. 7, Fires outside state forests in Auckland Fire districts 1961–63, ANZ.

186 F1, W3219, Box 113, 12/2/6A, Fire control and prevention/Mt White fire investigation, ANZ. Recommendations by the Combined Fire Committees for improvements in fire prevention and control, Wellington, 1973.

187 Gavin Wallace, pers. comm., April 2010.

188 Interviews: Roy Knight, Jack Barber, John Barnes and John Ward, RFP.

189 Jack Barber, Interview, RFP.

190 Team membership is now about 30, down from the original 90 or so members. The initiative was first mooted in 1970 (F1, W3219, Box 113, 12/2/6A, Fire control and prevention/Mt White fire investigation, ANZ, Report, Fire in Mt Cook National Park, 29–30 March 1970, Actions to stop such fires). For the extensive training see ABGE, 7498, Box 48, 12/1/3, Fires other than Marlborough–Nelson fire district, ANZ. However, by 1982 the frequency of training had changed from twice weekly to once every two months, with an outdoor exercise in the fire season. John Barnes (Interview, RFP) remembers the team being pointed out to him with some veneration. Reference to the insignia etc is in AJHR, 1959, C.–3, p. 94; AANI, W3219, Box 62, 12/6/2, Pt. 7, Fire-fighting and control equipment/General info 1976–79 (ANZ), Regular visits round conservancies, R. A. Collier, forest ranger, fire control, Report ending June 1976, 14 July 1976. Roy Knight, Interview, RFP.

191 ABGE, 7498, Box 48, 12/1/3, Fires other than Marlborough–Nelson fire district 1977–87, ANZ. This includes permits within SF/SF management or high country; Murray Dudfield, Interview, RFP.
192 Penny Walker, *Sound the Siren: A history of the Wainuiomata Volunteer Fire Brigade 1944–1994*, pp. 12, 26 and 35. It is worth noting that in the late 1960s in Wainuiomata and the Hutt Valley the Fire Service and the local city council planned for firebreaks, a fire permit system, and mandatory requirements on householders to have water laid on and a hose handy (F1, W3129, Box 108, 12/1/10, Pt. 2, Fire control and prevention/Co-ordination by government departments 1953–62, ANZ. Joint report on fire protection of the Eastern Hutt hills, undated (late 1960); Ibid., C. H. Lorden, personal assistant to Director-General of Forests, to chief fire officer, Lower Hutt Fire Brigade, 21 October 1960). However, if those intentions were implemented they do not seem to have affected the general perception that the local bodies were neither interested nor willing to act – see Interviews: Murray Ellis, Arnold and Jan Heine, Dave Thurley and Syd Moore, RFP.
193 'Scientist a man of diverse talents', obituary for William James (Bill) McCabe, *Dominion Post*, March 2009.
194 Murray Ellis, Interview, RFP.
195 Interviews: Arnold Heine and Dave Thurley, RFP. Syd Moore (Interview, RFP) gives a lively and more informal account of the group's activities.
196 *AJHR*, 1971, C.–3, p. 22.
197 AANI, W3219, Box 2, 1/1/4/0, Pt. 4, Revision of Forest and Rural Fires legislation 1977–77, ANZ. Submission from Hutt County Council, August 1977.
198 The Forest Service published the costs of fire prevention and suppression in its annual reports (*AJHR*, C.–3). The reports cover 'Fires, prevention, precautions, administering fire districts'.
199 Jiani Wu et al., *The Economic Cost of Wildfires*, pp. 5–6.
200 See, for instance, AANS, 828, W5491, Box 766, 1/1/4/0, Pt. 6, Administration/Revision of forest and rural fires legislation 1980–85, ANZ. Tangoio Soil Conservation Reserve, proposed rural fire district, 4 September 1950. See also papers 16 November and 21 December 1950 and 17 January 1951. The Department of Lands and Survey in 1950 made the Tangoio Soil Conservation Reserve and adjoining scenic reserves on the Napier–Wairoa highway a rural fire district to eliminate risk to adjacent properties. The Ministry of Works would be the fire authority, the equipment the responsibility of the Soil Conservation Council, and a committee of three farmers and a Ministry of Works representative would implement the fire-fighting policy. This operation, they hoped, would be on a voluntary basis, with monetary contributions only for labour and vehicle costs.
201 AANI, W3219, Box 1, 1/1/4/0, Pt. 3, Revision of Forest and Rural Fires Act 1968–74, ANZ. Notes on fire organisation, address by L. G. Ardell, Maniototo County, Rural fire control in Otago, pp. 5–6, undated (1968). Some 160 volunteers, all with good local knowledge, were organised into fire teams led by wardens who were themselves actively involved with burning applications. There were clear chains of command, along with equipment and means of alerting members; the relationships and responsibilities between the county fire officers and the wardens were clearly delineated; and participants' responsibilities to provide for their food were described. Provision was made for the property owners' input and for advice from the chief fire officer, as well as back-up from 10 wage workers from Ranfurly (the voluntary urban brigade was called on as little as possible because of its responsibility for structural fires).
202 Forest and Rural Fires Association of New Zealand, 'Management and legislation': www.ruralfirehistory.org.nz/4th.htm.
203 F1, Box 299, 12/3/4, Pt. 1, Fires outside state forests/Nelson region 1941–53, ANZ. Extract from diary notes, W. F. Wright, Fire report, fire No. 13C, Kaiteretere and Motueka, 10 December 1954, (5) Re comments Ngatimoti and Maoriland fires.
204 John Ward, Interview, RFP.
205 Don Geddes, Interview, RFP.
206 AANI, W3219, Box 1, 1/1/4/0, Pt. 3, Revision of Forest and Rural Fires Act 1968–74, ANZ. Report by Forest Service on evidence heard by Select Committee on Lands and Agriculture

on 4 June 1968 regarding Forest and Rural Fires Amendment Bill 1967–68 [signed by M. Buist and Girling-Butcher], 20 June 1968.

207 Ibid. Federated Farmers of N.Z. (Inc.) (undated). These ranged from objections to perceived build-up of rubbish due to forest and burning restrictions, resulting depreciation in farmland and farmers being forced to sell; logging trucks damaging country roads; foresters not providing adequate firebreaks and being reluctant to accept a 50:50 responsibility for boundary fences; erosion; neglect of noxious weeds and animal control; trees shading narrow valleys; acid soil conditions created by conifers, adversely affecting lower farmland; claims that the Forest Service was taking Crown land away from farming and allowing it to revert without protection measures in the hope it would regenerate into native bush; and forest interests encroaching on farmland. Extravagant claims were also made on value of forestry compared with agriculture – yet the Service wouldn't disclose costs to taxpayer or produce a long-term budget.

208 Ibid. M. Buist to chairman, Tokoroa Rural Fire Committee, 5 January 1972.

209 Ibid. Girling-Butcher to Director-General of Forests, 2 February 1976.

210 ABFK, 7494, W4948, Box 226, 50/16/9, Pt. 5, Forest and Rural Fires Act draft revision 1977–81, ANZ. Minister of Defence, Forest and Rural Fires Bill, 3 June 1977.

211 Ibid. Forest and Rural Fires Bill, Submission of Federated Farmers of New Zealand to Lands and Agriculture Select Committee on 1977 Bill, August 1977, p. 6.

212 AANI, W3219, Box 2, 1/1/4/0, Pt. 4, Revision of forest and rural fires legislation 1977–77, ANZ. Forest and Rural Fires Bill 1977, Second reading, Speech for Minister of Forests, undated.

213 Ibid. Director-General of Forests to Minister of Forests, Forest and Rural Fires Act commencement order 1979, Forest and Rural Fires Regulations 1979, Memo for Cabinet, 20 February 1979.

214 Ian Millman, Interview, RFP.

215 Don Geddes, Interview, RFP.

216 'Fire permits may be needed all year', *Northland Age*, 18 April 1975.

217 AANI, W3219, Box 2, 1/1/4/0, Pt. 5, Revision of forest and rural fires legislation 1979–79, ANZ. Memo, Girling-Butcher, 9 March 1979.

218 AANI, W3219, Box 2, 1/1/4/0, Pt. 4, Revision of forest and rural fires legislation 1977–77, ANZ. Director-General of Forests to Minister of Forests, Forest and Rural Fires Act commencement order 1979, Forest and Rural Fires Regulations 1979, Memo for Cabinet, 20 February 1979.

219 'In the line of fire', *Straight Furrow*, 16 February 1982; D. G. Taylor, letter, *Straight Furrow*, 16 February 1982.

220 National Forest and Rural Fire Seminar, Summary of proceedings, 1981, NZ Forest Service, Wellington.

221 G. Hensley, *A Review of Rural Fire Services in New Zealand: Report of a committee chaired by G. C. Hensley, Coordinator Domestic and External Security*, Rural Fires Review Committee, Wellington, August 1989, pp. 5–6; Murray Dudfield, Interview, RFP. The fund, derived from the fire insurance levy on buildings, was paid through the Fire Service Commission.

222 Michael Roche, 'Exotic forestry: Government restructuring', *Te Ara: The Encyclopedia of New Zealand*: www.TeAra.govt.nz/en/exotic-forestry/5 (accessed March 2009); Bill Studholme, Interview, RFP. In the course of my research I also heard the Forest Service criticised as incapable of either recognising the change in government mood or mounting an adequate defence. I cannot comment on the former, but it is hard for those not involved in government at the time to realise how difficult it was to argue against what rapidly became economic orthodoxy.

Chapter Seven: Fragmentation, re-formations and the future: 1987 to the present day

1 Morrie Geenty, Interview, RFP. The changed systems now mean that cut-over burning of harvested areas 'is just about extinct'. Nor do harvesting systems now leave much on the ground. Trees are not delimbed until dragged out to landings, and the slash is consolidated into 'birds' nests' so that fires, though not entirely eliminated, are dealt with more easily. The

potential for unperceived underground fires, too, is reduced by using infra-red equipment.
2 Department of Internal Affairs, 'Strengthening our fire services', Annual report 2009, p. 20.
3 Murray Dudfield, pers. comm., 2010.
4 For the meteorological measurements, see Dr Neil Cherry, Lincoln College meteorologist, in Bill Studholme, Report to Selwyn Plantation Board, February 1989; 'Severe fire risk in Canterbury', Christchurch *Press*, 4 February 1987, p. 3, comments on the dry forests.
5 Bill Studholme, Interview, RFP.
6 Michael Burke, 'Canterbury rural fires of 1987/88 and 1988/89', Paper presented at the Prevent Rural Fires Convention, Ministry of Forestry, Wellington, May 1989: www.ruralfirehistory.org.nz/documents/Burke1989.htm (accessed February 2012); 'Stock killed as fire destroys farmland', *Press*, 4 February 1987, p. 1; 'Severe fire risk in Canterbury', *Press*, 4 February 1987, p. 3; 'Farms evacuated, houses burned as fire spreads', *Press*, 5 February 1987, p. 1; 'Careful watch being kept on blaze area', and 'Quick work saves farms', *Press*, 6 February 1987, p. 1.
7 Cherry, in Studholme, Report to Selwyn Plantation Board.
8 Burke, 'Canterbury rural fires of 1987/88 and 1988/89'; 'Fire rages near homes', *Press*, 17 October 1988.
9 Studholme, Interview, RFP. Unless otherwise indicated, the rest of this account is taken from Studholme, Report to Selwyn Plantation Board.
10 Studholme, Interview, RFP.
11 Ibid.
12 Ibid.
13 Ibid.
14 Ibid.
15 'Rain relief to fire-fighters', *Press*, 23 December 1988, p. 6.
16 Burke, 'Canterbury rural fires of 1987/88 and 1988/89'.
17 Ibid.
18 AAUA, 6915, W4625, Box 37, 6/175, Government departments & general/Forest and rural fire matters working party 1982–82, *Forest and Rural Firefighter*, Ministry of Forestry newsletter, no. 1, April 1987.
19 Gerald Hensley, *A Review of Rural Fire Services in New Zealand: Report of a committee chaired by G. C. Hensley, Coordinator Domestic and External Security*, Rural Fires Review Committee, Wellington, August 1989, pp. 2–4; M. Dudfield, 'Rural fire legislation and local government', undated, National Rural Fire Authority (NRFA); Barbara Hedley, 'Ready, aim … FIRE', *New Zealand Forest Industries*, October 1987, pp. 48–53.
20 Gerald Hensley, 'Comments on the 1989 review', Rural Fire Service Seminar, 29 June 2000, Circ. 2000/14, NRFA.
21 Bill Studholme, Interview, RFP.
22 Hensley, *A Review of Rural Fire Services in New Zealand*, p. 9; Bill Studholme, Interview, RFP.
23 Hensley, ibid., pp. 8, 9. Speaking from memory, Peter Berg (Interview, RFP) gives a lower level of manpower loss and suggests that the 1200 wage workers the corporation retained as contractors represented 'a substantial proportion of the pre-1987 capability'. He also points out that some ex-Forestry Service men had gone to DOC and would have been involved in DOC's fire-fighting activity on its land. While that is certainly true, it ignores DOC's lack of fire-fighting capacity at the time and ForestryCorp's position that it was 'not prepared or equipped to underwrite cover for other authorities' (Hensley, p. 9).
24 Kerry Hilliard, Interview, RFP; Warren (Nobby) Reekie (Interview, RFP) also comments on DOC's difficulties in getting equipment.
25 Kerry Hilliard, Interview, RFP.
26 Hensley, *A Review of Rural Fire Services in New Zealand*, p. 4.
27 The quote is taken from Bill Studholme, Interview, RFP; Hensley, *A Review of Rural Fire Services in New Zealand*, pp. 9–10, makes the same point, but employs a different language.
28 Bill Studholme, Interview, RFP.
29 Hensley, *A Review of Rural Fire Services in New Zealand*, pp. 10, 13, 14; Hensley, 'The basic

principles we followed', Rural Fire Seminar, 29 June 2000, no. 1, NRFA Circ. 2000/14.

30 Hensley, *A Review of Rural Fire Services in New Zealand*, p. 37; Hensley, 'What we found', Rural Fire Seminar, 29 June 2000.

31 Hensley, *A Review of Rural Fire Services in New Zealand*, pp. 10–11; Burke, 'Canterbury rural fires of 1987/88 and 1988/89'.

32 Studholme, Report to Selwyn Plantation Board.

33 Peter Berg, Interview, RFP. Berg has recently retired as president of the New Zealand Forest Owners' Association.

34 Kerry Hilliard, Interview, RFP. Hensley (pers. comm., 2010) also stressed to me the difficulties of considering initiatives outside the Treasury's economic orthodoxy at that time. The final comment is Hensley's.

35 Burke, 'Canterbury rural fires of 1987/88 and 1988/89'.

36 *The Rural Firefighter*, February 1997; Morrie Geenty, Interview, RFP.

37 Doug Ashford, Interview, RFP.

38 Hensley, pers. comm., November 2010.

39 Hensley, Letter of transmittal, *A Review of Rural Fire Services in New Zealand*, 23 August 1989.

40 Hensley, 'The basic principles we followed'.

41 Hensley, *A Review of Rural Fire Services in New Zealand*, pp. 5–6, 13–14; Hensley, 'The basic principles we followed'.

42 Hensley, *A Review of Rural Fire Services in New Zealand*, pp. 16–20.

43 Hensley, pers. comm., November 2010.

44 Grant Pearce, Geoff Cameron, Stuart A. J. Anderson and Murray Dudfield, 'An overview of fire management in New Zealand forestry', *NZ Journal of Forestry*, 53 (3) pp. 7–11. The legislation referred to is the Resource Management Act 1991 (which defined fire as a 'natural hazard'), the Local Government Act 2002, and the Civil Defence and Emergency Management Act 2002.

45 Pearce et al., 'An overview of fire management in New Zealand forestry'; National Rural Fire Authority: www.nrfa.org.nz (accessed June 2010).

46 Hensley, *A Review of Rural Fire Services in New Zealand*, pp. 34–35. When he wrote the report Hensley expected that, as a result of local government organisation, significant amounts of rural-character land such as green belts would fall within urban districts. This has not happened on any significant basis, and these blocks are dealt with under s. 9 of the Forest and Rural Fires Act, which allows such land to be designated as rural land for the purposes of fire-fighting.

47 Forest and Rural Fires Association of New Zealand, 'Fire training': www.ruralfirehistory.org.nz/5th.htm (accessed June 2010).

48 Murray Dudfield, Interview, RFP. Kerry Hilliard (Interview, RFP) comments on the scheme's origin and on the use of the same principles and language.

49 Kerry Hilliard, Interview, RFP.

50 Interviews: Jock Darragh, Ross Hamilton and Kerry Hilliard, RFP.

51 Ross Hamilton, pers. comm.

52 John Barnes, Interview, RFP.

53 Jock Darragh, Interview, RFP.

54 Letter, Chris and Philippa Gorman to Bill Girling-Butcher, 16 June 1991, Papers, NRFA.

55 Morrie Geenty, Interview, RFP. Geenty comments on the 'wishy-washy' reports that do not name names or place responsibility, and the absence of clear recommendations. This follows Australian practice, he says.

56 Forest and Rural Fires Association of New Zealand: www.ruralfirehistory.org.nz/5th.htm; Murray Dudfield, Interview, RFP. Susan M. Timms, 'Wetland vegetation recovery after fire: Eweburg Bog, Te Anau', *NZ Journal of Botany*, 30, 1992, pp. 383–99, notes the absence of published papers on fire in wetland at that time. On the new directions in research see Dudfield, pers. comm., March 2010. In relation to the New Zealand work on human dimensions see, for instance, Mary Hart, Lisa Langer, Muriel McGlone and H. Grant

Pearce, *Review of Fire Recovery Planning in Two Regions of New Zealand*, SCION, 2009. For examples of the Australian work see the papers listed in the bibliography under Bushfire CRC, Fire notes, issues 41, 42, 44.

57 Peter Berg, Interview, RFP.

58 Morrie Geenty, Interview, RFP; Pearce et al, 'An overview of fire management in New Zealand'.

59 Grant et al., 'An overview of fire management in New Zealand'.

60 John Ward, Interview, RFP.

61 Interviews: John Barnes and Kerry Hilliard, RFP.

62 New Zealand Fire Service Commission, *The New Zealand Coordinated Incident Management System (CIMS)*, New Zealand Fire Service Commission, Wellington, 1998, pp. 8–10. The American Incident Management System was developed in 1970, and was followed by a national inter-agency version on which the Australian Inter-service Incident Management System (adopted in the early 1980s) is closely modelled (ibid., p. 7). Forest and Rural Fires Regulations 2005 (reprinted 2008), ss. 39–46, sets out what rural fire plans have to cover.

63 Officials' report on rural fire services, 15 March 1990, in Restructured rural fire-fighting fund, Minutes of Cabinet Committee, POL (90) 171, 11 July 1990. This report recommended adding the $300,000–400,000 DOC spent annually on rural fire-fighting to the fund. We have already identified the benefits the report noted – incentives for prompt suppression work, perception of value from the rural levy, and improving rural fire authorities' confidence. Cabinet approved the recommendation on 11 July 1990. Murray Dudfield, Interview, RFP, comments on DOC's previous parlous condition.

64 Hensley, *A Review of Rural Fire Services in New Zealand*, pp. 39–40; Murray Dudfield, Interview, RFP. The fund, derived from the fire insurance levy on buildings, continues to be paid through the Fire Service Commission. The Department of Internal Affairs once paid any shortfall, with a subsequent refund once the Fire Service and DOC levies were adjusted (R. L. Mark, 'Report: A proper and equitable base for the funding of the Fire Service', undated, NRFA) but now does not provide any funds towards the Fire Service Commission's annual budget.

65 Forest and Rural Fires Association of New Zealand: www.ruralfirehistory.org.nz/5th.htm. Bona fide recipients are rural fire-fighting groups registered with the National Rural Fire Authority and which have a constitution.

66 Murray Dudfield, pers. comm., June 20. The subsidy varies. It provides $3 for every $1 the authority provides for clothing and equipment. The subsidy for new tankers is dollar for dollar.

67 'The cost of wildfires in New Zealand', circular 2010/05, 17 March 2010, NRFA.

68 Jiani Wu, *The Economic Cost of Wildfires*, New Zealand Fire Services Commission, Wellington, 2009, pp. 39–40.

69 Wu, *The Economic Cost of Wildfires*, pp. 33–34, 38.

70 Department of Internal Affairs, *New Fire Legislation: The functions and structure of New Zealand's fire and rescue services: A discussion document*, Department of Internal Affairs, Wellington, 2004, p. 42.

71 Unless otherwise indicated, all the material on the DOC response is from Kerry Hilliard, Interview, RFP.

72 Lesley Porter, Interview, RFP.

73 Murray Ellis, Interview, RFP.

74 Ibid.

75 Lesley Porter, Interview, Programme 75, Open Country Sound Recordings, recorded March 1964, OHT 5-0064, OHC.

76 Lesley Porter, Interview, RFP.

77 Wu, *The Economic Cost of Wildfires*, p. 4. This figure is averaged from the costs of wildfire 2002–07.

78 Peter Berg, Interview, RFP.

79 Department of Internal Affairs, *New Fire Legislation discussion document*; Department of

Internal Affairs, *New Fire Legislation: A framework for New Zealand's fire and rescue services and their funding: A proposal to stakeholders*, Wellington, 2007; Department of Internal Affairs, *New Fire Legislation: A framework for New Zealand's fire and rescue services and their funding: Report on submissions*, Wellington, 2007.

80 Department of Internal Affairs, Annual report 2008/09, p. 20.

81 National Rural Fire Authority (NRFA), *Enlarged Rural Fire Districts*, National Rural Fire Authority, undated (2009?), p. 14.

82 Ibid., p. 9.

83 Doug Ashford, pers. comm., July 2010. Ashford notes that the Forest and Rural Fire Association does not have a firm view on this issue and members' views differ, so any estimate of the level of support or opposition is somewhat anecdotal. Opposition is said to come from, variously, big rather than small rural fire districts, from hands-on people, and from urban fire-fighting forces. The material in this section, unless otherwise referenced, reflects the issues raised by the people whom I interviewed.

84 Ross Hamilton, Interview, RFP. The research mentioned is Hart et al., *Review of Fire Recovery Planning in Two Regions of New Zealand*, pp. 9–13. Having reviewed the way that two regions approached fire recovery planning, the authors discuss the limited ways in which rural fire districts treat fire, the limited responsibilities of principal rural fire officers, and lack of clearly defined recovery procedures in rural fire plans. These all impinge on a range of issues involved in recovery.

85 Jock Darragh, Interview, RFP.

86 Kerry Hilliard, Interview, RFP. His view was shared by many other interviewees.

87 John Barnes, Interview, RFP.

88 Ibid.

89 2007 Census

90 Pearce et al., 'An overview of fire management in New Zealand'.

91 Murray Dudfield, pers. comm., June 2010.

92 Ian J. Payton and H. Grant Pearce, *Fire-induced Changes to the Vegetation of Tall-tussock (*Chionochloa rigida*) Grassland Systems*, Science for Conservation, Department of Conservation, Wellington, 2009, pp. 6, 33; Kerry Hilliard, Interview, RFP.

93 Department of Conservation, *The Value of Conservation*, Wellington, 2006: www.doc.govt.nz/publications/conservation/benefits-of-conservation/the-value-of-conservation (accessed February 2012).

94 Kerry Hilliard, Interview, RFP.

95 United Nations Development Programme, 'Changing attitudes to rural fire', International Wildland Fire Conference, Boston, 1989.

96 Kerry Hilliard, Interview, RFP.

97 Commissioner Piers Reid, Address to the FRFANZ conference, 2002: www.ruralfirehistory.org.nz/5th.htm.

98 Hart et al., *Review of Fire Recovery Planning*, p. 7.

Bibliography

A list of abbreviations used appears on page 279.

Books and book chapters

Allsop, R., *The First Fifty Years of New Zealand's Forest Service*, Government Printer, Wellington, 1973

Arnold, Rollo, *New Zealand's Burning: The settler's world in the mid 1880's*, Victoria University Press, Wellington, 1994

Atkinson, Gwenda, *Flaming Century: Thames Volunteer Fire Brigade 1887–1987*, Hauraki Publishers, 1987

Bain, W., *Greytown Volunteer Fire Brigade: 100 years of community service 1882–1982*, Greytown Volunteer Fire Brigade, Greytown, 1982

Baird, Donald, *The Story of Firefighting in Canada*, Boston Mills Press, Erin, Ontario, 1986

Barnett, Tim, *Aroha, Poha, Tikanga: Volunteering in Aotearoa/New Zealand*, Department of Labour, Wellington, 1996

Beaglehole, J. C. (ed.), *The Journals of Captain James Cook, Vol. 1, The Voyage of the Endeavour 1768–1771*, Cambridge University Press for the Hakluyt Society, Cambridge, 1955

Bellamy, A. C., *A Century of Service to Tauranga 1882–1982*, Tauranga Fire Brigade, Tauranga, 1982

Berry, Ken, *Duty Calls: The Picton Volunteer Fire Brigade 1872–1972*, Picton Volunteer Fire Brigade, Picton, 1997

Binney, Judith, *Encircled Lands: Te Urewera 1820–1921*, Bridget Williams Books, Wellington, 2009

Blewett, D. W. and Mark Hutton (compilers), *Alexandra Volunteer Fire Brigade 1903–2003: 100 Years*, Otago Daily Times Print, Alexandra, 2003

Bloomfield, G. T., *New Zealand: A handbook of historical statistics*, G. K. Hall and Co., Boston, 1984

Boast, Richard, *Buying the Land, Selling the Land: Governments and Maori land in the North Island 1865–1921*, Victoria University Press, Wellington, 2008

Bradley, W. I., *Lyttelton Volunteer Fire Brigade Centennial 1873–1973*, Lyttelton Volunteer Fire Brigade, Lyttelton, 1973

Bulls Volunteer Fire Brigade History, Bulls Volunteer Fire Brigade, Bulls, 1985

Cathcart, B. C., *Water On! The first hundred years of the United Fire Brigades' Association of New Zealand*, incorporating 'A brief history of the first 50 years of the UFBA', N. G. Buick, United Fire Brigades' Association, Te Awamutu, 1979

Cochrane, R., *Early forestry and sawmilling in Mid Canterbury*, Plain Words, Ashburton, 1996

Coughlan, Lanna, *The Growth of State Forestry in New Zealand: A brief review*, Wellington, 1964

Cross, J. C., *Fire and Flood: Pleasant Point Volunteer Fire Brigade 1944–1994*, Pleasant Point Volunteer Fire Brigade, 1994?

Crow, P. M. and D. G. Crow, *Waitara Volunteer Fire Brigade 75th Jubilee 1910–1985*, Waitara Volunteer Fire Brigade, Waitara, 1985

Cumberland, Kenneth B., ' "Climatic change" or cultural interference', in Murray McCaskill (ed.), *Land and Livelihood: Geographical essays in honour of George Jobberns*, New Zealand Geographic Society, Christchurch, 1962

Dalziel, Raewyn, *Julius Vogel: Business politician*, Auckland University Press/Oxford University Press, Auckland, 1986

Davidson, R. J. and V. I. Davidson, *Clyde Volunteer Fire Brigade: 75th Jubilee Programme and History 1906–1981*, Clyde Volunteer Fire Brigade, 1981

Doherty, J. J., S. A. J. Anderson and G. Pearce, *An analysis of wildfire records in New Zealand 1991–2007*, SCION report no. 12789, 2008

Dudfield, M., *Rural Fire Legislation and Local Government*, NRFA, undated

Dunsbee, Bruce, *Pukekohe Volunteer Fire Brigade 75th Anniversary 1911–1986*, New Zealand Fire Service, Auckland, 1986

Ellis and Burnand, *Ellis and Burnand: Fifty years of service in the timber industry of N.Z.*, Hamilton, 1953

Elson, Grahame, *Fifty Not Out: A brief history of the Kaikoura Volunteer Fire Brigade*, Kaikoura Volunteer Fire Brigade, Kaikoura, 1985

Foster, Ted, *Bushfire: History, prevention, control*, A. H. & A. W. Reed, Sydney, 1976

Galbraith, A. V., *Forest Fires: Cause and effect*, H. J. Green, Government Printer, Melbourne, 1926

Geraldine Volunteer Fire Brigade, *A History of the Geraldine Volunteer Fire Brigade, 1889–1989*, Geraldine, 1989

Gisborne Fire Brigade, *From the Ashes: 100 years of the Gisborne Fire Brigade*, 1980

Greymouth Fire Brigade, *Eighty Years of Fire Fighting*, Greymouth Evening Star Company, 1947

Grossman, J. P., *The Evils of Deforestation*, Brett Printing and Publishing Co., Auckland, 1909

Halkett, John, Peter Berg and Brian Mackrell, *Tree People: Forest Service memoirs*, New Zealand Forestry Corporation, GP Print, Wellington, undated (c. 1987/88)

Hawke Grimwade, Karen, *100 Years of Fire Fighting in Ohakune 1908–2008: A history, to commemorate the Centennial Jubilee of the Ohakune Volunteer Fire Brigade, Ohakune Volunteer Fire Brigade*, Ohakune, 2008

Healey, Brian, *A Hundred Million Trees: The story of NZ Forest Products Ltd*, Hodder & Stoughton, Auckland, 1982

Henderson, Alan, *Competition and Co-operation: The Insurance Council and the general insurance industry 1895–1995*, Insurance Council of New Zealand/ Historical Branch, Department of Internal Affairs, Wellington, c. 1995

Hitchcock, Walter M., *Reminiscences of a Volunteer Fireman in Australia and England 1854–1912*, 3rd edn, W. M. Hitchcock, Blackheath, England, 1913

Jemison, George M., *The Measurement of Forest Fire Danger in the Mountains of Eastern United States*, Department of Agriculture, Forest Service Appalachian Forest Experiment Station, Asheville, 1939

Kaiapoi Volunteer Fire Brigade Centennial 1870–1970, Kaiapoi Volunteer Fire Brigade, Kaiapoi, 1970

Kawakawa Volunteer Fire Brigade Golden Jubilee 1920–1980, Kawakawa Volunteer Fire Brigade, Kawakawa, 1980

Kear, G. R., *A Century of Service: A history of the Palmerston North Fire Brigade 1883–1983*, Dunmore Press, Palmerston North, 1983

Kingsbury, B. S., *The Cust Volunteer Fire Brigade 1948–2000: A record of the Brigade's formation and activities over its first 52 years*, Cust Volunteer Fire Brigade, 2000

Kirkland, A. and P. Berg, *A Century of State-honed Enterprise: 100 years of state plantation forestry in New Zealand*, Profile Books, Auckland, 1997

Knowles, A. G., *A Century of Service: The story of the first hundred years of the Carterton Volunteer Fire Brigade 1878–1978*, Carterton Volunteer Fire Brigade, Carterton, 1978

Krissansen, Grant (ed.), *Henderson Fire Brigade 1929–1979, Golden Jubilee: Fifty years' community service*, Henderson Fire Brigade, Auckland, 1979

Luke, R. H. and McArthur, A. G., *Bushfires in Australia*, Australian Government Publishing

Service, Canberra, 1978
Luke, R. H., *Bush Fire Control in Australia*, Hodder & Stoughton, Melbourne, 1961
Lyttelton Volunteer Fire Brigade Centennial 1873–1973, Lyttelton Volunteer Fire Brigade, Lyttelton, 1973
Marton Volunteer Fire Brigade Centennial 1879–1979: The first 100 years of the Marton Volunteer Fire Brigade: A brief history of the passing years 1879–1979, Marton Volunteer Fire Brigade, Marton, 1979
McCormick, John J., *Oamaru Volunteer Fire Brigade: Looking back 1879–2004*, Oamaru, 2004?
McIntosh, A. D. (ed.), *Marlborough: A provincial history*, Marlborough Provincial Historical Committee, Blenheim, 1940
McKinnon, Malcolm (ed.), *New Zealand Historical Atlas: Visualising New Zealand*, David Bateman/ Department of Internal Affairs, Auckland, 1997
Molloy, J. K., *Port Chalmers Volunteer Fire Brigade 1876–1976: Centennial History*, Port Chalmers Volunteer Fire Brigade, Port Chalmers
Nathan Simon and Mary Varnham (eds), *The Amazing World of James Hector*, Awa Science, Wellington, 2008
New Zealand Forestry League, *History of the New Zealand Forestry League: A remarkable record*, 1935?
New Zealand Forestry League, *New Zealand Forestry League: Reasons for its establishment, its aims and objectives*, 1917
New Zealand Forestry League, *Rules and Objects of the New Zealand Forestry League (Inc)*, 1916
NZ Perpetual Forests Ltd, *N.Z. Perpetual Forests*, Auckland, 1931
NZ Perpetual Forests Ltd, *Reports Covering Operations and Progress of N.Z. Perpetual Forests Ltd, 1924–1926*
Osborne, W. B., 'The lookout system', in *The Western Fire Fighters' Manual* (6th edn), Western Forestry and Conservation Association, Seattle, 1934 (Millman collection)
Pound, Francis, *The Invention of New Zealand Art and National Identity 1930–1970*, Auckland University Press, 2009
Putaruru Volunteer Fire Brigade 50th Jubilee 1943–1993
Pyne, Stephen J., *Fire in America: A cultural history of wildland and rural fire*, Princeton University Press, Princeton, 1982
Richards, Glenn, *The History of Firefighting at Rolleston: Written to celebrate 25 years as a fire brigade*, Rolleston Volunteer Fire Brigade, Rolleston, 1994
Riseborough, Hazel, *Ngamatea: The land and the people*, Auckland University Press, Auckland, 2006
Robb, Leo, *Balfour Volunteer Fire Brigade 50th Jubilee 1955–2005*, Balfour Volunteer Fire Brigade, 2005
Roche, Michael, 'Ellis, Leon MacIntosh', *Dictionary of New Zealand Biography*: http://www.teara.govt.nz/en/biographies (accessed February 2012)
Roche, Michael, *History of New Zealand Forestry*, New Zealand Forestry Corporation/GP Books, Wellington, 1990
Rule, S. M. (ed.), *Sumner Volunteer Fire Brigade 1894–1994*, Sumner Volunteer Fire Brigade, Christchurch, 1994
Scanlan, A. B., *100 Years of Firefighting: The New Plymouth Fire Brigade 1866–1966*, New Plymouth Fire Board, 1966
Simpson, Thomas E., *Kauri to Radiata: Origin and expansion of the timber industry of New Zealand*, Hodder & Stoughton, Auckland, 1973
Stanbury, Peter (ed.), *Bushfires: Their effect on Australian life and landscape*, Macleay Museum, University of Sydney, Sydney, 1981
Stringer, Marion J., *Wakefield Volunteer Fire Brigade 1953–1993: The first forty years, Wakefield Volunteer Fire Brigade*, Nelson, 1993
Taupo Totara Timber Company Ltd, *Sixty Years of Progress in the Taupo Totara Timber Company 1901–1961*, Taupo Totara Timber Company, undated
Taylor, Nancy, *The New Zealand People at War: The home front*, vol. II, Historical Publications Branch, Department of Internal Affairs, Wellington, 1986

Taylor, Rowan and Ian Smith, *The State of New Zealand's Environment 1997*, Ministry for the Environment, GP Publications, Wellington, 1997
Thomson, K. W. (ed.), *The Legacy of Turi: An historical geography of Patea County*, Dunmore Press, Palmerston North, 1976
Tindall, France J., *Stokes Valley Volunteer Fire Brigade: 50 flaming years*, Stokes Valley Volunteer Fire Brigade, Lower Hutt, 1990
Titirangi Volunteer Fire Brigade Jubilee 1949–1999, Titirangi Volunteer Fire Brigade, Auckland, 1999
Trass, D. G., *Nuhaka Fire Service: The first 30 years, 1960–1990*, Nuhaka Fire Service, Nuhaka, 1990
Turner, J. L., *Patea Volunteer Fire Brigade Centennial 1884–1984*, Patea, 1984]
United Nations Development Programme, *Changing Attitudes to Rural Fire*, International Wildland Fire Conference, Boston, 1989
Waipara Volunteer Fire Brigade: First 25 years, Waipara Volunteer Fire Brigade, Waipara, 1990
Walker, Penny, *Sound the Siren: A history of the Wainuiomata Volunteer Fire Brigade 1944–1994*, Wainuiomata Volunteer Fire Brigade, Wainuiomata, 1994
Ward, John and Cooper, Neil, *Seventy Years of Forestry: Golden Downs Forest, Nelson 1927–1997*, Forest History Trust, Richmond, 1997
Ward, John and the Kowhai Archives Society, *A History of Ashley Forest including Mt Thomas, Okuku and Omihi Forests*, Kowhai Archives Society, Rangiora, 2010
Watton, Graham, *100 Years of Community Service: The history of the Paeroa Volunteer Fire Brigade, October 1895–October 1995*, Paeroa Volunteer Fire Brigade, 1995
Whitianga Volunteer Fire Brigade 1952–2002: From buckets to high pressure hoses, Whitianga Volunteer Fire Brigade, Whitianga, 2002

Official/government publications

Appendices to the Journals of the House of Representatives, Government Printer, Wellington
'Conservation or planting of forests', 1876, H.–41
'Correspondence relative to the present condition of the forests of New Zealand', 1869, D.–22
'Crown lands', report of the Conservator of State Forests, 1877, C.–3
Crown Lands Department annual reports 1888–1891, C.–1
Department of Internal Affairs, 'Strengthening our fire services', annual report 2009
Department of Lands and Survey annual reports 1892–1920, C.–1, *AJHR*
Fire Service Council annual reports, H.–12
'Forests in New Zealand: Suggestions on colonial revenues derivable from New Zealand forests', 1881, H.–11
'Native forests and the state of the timber trade', Professor Kirk to Minister of Lands, 1886, C.–3A
Reports on the Fire Brigades of the Dominion to the Minister of Internal Affairs, 1909–1926, H.–6A; 1927–1949, H.–12
State Forest Department annual reports 1886–1887, C.–3D
State Forest Service annual reports 1921–1987, C.–3
'Suggestions on forests in New Zealand', Mr A. Lecoy, 1880, H.–3
'The forests of the colony', Papers relating to state forests/Parliamentary debates and resolutions 1874, H.–5
'The timber industry', William C. Kensington, Report to the Minister of Department of Lands, 1907, C.–4
Department of Conservation, *The Value of Conservation*, Wellington, 2006: www.doc.govt.nz/publications/conservation/benefits-of-conservation/the-value-of-conservation (accessed February 2012)
Department of Conservation, *Fire-induced Changes to the Vegetation of Tall-tussock (*Chionochloa rigida*) Grassland Systems*, by Ian J. Payton and H. Grant Pearce, Wellington, 2009
Department of Internal Affairs, *New Fire Legislation: A framework for New Zealand's fire and rescue services and their funding: A proposal to stakeholders*, Wellington, 2007
Department of Internal Affairs, *New Fire Legislation: A framework for New Zealand's fire and rescue services and their funding: Report on submissions*, Wellington, 2007

Department of Internal Affairs, *New Fire Legislation: The functions and structure of New Zealand's fire and rescue services: A discussion document*, Wellington, 2004
Hart, Mary, Lisa Langer, Muriel McGlone and H. Grant Pearce, *Review of Fire Recovery Planning in Two Regions of New Zealand*, SCION, 2009
Hensley, G. C., *A Review of Rural Fire Services in New Zealand: Report of a committee chaired by G. C. Hensley, Coordinator Domestic and External Security*, Rural Fires Review Committee, Wellington, 1989
Ministry of Agriculture and Fisheries, *Burning Scrub (or Bush Cover): Requirement and methods*, Ministry of Agriculture and Fisheries, Wellington, undated (mid-1980s)
Ministry of Forestry and the New Zealand Logging Industry Research Organisation, *Harvesting a Small Forest*, Ministry of Forestry, Wellington, 1996
National Rural Fire Authority, *A Farmers' Practical Guide to Rural Fire*, 2nd edn, 2007
National Rural Fire Authority, *A Landowners' Guide to Land Clearing by Prescribed Burning*, 2005
National Rural Fire Authority, *Air Operations Information and Checklist*, 2009
National Rural Fire Authority, *Enlarged Rural Fire Districts*, undated (2009?)
National Rural Fire Authority, *FireSmart: Home Owners' Manual*, 2006
National Rural Fire Authority, *FireSmart: Partners in protection*, 2004
National Rural Fire Authority, *Rural Fire Management Handbook*, 2005
New Zealand Fire Service Commission, *The New Zealand Coordinated Incident Management System (CIMS)*, Wellington, 1998
New Zealand Forest Service, *Basics of Fire Fighting*, 1980
New Zealand Forest Service, *Fire Engine Operators Manual*, 1983
New Zealand Forest Service, *Forest and Rural Fire Precautions: Notes for contractors and permittees*, 1973, 1979, 1981
New Zealand Forest Service, *Helicopters in Forest Fire Operations*, 1983
New Zealand Forest Service, *Rural Fire Fighting*, 1984, 1985
New Zealand Forest Service, *Safety and Survival in Forest Fires*, Information series no. 68, 1974, 1984
New Zealand Parliamentary Debates, Government Printer, Wellington
Regulations for New Zealand Forests and Forest Reserves subject to the provisions of 'The New Zealand State Forests Act 1885' and forests growing on any Crown land, Government Printer, Wellington, 1900
Statistics New Zealand, *Unpaid Work*, Wellington, 1998
Statutes of New Zealand, Government Printer, Wellington
US Department of Agriculture, Forest Service and the Resources Agency of California, Department of Conservation, Division of Forestry, *Forest Fire Fighting Fundamentals*, California, undated
US Department of Agriculture, Forest Service, *Water vs. Fire*, California, 1950

Articles and monographs

Anon., Institute news, 'Inaugural members', *NZJF*, vol. 7, no. 2, 1955, pp. 107–08
Anon., Obituary, A. D. McGavock, *NZJF*, vol. 8, no. 1, 1959, pp. 167–69
Anon., Obituary, Arnold Hansson, *NZJF*, vol. 27, no. 1, 1982, pp. 19–21
Anon., Obituary, Norman James Dolamore', *NZJF*, vol. 7, no. 3, 1956, pp. 9–10
Anon., Obituary, S. A. C. Darby, *NZJF*, vol. 30, no. 1, 1985, pp. 16–17
Berg, Peter, Obituary, Bill Girling-Butcher, *NZJF*, vol. 44, no. 4, 2000, p. 34
Brooking, Tom, 'Use it or lose it: Unravelling the land debate in late nineteenth-century New Zealand', *NZ Journal of History*, 30 (2), October 1996, pp. 141–62
Bushfire CRC and AFAC, 'Investigating perceived teamwork effectiveness in incident management teams', *Fire Note*, no. 41, October 2009
Bushfire CRC and AFAC, 'Observing teamwork in emergency management', *Fire Note*, no. 42, October 2009
Bushfire CRC and AFAC, 'How human factors drive decisions at fire ground level', *Fire Note*, no. 44, October 2009

Cameron, R. J., 'Maori impact upon the forests of New Zealand', *Bay of Plenty Journal of History*, vol. 9, no. 3, September 1961, pp. 131–41

Campbell, D. A., 'Aerial fire fighting', *NZ Science Review*, December 1959

Cooper, A. N., *Forest fires in New Zealand: History and status 1880–1980*, NRFA, undated

Cooper, A. N., 'Training using simulated forest fires', *Commonwealth Forestry Review*, vol. 62, no. 2, 1986, pp. 131–39

Guild, Dennys and Murray Dudfield, 'A history of fire in the forest and rural landscape in New Zealand, Part 2: Post-1830 influences, and implications for future fire management', *NZJF*, vol. 54, no. 4, 2008, pp. 31–38

Guild, Dennys and Murray Dudfield, 'A history of fire in the forest and rural landscape in New Zealand, Part 1: Pre-Maori and pre-European influences', *NZJF*, vol. 54, no. 1, 2009, pp. 34–38

Hedley, Barbara, 'Ready, aim ... FIRE', *New Zealand Forest Industries*, October 1987, pp. 48–53

Holloway, J. T., 'Forests and climates in the South Island of New Zealand', *Transactions of the Royal Society of NZ*, vol. 82, no. 2, 1954, pp. 329–410

Holloway, J. T., 'The forests of the South Island: The status of the climate change hypothesis', *NZ Geographer*, vol. 20, no. 1, April 1964

Johnston, Judith A., 'New Zealand bush: Early assessments of vegetation', *NZ Geographer*, vol. 37, no. 1, April 1981

Kenneally, Christine, 'The inferno', *New Yorker*, October 2009

McGlone, M. S., 'The Polynesian settlement of New Zealand in relation to environmental and biotic changes', *NZ Journal of Ecology*, vol. 12 (supplement), 1989, pp. 115–19

McGlone, Mike, 'Interpreting the pollen record: Reconstructing the deforestation of New Zealand': www.landcareresearch.co.nz/research/ecosystems/past_env/pollen_interpretation.asp, 2009 (accessed February 2012)

McGlone, Mike, 'The origin of the indigenous grasslands of southeastern South Island in relation to pre-human woody systems', *NZ Journal of Ecology*, vol. 1, no. 25, 2001, pp. 1–15

McKelvey, Peter, 'Macintosh Ellis in France', *NZ Forestry*, no. 34, vol. 2, pp. 15–18

McWethy, David B., Cathy Whitlock, Janet Wilmshurst, Matt McGlone and Xun Li, 'Rapid deforestation of South Island, New Zealand, by early Polynesian fires', *The Holocene*, vol. 19, no. 6, 2009, pp. 883–87.

Molloy, P. J., 'Distribution of subfossil forest remains, eastern South Island, New Zealand', *NZ Journal of Botany*, vol. 1, 1963, pp. 68–77

'Monk family meets to celebrate 150 yrs', *Rodney and Waitemata Times*, 29 April 1980

Monk, Richard, *Our Forests and How to Conserve them: A paper on the New Zealand State Forests Act 1885, read before the Waitemata County Council, 5 February 1886*. Printed at the Star Office, Auckland

Morgan, Harold C., 'Fighting an Australian bush-fire', *Chambers Journal*, June 1937, pp. 445–58

New Zealand Forestry League, *The Forest Magazine of New Zealand: Official journal of the New Zealand Forestry League*, 1922

Odgen, John, Les Basher and Matt McGlone, 'Fire, forest regeneration and links with early human habitation: Evidence from New Zealand', *Annals of Botany*, vol. 81, 1998, pp. 687–96

O'Donnell, James, 'Forest fire control in Western Australia', repr. from *Australian Forestry*, vol. IV, no. l, undated (c. 1939)

Pearce, Grant, Cameron, Geoff, Anderson, Stuart A. J. and Dudfield, Murray, 'An overview of fire management in New Zealand forestry', *NZJF*, vol. 53, no. 3, pp. 7–11

Roche, Michael, 'Latter day 'Imperial Careering': L. M. Ellis, A Canadian forester in Australia and New Zealand 1920–1941': http://fennerschool-associated.anu.edu.au/environhist/newzealand/journal/2009/april/roche.php (accessed 2010)

The Rural Firefighter, Newsletter of the Forest and Rural Fire Association of New Zealand, no. 31, February 1997

Studholme, Bill, 'Dunsandel fire', Report to the Selwyn Plantation Board, undated [February 1989]: www.ruralfirehistory.org.nz/documents/Studholme1989.htm (accessed February 2012)

Timms, Susan M., 'Wetland vegetation recovery after fire: Eweburg Bog, Te Anau', *NZ Journal of*

Botany, *30*, 1992, pp. 383–99
V. T. F., Book review, *The First Fifty Years of New Zealand's Forest Service* by F. Allsop, *NZJF*, vol. 19, no. 1, 1974, pp. 146–48.
Walsh, P. (Rev.), 'The effect of deer on the New Zealand bush: A plea for protection of our forest reserves', *Transactions and Proceedings of the Royal Society of New Zealand 1868–1912*, vol. 25, 1892, pp. 435–39
Wheeler, P. B., *The Communal Character of Scattered Rural Settlements in New Zealand*, Geography discussion papers, no. 19, University of Otago, 1980?

Reports, circulars, papers, conference proceedings

Amner, Peter, 'New Zealand Forest Service fire-control staff 1947 until 1 April 1987', MS papers 9564, ATL
Burke, Michael, 'Canterbury rural fires of 1987/88 and 1988/89', paper presented at the Prevent Rural Fires Convention, Ministry of Forestry, Wellington, May 1989: www.ruralfirehistory.org.nz/documents/Burke1989.htm (accessed February 2012)
Entrican, Alex R., 'Influence of forestry on forest policy and forest products trade in Australia and New Zealand', MacMillan Lecture, Vancouver, January 1963, *NZJF*, vol. 41, no. 2, 1996, pp. 46–47
Gisborne, H. T., *Measuring Fire Weather and Forest Inflammability*, US Department of Agriculture, circular no. 398, Washington DC, 1936
Hensley, Gerald, 'Comments on the 1989 review', Rural Fire Service Seminar, 29 June 2000, circular 2000/14, NRFA
Johnstone, W., *The New Zealand Volunteer Fire Service in 3 Rural Communities in Northland*, Fire Service Commission Report no. 3, Fire Service Commission, Wellington, 2002
Mark, R. L., 'Report: A proper and equitable base for the funding of the Fire Service' NRFA, undated
National Forest and Rural Fire Seminar, Summary of proceedings, 1981, Forest Service, Wellington
O'Loughlin, Colin, 'Historical perspectives of indigenous forests', in *Native Trees for the Future: Potential, possibilities, problems of planting and managing New Zealand forest trees*, proceedings of the forum held at the Centre for Continuing Education and Department of Biological Sciences, University of Waikato, Hamilton, 8–10 October 1999
Roche, M., 'Edward Phillips Turner: The development of a "Forest Consciousness" in New Zealand 1890s to 1930s', proceedings of the 6th National Australian Forest History Conference, Augusta, WA, 12–17 September 2004
Wu, Jiana, William Kaliyati and Kerl Sanderson, *The Economic Cost of Wildfires: Report to the New Zealand Fire Service Commission*, BERL Economics, Wellington, 2009

Government archives (Archives New Zealand)

Department of Conservation: AANS series
Department of Internal Affairs: AD1 series, files 270; AAUA series; AANX series
Local Government Association: AAUA series
Local Government Commission: AANX series
Ministry of Agriculture and Forestry head office: ACCQ series
Ministry of Agriculture and Forestry: ACCQ series
Ministry of Works and Development residual unit: AATE series
Ministry of Works: AATE series
New Zealand Defence Force: ABFK series
New Zealand Forest Service: F1 series, record group ADSQ series, head office AADY series, residual unit AANI series, Nelson forestry BBQI
New Zealand Timberlands: head office ABHR series, Nelson district ABGE, northern region, Hawke's Bay ABDT series
Te Puni Kokiri: head office ABJZ series

Interviews

Programme 75, Open Country Sound Recordings, recorded March 1964, OHT 5-0064, Oral History Centre, ATL. Includes: Cassie Alex, Ernest Myer and Trevor Myer, 'The flame in the forest' and 'Bushwackers and bushfires'; and Francis Guy, 'How we fought the bush fires'.

Rural Firefighting Project, recorded 2009–10, Oral History Centre, ATL. Interviewees: Doug Ashford, Jack Barber, John Barnes, Peter Berg, Jock Darragh, Murray Dudfield, Murray Ellis, Rod Farrow, Don Geddes, Morrie Geenty, Arnold Heine, Jan Heine, Kerry Hilliard, Mike Hockey, Roy Knight, Ian Millman, Syd Moore, Lesley Porter, Warren (Nobby) Reekie, Bill Studholme, Dave Thurley, Gavin Wallace, John Ward

Websites

www.nzbirds.com/birds/moa.html (accessed August 2012)

www.insights.co.nz/story_behind_d.aspx – Topic: Deforestation (last accessed October 2008)

www.ruralfirehistory.org.nz

INDEX

Italics indicate illustrations